An Inquiry into Modes of Existence

An Inquiry

into Modes of Existence

An Anthropology of the Moderns

·

BRUNO LATOUR

·

Translated by Catherine Porter

HARVARD UNIVERSITY PRESS
CAMBRIDGE, MASSACHUSETTS
LONDON, ENGLAND
2013

The book was originally published as *Enquête sur les modes d'existence: Une anthropologie des Modernes*, copyright © Éditions La Découverte, Paris, 2012.

The research has received funding from the European Research Council under the European Union's Seventh Framework Programme (FP7/2007-2013)
ERC Grant 'IDEAS' 2010 n° 269567

Typesetting and layout:	Donato Ricci
This book was set in:	Novel Mono Pro; Novel Sans Pro; Novel Pro
	(CHRISTOPH DUNST | BÜRO DUNST)

Library of Congress Cataloging-in-Publication Data
Latour, Bruno.
[Enquête sur les modes d'existence. English]
An inquiry into modes of existence : an anthropology of the
moderns / Bruno Latour ; translated by Catherine Porter.
pages   cm
"The book was originally published as Enquête sur les modes
d'existence : une anthropologie des Modernes."
ISBN 978-0-674-72499-0 (alk. paper)
1. Civilization, Modern—Philosophy.
2. Philosophical anthropology.
I. Title.
CB358.L27813   2013
128—dc23   2012050894

"Si scires donum Dei."

·Contents·

·

·

·Introduction·

A shocking question addressed to a climatologist (02) that obliges us to distinguish values from the accounts practitioners give of them (06).

Between modernizing and ecologizing, we have to choose (08) by proposing a different system of coordinates (10).

Which leads us to define an imaginary diplomatic scene (13): in the name of whom to negotiate (13) and with whom to negotiate? (15)

The inquiry at first resembles the one involving speech acts (17) while we learn to identify different modes of existence (19).

The goal is, first, to accompany a people vacillating between economy and ecology (22).

·

Part One

How to Make an Inquiry into the Modes of Existence of the Moderns Possible

·Chapter 1·

An investigator goes off to do fieldwork among the Moderns (28) without respecting domain boundaries, thanks to the notion of actor-network (30), which makes it possible to distinguish networks as result from networks as process (31).

The inquiry defines a first mode of existence, the network [NET], through a particular "pass," or passage (33).

But networks [NET] have a limitation: they do not qualify values (35).

Law offers a point of comparison through its own particular mode of displacement (38).

There is thus a definition of "boundary" that does not depend on the notions of domain or network (38).

The mode of extension of objective knowledge can be compared with other types of passes (39).

Thus any situation can be defined through a grasp of the [NET] type plus a particular relation between continuities and discontinuities (41).

Thanks to a third type of "pass," the religious type, the investigator sees why values are difficult to detect (42) because of their quite particular ties to institutions (43), and this will oblige her to take into account a history of values and their interferences (45).

language and existence (88). This becomes particularly clear in the prime example of the laboratory (89). Hence the salience of a new mode of existence, [REP], for reproduction (91) and of a crossing [REP · REF] that is hard to keep in sight (93) especially when we have to resist the interference of Double Click (93).

To give the various modes enough room (98) we must first try to grasp existents according to the mode of reproduction [REP] (99) by making this mode one trajectory among others (100) in order to avoid the strange notion of an invasive material space (103).

If those who have occupied all the space nevertheless lack room (104) it is because they have been unable to disamalgamate the notion of matter (105) by the proper use of the [REP · REF] crossing (106).

Now, as soon as we begin to distinguish two senses of the word "form" (106), the form that maintains constants and the form that reduces the hiatus of reference (107), we begin to obtain a nonformalist description of formalism (108), which turns out, unfortunately, to be wiped out by a third sense of the word "form" (109).

At this point we risk being mistaken about the course followed by the beings of reproduction (110) in that we risk confusing two distinct courses in the idea of matter (111).

A formalist description of the outing on Mont Aiguille (112) generates a double image through a demonstration per absurdum (114) that would lead to a division into primary and secondary qualities (115).

But once the origin of this Bifurcation into primary and secondary qualities has been accurately identified (115) it becomes a hypothesis too contrary to experience (116) and the magic of rationalism vanishes (117) since we can no longer confuse existents with matter (118), a matter that would no more do justice to the world than to "lived experience" (120).

If we had to begin with the hardest part (124) it was because of an insistence on "straight talk" that connects formalism with closing off discussion (125).

Although this straight talk cannot rely on the requirements of reference [REF] (126), it leads to the disqualification of all the other modes (127) by creating a dangerous amalgam between knowledge and politics [REF · POL] (128), which makes it necessary to abandon the thread of experience in order to put an end to debates (129).

Fortunately, the method that allows us to recognize a crossing (131) will succeed in identifying a veridiction proper to politics [POL] (132), which has to do with the continual renewal of a Circle (133) that the course of reference cannot judge properly (134).

Thus we have to acknowledge that there is more than one type of veridiction (136) to foil the strange amalgam of "indisputable facts" (137) and thus to restore to natural language its expressive capacities (138).

The most difficult task remains: going back to the division between words and things (139) while liberating ourselves from matter, that is, from the res ratiocinans (140) and giving ourselves new capacities of analysis and discernment (142) in order to speak of values without bracketing reality (143).

Language is well articulated, like the world with which it is charged (144), provided that we treat the notion of sign with skepticism (145).

Modes of existence are indeed at stake, and there are more than two of these (146), a fact that obliges us to take the history of intermodal interferences into account (148).

·Chapter 6·

CORRECTING A SLIGHT DEFECT IN CONSTRUCTION\dotfill

The difficulty of inquiring into the Moderns (152) comes from the impossibility of understanding in a positive way how facts are constructed (153), which leads to a curious connivance between the critical mind and the search for foundations (155).

Thus we have to come back to the notion of construction and distinguish three features (157): 1. the action is doubled (157); 2. the direction of the action is uncertain (158); 3. the action is qualified as good or bad (159).

Now, constructivism does not succeed in retaining the features of a good construction (159).

We thus have to shift to the concept of instauration (160), but for instauration to occur, there must be beings with their own resources (161), which implies a technical distinction between being-as-being and being-as-other (162) and thus several forms of alterity or alterations (163).

We then find ourselves facing a methodological quandary (164), which obliges us to look elsewhere to account for the failures of constructivism (164): iconoclasm and the struggle against fetishes (165).

It is as though the extraction of religious value had misunderstood idols (166) because of the contradictory injunction of a God not made by human hands (167), which led to a new cult, antifetishism (168), as well as the invention of belief in the belief of others (169), which turned the word "rational" into a fighting word (170).

We have to try to put an end to belief in belief (171) by detecting the double root of the double language of the Moderns (172) arising from the improbable link between knowledge and belief (174).

Welcome to the beings of instauration (176).

Nothing but experience, but nothing less than experience (178).

•

PART TWO

HOW TO BENEFIT FROM THE PLURALISM OF MODES OF EXISTENCE

·Chapter 7·

Its mode of existence depends on the [MET · TEC] ruse (224) as much as on the persistence of the beings of reproduction [REP · TEC] (225).

The veridiction proper to [TEC] (226) depends on an original folding (227) detectable thanks to the key notion of shifting (228).

The unfolding of this mode gives us more room to maneuver (230).

Multiplying the modes of existence implies draining language of its importance (234), which is the other side of the Bifurcation between words and the world (235).

To avoid confusing sense with signs (236) we have to come back to the experience of the beings of fiction [FIC] (238).

Beings overvalued by the institution of works of art (238) and yet deprived of their ontological weight (239).

Now, the experience of the beings of [FIC] invites us to acknowledge their proper consistency (240) an original trajectory (241) as well as a particular set of specifications (242).

These beings arise from a new alteration: the vacillation between raw material and figures (243), which gives them an especially demanding mode of veridiction (245).

We are the offspring of our works (246).

Dispatching a work implies a shifting (246) different from that of the beings of technology [TEC · FIC] (248).

The beings of fiction [FIC] reign well beyond the work of art (249); they populate a particular crossing, [FIC · REF] (250), where they undergo a small difference in the discipline of figures (251) that causes the correspondence to be misunderstood (252).

We can then revisit the difference between sense and sign (254) and find another way of accessing the articulated world (256).

To remain sensitive to the moment as well as to the dosage of modes (260) the anthropologist has to resist the temptations of Occidentalism (261).

Is there a mode of existence proper to essence? (262)

The most widespread mode of all, the one that starts from the prepositions while omitting them (264), habit [HAB], too, is a mode of existence (265) with a paradoxical hiatus that produces immanence (266).

By following the experience of an attentive habit (267) we see how this mode of existence manages to trace continuities (268) owing to its particular felicity conditions (268).

Habit has its own ontological dignity (270), which stems from the fact that it veils but does not hide (272). We understand quite differently, then, the distance between theory and practice (273), which allows us to define Double Click more charitably [HAB · DC] (274).

Each mode has its own way of playing with habits (275).

This mode of existence can help define institutions positively (277), provided that we take into account the generation to which the speaker belongs (278) and avoid the temptation of fundamentalism (280).

·Conclusion, Part Two·

·

PART THREE
HOW TO REDEFINE THE COLLECTIVES

·Chapter 11·

Rationalization is what produces belief in belief (319) and causes the loss of both knowledge and faith (321), leading to the loss of neighboring beings and remote ones alike (322) as well as to the superfluous invention of the supernatural (323).

Hence the importance of always specifying the terms of the metalanguage (324).

contributions: first, beings of politics [POL] (373), then beings of law [LAW] (373), and finally beings of religion [REL] (374).

Quasi subjects are all regimes of enunciation sensitive to tonality (375).

Classifying the modes allows us to articulate well what we have to say (376) and to explain, at last, the modernist obsession with the Subject/Object difference (378).

New dread on the part of our anthropologist: the fourth group, the continent of The Economy (379).

By returning to the experience of what sets scripts in motion (422) we can measure what has to be passed through in order for beings to subsist (423) while discovering the beings of passionate interest [ATT] (424).

But several obstacles to the depiction of this new experience have to be removed: first, the notion of embedding (426); then the notion of calculating preferences (427); then the obstacle of a Subject/Object relation (427); fourth, the obstacle of exchange (429); and fifth and last, the cult of merchandise (430).

Then a particular mode of alteration of being appears (432) with an original pass: interest and valorization (434) and specific felicity conditions (435).

This kneading of existents (437) leads to the enigma of the crossing with organization [ATT · ORG] (438), which will allow us to disamalgamate the matter of the second Nature (439).

Detecting the [ATT · ORG] crossing (444) ought to lead to praise for accounting devices (445).

However, economics claims to calculate values via value-free facts (447), which transforms the experience of being quits (448) into a decree of Providence capable of calculating the optimum (449) and of emptying the scene where goods and bads are distributed (450).

While the question of morality has already been raised for each mode (452), there is nevertheless a new source of morality in the uncertainty over ends and means (454).

A responsible being is one who responds to an appeal (456) that cannot be universal without experience of the universe (457).

We can thus draw up the specifications for moral beings [MOR] (458) and define their particular mode of veridiction: the taking up of scruples (459) and their particular alteration: the quest for the optimal (461).

The Economy is transformed into a metaphysics (462) when it amalgamates two types of calculations in the [REF · MOR] crossing; (462) this makes it mistake a discipline for a science (464) that would describe only economic matter (466).

So The Economy puts an end to all moral experience (466).

The fourth group, which links quasi objects and quasi subjects (467), is the one that the interminable war between the two hands, visible and invisible, misunderstands (469).

Can the Moderns become agnostic in matters of The Economy (470) and provide a new foundation for the discipline of economics? (472)

To avoid failure, we must use a series of tests to define the trial that the inquiry must undergo (476): First test: can the experiences detected be shared? (477)

·

To the Reader:

User's Manual for the Ongoing Collective Inquiry

This book summarizes an inquiry that I have been pursuing rather obstinately for a quarter century now. Thanks to a generous subsidy from the European Union, I have been able to create a platform that will allow you not only to read this *provisional report* but also to extend the inquiry by using the research apparatus available at MODESOFEXISTENCE.ORG.

Begun in solitude, my work is being extended here with the help of a small team brought together under the code name AIME: An Inquiry into Modes of Existence, the English translation of the French EME: Enquête sur les Modes d'Existence. If all goes well, our platform will allow us to mobilize a considerably larger research community.

Once you have registered on the site, you will have access to the digital version of this book, and thus to the notes, bibliography, index, glossary, and supplementary documentation that we have provided. The digital interface as such is now well anchored in contemporary working practices; ours is flexible enough to multiply ways of reading while providing a constantly evolving critical apparatus benefiting from the commentary that you and other readers will not fail to add. (The boldface terms in the text refer to the digital glossary.)

What makes this project so interesting, and also, of course, so difficult, is that you are going to find yourself invited not only to read the work but to explore a somewhat unfamiliar environment. The digital interface is designed to provide you with enough handholds to let you retrace a certain number of experiences that lie, as I see it, at the heart of the history of the Moderns, whereas the Moderns' own accounts of these experiences do very little to make them understandable. In my view, this contradiction between the experiences themselves and the accounts of them authorized by the available metaphysics is what makes it so hard to describe the Moderns empirically. It is in order to move beyond this contradiction that I invite you to join me in paying close attention to the conflicts of interpretation surrounding the various truth values that confront us every day. If my hypothesis is correct, you will find that it is possible to distinguish different *modes* whose paired intersections, or *crossings*, can be defined empirically and can thus be *shared*. I encourage you to participate in this sharing via the digital environment we have developed for the purpose.

For I am convinced that, once you have discovered new ways of familiarizing yourself with the arguments of the inquiry, you will be able to propose quite different answers to the *questionnaire* around which it is structured. Thanks to the digital interface, you will be able to navigate in each mode and at each crossing where two modes intersect. After you examine the documents that we have begun to assemble, you may be prepared to contribute others. The entire interest of the exercise lies in the possibility that you and other participants, whether or not they have read the book, will extend the work begun here with new documents, new sources, new testimonies, and, most important, that you will modify the questions by correcting or modulating the project in relation to the results obtained. The laboratory is now wide open for new discoveries.

In a final phase, if you wish, you will even be able to participate in an original form of *diplomacy* by proposing accounts different from mine as interpretations of the experiences that we shall have revisited together. Indeed, in a planned series of encounters, with the help of mediators, we shall try to propose other versions, other metaphysics, beyond the ones proposed in this provisional report. We may even be able to sketch out the lineaments of other institutions better suited to shelter the values we shall have defined.

This project is part of the development of something known by the still-vague term "digital humanities," whose evolving style is beginning to supplement the more conventional styles of the social sciences and philosophy. Although I have learned from studying technological projects that innovating on all fronts at once is a recipe for failure, here we are determined to explore innovations in method, concept, style, and content simultaneously. Only experience will tell us whether this hybrid apparatus using new techniques of reading, writing, and collective inquiry facilitates or complicates the work of empirical philosophy that it seeks to launch. Time is short: we are obliged to conclude this attempt to describe the Moderns' adventure differently by August 2014—a century after another, tragically memorable August '14. You can already see why there is no question of my succeeding in this enterprise on my own!

Acknowledgments

The publication of this work as well as the development of the AIME platform were made possible by a research grant from the European Research Council, ERC no. 269567. I thank the European Research Mission of Sciences Po for supporting this project from start to finish, and also the École des Mines for the two sabbatical years I was granted, in 1995 and again in 2005.

OVERVIEW

Since I cannot disguise the difficulty of the exercise I am going to ask readers to undertake, I am going to try to give them the overall thrust at the start, so they will know where I want to lead them—which may help them hang on during the rough passages. A guide leading the way can announce the trials to come, extend a hand, multiply the rest stops, add ramps and ropes, but it is not within his power to flatten the peaks that his readers have agreed to cross with him.

I have divided the report on this investigation into three parts. In Part One, I seek first of all to establish the object (Chapter 1) and then the data needed for this rather unusual investigation (Chapter 2). I must also remove the two principal obstacles that would make all our efforts to advance our understanding of the Moderns incomprehensible and even absurd. These two obstacles are obviously related, but I have nevertheless distinguished them by devoting two chapters (3 and 4) to the key question of objective knowledge—why has the advent of Science made it so difficult to grasp the other modes?—and two more chapters (5 and 6) to the question of how construction and reality are connected— why can't we say of something that it is "true" and "made," that is, both "real" and manufactured, in a single breath? At the end of Part One, we shall know how to speak appropriately about a plurality of types of beings by relying on the guiding thread of experience, on empiricism as William James defined it: nothing but experience, yes, but *nothing less* than experience.

Thus when the ground has been cleared, when experience has become a reliable guide once more, when speech has been freed from the awkward constraints peculiar to Modernism, we shall be in a position to profit, in Part

Two, from the pluralism of modes of existence and to get ourselves out of the prison, first, of the Subject/Object division. The first six modes that we are going to identify will allow us to offer an entirely different basis for comparative anthropology, these being the contrasts on which other cultures have focused particular attention. We shall then be able to understand the emergence of the modes, the fluctuation of their values, the adverse effects that the emergence of each one has had on our ability to grasp the others. I shall take advantage of this analysis, too, in order to arrange them in a somewhat more systematic way by proposing a different system of coordinates.

This system will allow us, in Part Three, to identify six additional modes, more regional ones, closer to the habits of the social sciences; these modes will help us get around the last major obstacles to our investigation: the notion of Society and above all that of The Economy, that second nature that defines probably better than all the other modes the anthropological specificity of the Moderns.

Just as, at the beginning of Fellini's *Orchestra Rehearsal*, each instrumentalist, speaking in front of the others, tells the team who have come to interview them that his or her instrument is the only one that is truly essential to the orchestra, this book will work if the reader feels that each mode being examined in turn is the best one, the most discriminating, the most important, the most rational of all . . . But the most important test is that, for each mode, the experience whose guiding thread I claim to have found can be clearly distinguished from its institutional report. This is the only way to be able to propose more satisfying reports in the next phase. At the end of these two Parts, we shall at last be able to give a positive, rather than merely a negative, version of those who "have never been modern": "Here is what has happened to us; here is what has been passed along to us; now, what are we going to do with this historical anthropology or, better, with this regional ontology?"

What to do? This is the object of the general conclusion, necessarily very brief because it depends on the fate of the collaborative research platform in which this text, a simple summary report on an inquiry, aspires to interest the reader. Here the anthropologist turns into a chief of protocol to propose a series of "diplomatic representations" that would allow us to inherit the set of values deployed in Parts Two and Three—all of which define the very local and particular history of the Moderns—but within renewed institutions and according to renewed regimes of speech.

Then, but only then, might we turn back toward "the others"—the former "others"!—to begin the negotiation about which values to institute, to maintain, perhaps to share. If we were to succeed, the Moderns would finally know what has happened to them, what they have inherited, the promises they would be ready to fulfill, the battles they would have to get ready to fight. At the very least, the others would finally know where they stand in this regard. Together, we could perhaps better prepare ourselves to confront the emergence of the global, of the Globe, without denying any aspect of our history. The universal would perhaps be within their grasp at last.

TRUSTING

INSTITUTIONS AGAIN?

....

A shocking question addressed to a climatologist ⊙ that obliges us to distinguish values from the accounts practitioners give of them.

Between modernizing and ecologizing, we have to choose ⊙ by proposing a different system of coordinates.

Which leads us to define an imaginary diplomatic scene: ⊙ in the name of whom to negotiate ⊙ and with whom to negotiate?

The inquiry at first resembles the one involving speech acts ⊙ while we learn to identify different modes of existence.

The goal is, first, to accompany a people vacillating between economy and ecology.

B EFORE THE READER CAN UNDERSTAND HOW WE ARE GOING TO WORK TOGETHER—I HOPE!—BY EXPLORING THE NEW MEANS THAT THE DIGITAL ENVIRONMENT MAKES AVAILABLE TO US, I need to offer a foretaste of what is at stake in such an inquiry. Since the smallest elements can lead step-by-step to the largest, let me begin with an anecdote.

A SHOCKING QUESTION ADDRESSED TO A CLIMATOLOGIST ⊙ They're sitting around a table, some fifteen French industrialists responsible for sustainable development in various companies, facing a professor of climatology, a researcher from the Collège de France. It's the fall of 2010; a battle is raging about whether the current climate disturbances are of human origin or not. One of the industrialists asks the professor a question I find a little cavalier: "But why should we believe *you*, any more than the others?" I'm astonished. Why does he put them on the same footing, as if it were a simple difference of opinion between this climate specialist and those who are called climate skeptics (with a certain abuse of the fine word "skeptic")? Could the industrialist possibly have access to a measuring instrument superior to that of the specialist? How could this ordinary bureaucrat be in a position to weigh the positions of the experts according to a calculus of more and less? Really, I find the question almost shocking, especially coming from someone whose job it is to take particular interest in ecological matters. Has the controversy really degenerated to the point where people can talk about the fate of the planet as if they were on the stage of a televised jousting match, pretending that the two opposing positions are of equal merit?

I wonder how the professor is going to respond. Will he put the meddler in his place by reminding him that it's not a matter of belief but of fact? Will he once again summarize the "indisputable data" that leave scarcely any room for doubt? But no, to my great surprise, he responds, after a long, drawn-out sigh: "If people don't *trust the institution of science,* we're in serious trouble." And he begins to lay out before his audience the large number of researchers involved in climate analysis, the complex system for verifying data, the articles and reports, the principle of peer evaluation, the vast network of weather stations, floating weather buoys, satellites, and computers that ensure the flow of information—and then, standing at the blackboard, he starts to explain the pitfalls of the models that are needed to correct the data as well as the series of doubts that have had to be addressed on each of these points. "And, in the other camp," he adds, "what do we find? No competent researcher in the field who has the appropriate equipment." To answer the question raised, the professor thus uses the notion of institution as the best instrument for measuring the respective weight of the positions. He sees no higher court of appeals. And this is why he adds that "losing trust" in this resource would be, for him, a very serious matter.

His answer surprises me as much as the question. Five or ten years ago, I don't believe that a researcher—especially a French researcher—would have spoken, in a situation of controversy, about "trust in the institution of science." He might possibly have pointed to "confidence intervals," in the scientific sense of the term, but he would have appealed to certainty, a certainty whose origin he would not have had to discuss in detail before such an audience; this certainty would have allowed him to treat his interlocutor as an ignoramus and his adversaries as irrational. No institution would have been made visible; no appeal to trust would have been necessary. He would have addressed himself to a higher agency, Science with a capital S. When one appeals to Science, there is no need for debate, because one always finds oneself back in school, seated in a classroom where it is a matter of learning or else getting a bad grade. But when one has to appeal to trust, the interlocutory situation is entirely different: one has to share the concern for a fragile and delicate institution, encumbered with terribly material and mundane elements—oil lobbies, peer evaluation, the constraints of model-making, typos in thousand-page-long reports, research contracts, computer bugs, and so on. Now, such a concern—and this is the essential point—does not aim to cast doubt on research results; on the contrary, it is what ensures that they are going to become valid, robust, and shared.

Whence my surprise: How can this researcher at the Collège de France abandon the comfort provided by the appeal to indisputable certainty and lean instead on trust in science as an institution? Who still has confidence in institutions today? Is this not the worst moment to set forth in full view the frightful complexity of the countless offices, meetings, colloquia, summits, models, treatises, and articles by means of which our certainties about the anthropic origin of climate disruption are milled? It is a little as though, responding to a catechumen who doubts the existence of God, a priest were to sketch out the organizational chart of the Vatican, the bureaucratic history of the Councils, and the countless glosses on treatises of canon law. In our day, it seems that pointing one's finger at institutions might work as a weapon to criticize them, but surely not as a tool for reestablishing confidence in established truths. And yet this is actually how the professor chose to defend himself against these skeptical industrialists.

And he was right. In a situation of heated controversy, when it is a matter of obtaining valid knowledge about objects as complex as the whole system of the Earth, knowledge that must lead to radical changes in the most intimate details of existence for billions of people, it is infinitely safer to rely on the institution of science than on indisputable certainty. But also infinitely riskier. It must have taken a lot of nerve for him to shift his argumentative support that way.

Still, I don't think the professor was quite aware of having slipped from one philosophy of science to another. I think, rather, that he no longer had the choice of weapons, because his climate-skeptic adversaries were the ones talking about waiting to act until they had achieved total certainty, and who were using the notion of institution only to put him in a bind. Weren't they in effect accusing climatologists of being a "lobby" like any other, the model-makers' lobby? Weren't they taking pleasure in tracing the monetary circuits necessary to their research as well as the networks of influence and complicity that were attested by the e-mails these skeptics had managed to get hold of? And how did they come by *their* knowledge? Apparently, they could boast of being right where everyone else was wrong because Certainty is "never a question of numbers." Every time someone alluded to the throng of climatologists and the scope of their equipment and their budgets, the skeptics raised indignant voices against what they called "an argument from authority." And repeated the lofty gesture of Certainty against Trust by appealing to Truth with a capital T, which no institution can corrupt. And wrapped themselves in the folds of the Galileo affair: didn't Galileo triumph all by himself against institutions, against the Church, against religion, against the scientific bureaucracy of the period? Caught in such a vise, the professor had little choice. Since Certainty had been commandeered by his enemies and the public was beginning to ask rude questions; since there was a great risk that science would be confused with opinion, he fell back on the means that seemed to be at hand: trust in an institution that he had known from the inside for twenty years and that he ultimately had no reason to doubt.

But about which no one ever speaks. Here is where we find the fragility of the buttress on which he chose to lean. If he found me looking at him a bit wryly as he struggled to respond, he will have to forgive me, for I belong to a field, SCIENCE STUDIES, which has been working hard to give a positive meaning to the term "scientific institution." Now, in its early days, in the 1980s, this field was perceived by many scientists as a critique of scientific Certainty—which it was—but also of reliable knowledge— which it most certainly was not. We wanted to understand how—with what instruments, what machinery, what material, historical, anthropological conditions—it was possible to produce objectivity. And of course, without appealing to any transcendent Certainty that would have all at once and without discussion raised up Science with a capital S against public opinion. As we saw it, scientific objectivity was too important to be defended solely by what is known by the umbrella term "RATIONALISM," a term used too often to bring debate to a halt when an accusation of irrationality is hurled against overly insistent adversaries. Well before questions of ecology came to the forefront of politics, we already suspected that the distinction between the rational and the irrational would not suffice to settle the debates over the components of the COMMON WORLD. As we saw it, the question of the sciences was rather more complicated than that; we sought to investigate the manufacture of objectivity in a new way. And that is why we are always astonished, my colleagues in the history or sociology of the sciences and I, at the hostility of certain researchers toward what they call the "relativism" of our inquiries, whereas we have only been trying to prepare scientists for a finally realistic defense of the objectivity to which we are just as attached as they are—but in a different way.

So my mild surprise at the climatologist's response will be understandable. "Well, well, here you are, speaking positively of the trust one must have in the institution of science . . . But, my dear colleague, when have you ever publicly invoked the necessity of such trust? When have you agreed to share your manufacturing secrets? When have you pleaded loud and long that scientific practice must be understood as a fragile institution that has to be carefully maintained if people are to trust the sciences? Are we not the ones, on the contrary, who have done this work all along? We, whose help you have spurned somewhat gruffly by calling us relativists? Are you really ready for such a change in epistemology? Are you really going to give up the comforting accusation of irrationality, that masterful way of shutting up everyone who picks a quarrel with you? Isn't it a little late to take refuge suddenly in the notion of 'trust,' without having prepared yourselves for this in any way?" If I did not raise such questions with the climatologist that day, it was because the time to debate the "relativism" of "science studies" had passed. This whole affair has become too serious for such squabbles. We have the same enemies, and we have to respond to the same emergencies.

⊙ THAT OBLIGES US TO
DISTINGUISH VALUES
FROM THE ACCOUNTS
PRACTITIONERS
GIVE OF THEM.

This anecdote should help the reader understand why we have to inquire into the role to be given to the key notion of INSTITUTION, and more especially the institution of science, since we find ourselves facing ecological crises that are unprecedented in kind and in scale. If I have committed myself to such an inquiry, it is because, in the professor's response, one can readily discern, if not a contradiction, at least a powerful tension between the VALUE that he wants to defend—objectivity—and the *account* he proposes to define this value. For he seems to hesitate, in effect, between an appeal to Certainty and an appeal to Trust, two things that involve, as we shall discover, entirely different philosophies, or rather metaphysics, or, better still, ontologies.

I am well aware that he had not had the time to get a good handle on this difference; it is not the sort of detail that one expects of a climatologist. But my own work as a sociologist, or philosopher, or anthropologist (the label hardly matters) is to explore this disconnect in as much depth as possible, for as long as it takes, and thus to propose—and for me this is the whole point of the project—a solution that will make such a value shareable and sustainable. As we shall see, the proposition I am exploring through this inquiry consists in using a series of CONTRASTS to distinguish the values that people are seeking to defend from the account that has been given of them throughout history, so as to attempt to establish these values, or better yet to install them, in institutions that might finally be designed for them.

I am all too well aware that the words "value" and "institution" can be frightening, can even sound terribly reactionary. What! Go back to values? Trust institutions? But isn't this what we've finally gotten away from, what we've done away with, what we've learned to fight and even to dismiss with scorn? And yet the anecdote analyzed above shows that we may have actually entered a new era. The scope of the ecological crises obliges us to reconsider a whole set of reactions, or rather conditioned reflexes, that rob us of all our flexibility to react to what is coming. This at least was the hypothesis with which I began. For a researcher at the Collège de France to shift from Certainty to Trust, something truly "serious" has to happen. This is the seriousness that weighs on our work together.

My goal for this inquiry is to create an arrangement that I call DIPLOMATIC, one that would make it possible, if I could make it work (but I can't do it alone), to help our researcher who has been attacked in the name of "rationalism" by offering him an alternative definition of what he holds dear. Can I succeed in redefining objectivity through trust in a scholarly institution without leaving him with the sense that he has lost the value for which he has been fighting? Even if, once the work has been done, he will have to rely on a totally different philosophy of science? And can I do this *with him*? Such are the stakes of this research: to *share* the experience of the values that my informants seem to hold dear, but by offering to *modify* the account, or more accurately the metaphysics, through which they seek to express the experience in the overly conflictual cases where they risk losing it while defending it clumsily. Can certain of the concepts that we have learned to cherish be offered the opportunity for a type of development that the much too narrow framework of modernization has not given them? After all, the notions of "sustainable development" and "protected species" can also apply to concepts!

BETWEEN MODERNIZING AND ECOLOGIZING, WE HAVE TO CHOOSE ⊙

Why can so many values no longer hold up against attacks? Because of another phenomenon that I have been seeking to document ever since I was initiated into fieldwork, in Africa, in the early 1960s, and that can be designated as the "end of the modernist parenthesis." In everything that follows, the terms "modernization" or "MODERNS" are opposed to "ECOLOGY." Between modernizing and ecologizing, we have to choose.

In a book published some twenty years ago, *We Have Never Been Modern*, I sought to give a precise meaning to the overly polysemic word "modern" by using as a touchstone the relationship that was beginning to be established in the seventeenth century between two worlds: that of Nature and that of Society, the world of nonhumans and the world of humans. The "we" of the somewhat grandiloquent title did not designate a specific people or a particular geography, but rather all those who expect Science to keep a radical distance from Politics. All those people, no matter where they were born, who feel themselves pushed by time's arrow in such a way that *behind them* lies an archaic past unhappily combining Facts and Values, and *before them* lies a more or less radiant future in which the distinction between Facts and Values will finally be sharp and clear. The modern ideal type is the one who is heading— who was heading—from that past to that future by way of a "MODERNIZATION FRONT" whose advance could not be stopped. It was thanks to such a pioneering front, such a Frontier, that one could allow oneself to qualify as "irrational" everything that had to be torn away, and as "rational" everything toward which it was necessary to move in order to progress. Thus the Moderns were those who were freeing themselves of attachments to the past in order to advance toward FREEDOM. In short, who were heading from darkness into light—into ENLIGHTENMENT. If I used Science as my touchstone for defining this singular system of coordinates, it was because any disruption in the way the sciences were conceived threatened the entire apparatus of modernization. If people began to mix up Facts and Values again, time's arrow was going to interrupt its flight, hesitate, twist itself around in all directions, and look like a plate of spaghetti—or rather a nest of vipers.

One didn't have to be a genius, twenty years ago, to feel that modernization was going to end, since it was becoming harder and harder by the day—indeed, by the minute—to distinguish facts from values because of the increased intermixing of humans and nonhumans. At the time, I offered a number of examples, referring to the multiplication of "hybrids" between science and society. For more than twenty years, scientific and technological controversies have proliferated in number and scope, eventually reaching the climate itself. Since geologists are beginning to use the term "ANTHROPOCENE" to designate the era of Earth's history that follows the Holocene, this will be a convenient term to use from here on to sum up the meaning of an era that extends from the scientific and industrial revolutions to the present day. If geologists themselves, rather stolid and serious types, see humanity as a force of the same amplitude as volcanoes or even of plate tectonics, one thing is now certain: we have no hope whatsoever—no more hope in the future than we had in the past— of seeing a definitive distinction between Science and Politics.

As a result, the touchstone that served to distinguish past from present, to sketch out the modernization front that was ready to encompass the planet by offering an identity to those who felt "modern," has lost all its efficacy. It is now before GAIA that we are summoned to appear: Gaia, the odd, doubly composite figure made up of science and mythology used by certain specialists to designate the Earth that surrounds us and that we surround, the Möbius strip of which we form both the inside and the outside, the truly global Globe that threatens us even as we threaten it. If I wanted to dramatize—perhaps overdramatize—the ambience of my investigative project, I would say that it seeks to register the aftershocks of the modernization front just as the confrontation with Gaia appears imminent.

It is as though the Moderns (I use the capitalized form to designate this population of variable geometry that is in search of itself) had up to now defined values that they had somehow sheltered in shaky institutions conceived on the fly in response to the demands of the modernization front while continuing to defer the question of how they themselves were going to last. They had a future, but they were not concerned with what was to come—or rather, what was coming. What is coming? What is it that is arriving unexpectedly, something they seem not to have anticipated? "Gaia," the "Anthropocene" era, the precise name hardly matters, something in any case that has deprived them forever of the fundamental distinction between Nature and Society by means of which they were establishing their system of coordinates, one step at a time. Starting from this event, everything has become more complicated for them. "Tomorrow," those who have stopped being resolutely modern murmur, "we're going to have to take into account even more entanglements involving beings that will conflate the order of Nature with the order of Society; tomorrow even more than yesterday we're going to feel ourselves bound by an even greater number of constraints imposed by ever more numerous and more diverse beings." From this point on, the past has an altered form, since it is no more archaic that what lies ahead. As for the future, it has been shattered to bits. We shall no longer be able to emancipate ourselves the way we could before. An entirely new situation: behind us, attachments; ahead of us, ever more attachments. Suspension of the "modernization front." End of emancipation as the only possible destiny. And what is worse: "we" no longer know *who* we are, nor of course *where* we are, we who had believed we were modern . . . End of modernization. End of story. Time to start over.

⊙ BY PROPOSING A DIFFERENT SYSTEM OF COORDINATES. Is there another system of coordinates that can replace the one we have lost, now that the modernist parenthesis is closing? This is the enterprise that I have been doggedly pursuing, alongside other endeavors, for a quarter of a century, and that I would like to share and extend through this book and its accompanying digital apparatus. I believe that it is actually possible to complement the starkly negative title *We Have Never Been Modern* with a *positive* version of the same assertion. If we have never been modern, then what has happened to us? What are we to inherit? Who have we been? Who are we going to become? With whom must we be connected? Where do we find ourselves situated from now on? These are all questions of historical and comparative anthropology that we cannot begin to approach without a thorough inquiry into the famous modernity that is in the process of shutting down.

Why do I believe I am capable of proposing such an inquiry and offering such an alternative? Simply because, by suspending the theme of modernity in order to characterize the adventure of the Moderns, I think I have localized the experience of a certain number of values that can be presented, I believe, in alternative versions.

I am convinced, for example, that the experience of objectivity did not seem to protect Science with a capital S very well because no one had ever really felt a need to defend it. As soon as objectivity is seriously challenged, as it was in the anecdote related above, it becomes desirable to describe the practice of researchers quite differently, offering scientists a different representation of themselves, one that would make it possible to regain trust at last in a profoundly redefined scientific institution. As we shall soon see, the work of redescription may be of value in that it may allow us to give more space to *other values* that are very commonly encountered but that did not necessarily find a comfortable slot for themselves within the framework offered by modernity: for example, politics, or religion, or law, values that the defense of Science in all its majesty had trampled along its way but which can now be deployed more readily. If it is a question of ecologizing and no longer of modernizing, it may become possible to bring a larger number of values into cohabitation within a somewhat richer ecosystem.

In all that follows, I am thus going to offer readers a double dissociation: first, I shall try to tease out an EXPERIENCE proper to each value from the account traditionally provided for it; next, I shall take it upon myself to give this experience an entirely provisional alternative formulation that I shall put on the bargaining table and submit to critique. Why proceed this way? Because it seems to me that an experience, provided that it is pursued with care, can be *shared*, whereas the alternative formulation that I offer of that experience cannot be—in any case, not at the outset.

The study of scientific practices that I have been carrying out for so long can serve as an example: I have rarely heard critiques of the *descriptions* that "science studies" has given of scientific networks (on the contrary, the veracity of these descriptions has always been recognized, as if, after Harvey, we had discovered the veins and arteries of the scholarly bloodstream). And yet the alternative *versions* my colleagues and I have proposed in order to account for the fabrication of objectivity have been hotly contested by some of the very researchers whose values we were trying to make comprehensible, at last, to others. The very words "network" and "fabrication" are sometimes enough to shock our interlocutors, which only shows how badly we have gone about it. What poor diplomats we have been!

Since the goal is take an inventory of the Moderns in order to know what we are to inherit, it would be tragic to confuse three ingredients in particular: the accounts the Moderns have invented in the course of their various struggles; the values they have held to during this same history, through experiences that can be shared; finally, my own formulation, overly particular or overly polemical, of this same experience.

This is why the apparatus I want to offer readers is presented in two sequences; the report of an inquiry to which they are welcome to add, or subtract, whatever seems to them to correspond, or not, to what is given in the experience; second, a procedure that really has to be called negotiation by means of which the author and some readers—who will have become COINVESTIGATORS—can envisage participating in a shared reformulation of these same experiences.

Such is my attempt. To put it bluntly, I think I am right in the detection of the experiences that I am going to try to bring to the reader's attention; I am sure that I am often wrong in the expression that I have proposed for each of them while seeking to offer an alternative to modernism. And if I am wrong, it is by construction, since a diplomat cannot be right *all alone*. He can only offer a formula for peace and send it out to be picked apart both by those whom he represents and by those who are on the receiving end. The object of this book is thus to serve as the report on an inquiry and also, perhaps, as a preliminary step in a peacemaking process.

The strangeness of this diplomatic situation does not stem solely from the procedure chosen (a digital environment!) or from the nature of those that I claim to represent (obviously, without the slightest mandate!); it also stems from the conflict itself, for which the intervention of diplomats is finally required. For modernization never takes the form of a war that could appropriately be brought to an end. What conflict, then, has so exhausted the parties that they now dream of holding peace talks? A strange conflict in which none of the protagonists can be defined: neither the aforementioned Moderns, since they "have never been" modern, nor of course the "others," since they were "others" only by comparison with a modernity maintained in vagueness. The diplomatic scene—a perfectly imaginary one, I confess—that I seek to set forth through this inquiry is one that would reunite the aforementioned Moderns with the aforementioned "others" as Gaia approaches. The situation that I would like to sketch out is one in which the Moderns *present themselves once again* to the rest of the world, but this time finally knowing, for real, what they value!

WHICH LEADS US TO DEFINE AN IMAGINARY DIPLOMATIC SCENE: ⊙

This may appear astonishing and even somewhat backward-looking, but it is in the Moderns, in "Occidentals," yes even in "Europeans," that we are going to have to take an interest, *at last*, in this inquiry. Not to worry: there is no narcissism here, no nostalgic search for identity. It is just that for a long time ANTHROPOLOGY has taken it for granted that it has had to set up a contrast between "the other CULTURES" and a process of modernization that was European, or in any case Western, in origin, a process that no one had tried to specify further, and that anthropologists did not see it as their job to study. Nevertheless, it was always in relation to that standard, defined by default, that the irrationality, or, more charitably, the alternative rationalities manifested by other cultures were judged. As respectful as anthropologists wanted to be of "the savage mind," it was from the starting point of "cultivated" or "learned" minds that they had to conceive of the difference. This is the ideal of modernity that has been used to identify the "cultural," "archaic," or "reactionary" elements

⊙ IN THE NAME OF WHOM TO NEGOTIATE ⊙

with which "modernity" itself has remained imbued. Moreover, it is in relation to this modernization front that some are still trying to penetrate the secrets of the future (are cultures going to converge, diverge, enter into conflict? and so on). The result of an approach like this is that we still lack an anthropology of the Moderns.

The fact is that these populations with elastic borders have always posed a real problem of description for themselves and for others, since, even if they have never been modern, they have certainly *thought* of themselves as such. The Moderns have never been modern, but they have *believed* they were modern, and this belief too is crucial, for it has made them act in a thousand contradictory ways that we must learn to sort out—while we may have to abandon, for our part, the very notion of "BELIEF." In other words, there is an opacity proper to the Moderns with which comparative anthropology will have to reckon sooner or later. This opacity is all the more enigmatic in that it contrasts with the Moderns' claim to practice self-awareness, self-analysis, critique, lucidity—and also with the odd idea that the "other cultures" would be the ones that are opaque and in great need of ethnography. It is to combat this opacity—or this false transparency, perhaps—that I have needed to develop a special protocol for inquiry. As we shall see, the anthropology of the Moderns is in no way easier than that of the "others"—who, moreover, having ceased to be "the others," have thereby probably become easier to analyze than the Moderns, who remain as opaque as ever!

I am going to proceed as if the Moderns had discovered during their history—most often as borrowings from other civilizations, moreover—a number of values that they hold dear and that constitute, as it were, their very self-definition, even though they have never had an entirely firm grasp on these values. Because of this lack of assurance, I am going to proceed as if they have not managed, this time on the theoretical level, to find a way to *respect* their own values—and still less, then, to respect those of others. In other words, in this hypothesis the Moderns did not invest as much energy in the overall design of their values as they put into discovering them in practice, one after another. It is not simply that the Moderns are two-faced, like "white men with forked tongues." It is rather that, encumbered by their treasures, they have never had the occasion to specify clearly what it is that they really hold dear. A matter of excessive gourmandise, or greed? Who knows! In any case, thanks to this sort of *charitable fiction*, I shall be able to extend my ethnographic investigations even as I acknowledge the immense gap between the official and the unofficial versions, without seeking to criticize them for all that. It is in this sense that I can claim to be offering a positive rather than a purely negative version of modernization—at the risk of appearing ever so slightly positivist and of being accused, basically, of connivance with my subject (but after all, isn't connivance another name for the attribute of ethnologists that we call empathy?).

This fiction of an embarrassment of riches, as will soon become clear, is not at all aimed at rendering the Moderns innocent by washing them clean of all their excesses. Its goal is above all to propose at last a somewhat realistic description of what could be called the modern adventure, while no longer confusing it all—for better and for worse—with the advent of a modernization front. If it is really a matter of war, then let war be declared; in particular, let its objectives be defined so that we can finally figure out how to end it. This descriptive project is useful in itself, since if it were to be successfully completed, it would allow us to provide comparative anthropology with a standard that would no longer be a fantasy (as was the advent of Reason) and that would not be, either, a negative or simply critical version of the goal of modernization. But it is useful in another way, too: if we were finally to learn what "we" Moderns have really been, we could renegotiate that "we" from top to bottom—and thus also renegotiate what we might become with the "others," as we face the new horizon of Gaia.

⊙ AND WITH WHOM TO NEGOTIATE?

The fact is that comparative anthropology remains hanging in the air as long as we do not have access to an alternative version of the point of comparison that always remains in the background: the "WEST" (a frightfully vague term to which we ought to be able to give a precise meaning at last). As long as we have not taken the inventory of the Moderns' legacy, we cannot undertake an authentic comparative anthropology, nor can we—and this is perhaps even more serious—come up with any long-term hypotheses about the future of their relations with the rest of the world. A "rest of the world," a "remainder" whose definition obviously varies depending on whether "we" have been modern or something else entirely; to begin with (but this was already becoming clear), this world no longer remains by any means a "remnant"!

After the terrifying scenes of empires in which all the other populations watched with alarm the downfall of the brilliant madmen who were overturning their own values along with those of others in an indescribable disorder while chopping up the planet in a sort of juvenile fury, their eyes fixed on the past as if they were fleeing backward away from some dreadful monster before covering everything over with the cloak of an inevitable modernization and the irreversible reign of Reason, I would like to proceed as if the madmen could calm down, go home, get a grip, chill out, and then come back to present themselves, not in order to apologize (who is weak enough to demand apologies?) but to explain what they were looking for, and to discover at last, on their own, what they were ultimately holding onto. It is not totally fanciful to imagine that the "others" might then take an interest, in part, in the "Western" project—at last.

This *recalling* of modernity, in all senses of the word "recall" (including the meaning it has in the automobile industry)—is more useful for preparing "Occidentals" for their future than their strange claim to be extending the modernization front to the antipodes. It is entirely possible—indeed, it is already largely the case—that the West (Europe, at least, unquestionably) is *finally* in a situation of relative weakness. No more question of hubris; no more question of repentance. It is high time to begin to spell out not only what happened in the name of "modernity" in the past (a patrimonial interest, as it were) but also and especially what this word will be able to mean in the near future. When the incontrovertible authority of force is lacking, when it has become impossible to "steal history," might the diplomats' moment finally be at hand?

This inquiry into values, as they have been extracted, cherished, misunderstood, mistreated, patched back together, and appropriated by the West as its patrimony, seeks to contribute to the planetary negotiation that we are going to have to undertake in preparation for the times when we shall no longer be in a position of strength and when the others will be the ones purporting to "modernize"—but in the old way and, as it were, without us! We shall claim, even so, that we have something to say about our values—and perhaps also about those of the others (but with none of the privileges of the old European history). In other words, "Occidentals" will have to be made present in a completely different way, first to themselves, and then to the others. To borrow the remarkable expression used in chancelleries, it is a matter of making "diplomatic representations" in order to renegotiate the new frontiers of self and other.

But if there is to be diplomacy there have to be diplomats, that is, people capable—unlike those who dispatch them—of discovering, finally, what their principals really cherish—at the price of some sacrifices that they learn to detect during often interminable negotiations. A delicate exploration that has to proceed by feeling one's way in the dark, efforts that accusations of treachery must not interrupt and that will occupy a privileged place in this inquiry.

The two questions that justify this work, then, are these: can we finally offer a realistic description of the modern adventure, one that will allow us to give comparative anthropology a more credible basis for comparison? Can this comparative anthropology serve as a preliminary to the planetary negotiation that is already under way over the future of the values that the notion of modernization had at once revealed and compromised? We shall be told that it is too late to plunge into such an exploration. Too late because of the crimes committed; too late because Gaia is irrupting too urgently. "Too little, too late." I believe, on the contrary, that it is because of the urgency that we must begin to reflect slowly.

How are we going to proceed? To use expressions that would be more suitable to an analytic philosopher, let us say that the inquiry will allow us to clarify, fairly systematically, for a large number of unexpected subjects, CATEGORY MISTAKES bearing on what I have called the various MODES OF EXISTENCE.

THE INQUIRY AT FIRST RESEMBLES THE ONE INVOLVING SPEECH ACTS ⊙

By comparing conflicts of values in pairs—scientific versus religious, for example, or legal versus political, or scientific versus fictional, and so on—we shall observe very quickly that a large proportion of the tensions (tensions that explain in part the opacity I mentioned above) stem from the fact that the *veracity* of one mode is judged in terms of the conditions of veridiction of a different mode. We shall have to spend a good deal of time on this essential issue; it clearly presupposes that we accept the *pluralism* of modes and thus the *plurality* of keys by means of which their truth or falsity is judged.

But the difficulty is not so great, after all, if we turn to the work done by J. L. AUSTIN and his successors on "speech acts." The notions of FELICITY AND INFELICITY CONDITIONS, now solidly established in our intellectual traditions, make it possible to contrast very different types of veridiction without reducing them to a single model. The difficulty will come later, when we shall need to go beyond the linguistic or language-bound version of the inquiry to make these modes more substantial realities. But in the meantime, the heart of the investigation will involve an effort to *clarify* assertions bearing on the truth or falsity of an experience. This is the only test that is worthwhile, it seems to me, for the reader: does the redescription of a mode of existence make it possible to clear up conflicts between values—conflicts that had previously given rise to more or less violent debates—or not? Thus the truth and falsity of distinct forms of experience will be our first concern. It turns out, though, that there are several types of truth and falsity, each dependent on very specific, practical, experiential conditions. Indeed, it can't be helped: there is more than one dwelling place in the Realm of Reason.

When I speak of several types, I am not making a relativist argument (in the sense given this term by the papacy) about the impossibility of reaching any truth whatsoever, but only an argument about the fact that there are incompatible felicity conditions that nevertheless allow us, each in its own way, to reach incontrovertible judgments (in practice, of course, they always generate controversy) on the truth and falsity (relative and not relativist) of what they are to judge. This is, for example, the case for LAW [LAW] (a topic on which we shall spend a good deal of time), which manages to persevere in its own system of truth and falsity, even though this value in no way resembles any of the ones that might be applied, with just the same taste for discrimination, to judgments said to be "scientific." And when we show how fragile these truths are, each in its own mode, the point will not be to stress their deficiencies, as skeptics do, but to invite attention to the institutions that would allow them to maintain themselves in existence a little longer (and it is here, as we have already seen, that the notion of trust in institutions comes to the fore).

Our project is thus in fact a rational project (if not rationalist) from start to finish, provided that we agree to define reason as what makes it possible to follow the various types of experience step-by-step, tracking down truth and falsity in each mode after determining the practical conditions that allow us to make such a judgment in each case. I have always thought that the metaphor of Occam's razor is misapplied when it is used to support a claim that one must empty the world of everything that is not rational; the metaphor is confused, it seems to me, with the metaphor of the Gordian knot, which Alexander sliced through with his sword rather than going to the trouble of untying it. I have always imagined for my part that the story of Occam's razor alluded to a little case made of precious wood like those once used by surgeons, in which a great many tools adapted to all the delicate operations of reason lie nestled in green felt compartments. Shouldn't even the most hardheaded rationalists rejoice that there are several types of instruments, as long as each one is well honed? Especially if this allows them to reconnect with the other cultures to prepare themselves for what lies ahead.

But why speak of an inquiry into modes of existence? It is because we have to ask ourselves why rationalism has not been able to define the adventure of modernization in which it has nevertheless, at least in theory, so clearly participated. To explain this failure of theory to grasp practices, we may settle for the charitable fiction proposed above, to be sure, but we shall find ourselves blocked very quickly when we have to invent a new system of coordinates to accommodate the various experiences that the inquiry is going to reveal. For language itself will be deficient here. The issue—and it is a philosophical rather than an anthropological issue—is that language has to be made capable of absorbing the pluralism of values. And this has to be done "for real," not merely in words. So there is no use hiding the fact that the question of modes of existence has to do with METAPHYSICS or, better, ONTOLOGY—regional matters, to be sure, since the question concerns only the Moderns and their peregrinations.

⊙ WHILE WE LEARN TO IDENTIFY DIFFERENT MODES OF EXISTENCE.

In fact, as will quickly become clear, to deploy the diversity of felicity conditions it would do no good to settle for saying that it is simply a matter of different "language games." Were we to do so, our generosity would actually be a cover for extreme stinginess, since it is to LANGUAGE, but still not to being, that we would be entrusting the task of accounting for diversity. Being would continue to be expressed in a single, unique way, or at least it would continue to be interrogated according to a single mode—or, to use the technical term, according to a single CATEGORY. Whatever anyone might do, there would still be only one mode of existence—even if "manners of speaking"—which are not very costly, from the standpoint of ordinary good sense—were allowed to proliferate.

"Keep talking: I'm interested!" wouldn't be too unfair a way of qualifying this curious mix of open- and closed-mindedness that has made it possible, in the West, to welcome the diversity of cultures. It is true that they interest us; but it is also true that these are "just manners of speaking." Through a somewhat perverse mental restriction, on the one hand we acknowledge the most extreme diversity among these representations, while on the other we deny them any access to reality. Relativism, in other words, never traffics in hard cash. All the weaknesses of the aborted dialogues about the diversity of cultures, the plurality of worlds, the future composition of a common world, the universals to be extended, can be explained by mental restrictions of this sort, by a bizarre mix of irenicism and condescension. In circles like this, no one pays the ontological price for open-mindedness. Different words, a single reality. Pluralism of representations, monism of being. And, consequently, no use for diplomacy, because every representative is convinced that at bottom the arbitration has already occurred, elsewhere, at a higher level; each party is convinced that there is an optimal distribution, an unchallengeable arbiter and thus, somewhere, a Game Master. In the final analysis, there is nothing to negotiate. Violence resumes under the benign appearance of the most accommodating reason. We haven't advanced an iota since the era of Divine Judgment: "Burn them all; the Real will recognize its own!"

To speak of different modes of existence and claim to be investigating these modes with a certain precision is thus to take a new look at the ancient division of labor between words and things, language and being, a division that depends necessarily on a history of philosophy that we shall have to confront, I am afraid, along with everything else. The goal will be to obtain *less* diversity in language—we shall have to pay in cash and not on credit—but *more* diversity in the beings admitted to existence—there is more than one category, or rather, the will to knowledge is not the only category that allows us to interrogate the diversity of being (we shall spend a great deal of time addressing this difficulty). Conditions of felicity or infelicity do not refer simply to manners of speaking, as in speech act theory, but also to modes of being that involve decisively, but differently in each case, one of the identifiable differences between what is true and what is false. What we say commits us much more extensively than we would like to think—enough to make us slow down and ponder before we speak.

Conversely, though, we may benefit from an ontological pluralism that will allow us to populate the cosmos in a somewhat richer way, and thus allow us to begin to compare worlds, to weigh them, on a more equitable basis. It should not be surprising if I speak throughout what follows about "the beings" of science, of technology, and so on. Basically, we have to go back to the old question of "what is X?" (what is science? what is the essence of technology?), but in the process we shall be discovering new beings whose properties are different in each case. What we shall lose in freedom of speech—words bear their weight of being—we shall regain through the power to enter into contact with types of entities that no longer had a place in theory and for which a suitable language will have to be found in each case. A perilous enterprise if ever there was one.

It would have been more reasonable, I admit, to limit the inquiry to its ethnographic dimension alone. But since it is a matter of finally bridging the abyss that separates what the Moderns say from what they do, I have not been able to see any way this inquiry into modes of existence could do without PHILOSOPHY. I am turning to philosophy, then, not in the vain hope of finding in the "foundations" what field study is unable to provide but, on the contrary, in the hope of forging a METALAN-GUAGE that will allow us finally to do justice, in theory, to the astounding inventions that the fields reveal at every step—among the Moderns too.

We shall see, moreover, that we cannot recover the notion of institution without addressing questions that might seem overly basic. If it is true that "there are more things in heaven and earth than in all of our philosophy," it is also true that, without philosophical exploration, we would not succeed in expressing very much about what is on earth, let alone in heaven. In any case, I have no choice: the Moderns are the people of Ideas; their dialect is philosophy. We shall have to concentrate first of all on their curious regional ontology if we want to have the slightest chance of confronting the "others"—the former others—and Gaia—the truly other Other.

THE GOAL IS, FIRST, TO ACCOMPANY A PEOPLE VACILLATING BETWEEN ECONOMY AND ECOLOGY.

At all events, we shall not cure the Moderns of their attachment to their cherished theme, the modernization front, if we do not offer them an alternate narrative made of the same stuff as the Master Narratives whose era is over—or so some have claimed, perhaps a bit too hastily. We have to fight trouble with trouble, counter a metaphysical machine with a bigger metaphysical machine. Diplomats, too, need a "narrative," as advocates of "storytelling" say in the American press. Why would the Moderns be the only ones who have no right to a dwelling place, a habitat, city planning? After all, they have cities that are often quite beautiful; they are city-dwellers, citizens, they call themselves (and are sometimes called) "civilized." Why would we not have the right to propose to them a form of *habitation* that is more comfortable and convenient and that takes into account both their past and their future—a more *sustainable* habitat, in a way? Why would they not be at ease there? Why would they wander in the permanent utopia that has for so long made them beings without hearth or home—and has driven them for that very reason to inflict fire and bloodshed on the planet?

The hypothesis is ludicrous, as I am very well aware, but it is no more senseless than the project of an architect who offers his clients a house with a new form, a new arrangement of rooms and functions; or, better still, an urbanist imagining a truly new city by redistributing forms and functions: why would we not put factories here, run subways there, ban cars in these zones? And so on. It would not be a matter of diplomacy—for the others—but of convenience—for oneself. "And if you were to put science over there, while relocating politics over here, at the same time that you run the law underneath and move fiction to this spot, wouldn't you be more at ease? Wouldn't you have, as people used to say, more *conveniences?*" In other words, why not transform this whole business of recalling modernity into a grand question of *design?*

Such castles in the air have to be judged by the only test worth its salt: would the potential inhabitant feel more comfortable there? Is it more habitable? And this is the test I was speaking of earlier under the name of diplomacy and under the still obscure notion of institution. At bottom, that is what this is all about: can one institute the Moderns in habitats that are, if not stable, at least sustainable and reasonable? More simply, more radically: can one offer them a dwelling place at long last? After all these years of wandering in the desert, do they have hope of reaching not the Promised Land but Earth itself, quite simply, the only one they have, at once underfoot and all around them, the aptly named Gaia?

The question is not as idle as one might think, if we remember that the adventure of these last three centuries can be summed up by the story—yes, I admit it, the Master Narrative—of a double displacement: from *economy* to *ecology.* Two forms of familiar habitats, *oikos:* we know that the first is uninhabitable and the second not yet ready for us! The whole world has been forced to move into "The Economy," which we now know is only a utopia—or rather a dystopia, something like the opium of the people. We are now being asked to move suddenly with our weapons and our baggage into the new dwelling place called "ecology," which was sold to us as being more habitable and more sustainable but which for the moment has no more form or substance than The Economy, which we are in such a hurry to leave behind.

It's hardly surprising, then, that the modernizers are gloomy. They are refugees twice over, twice driven out of artificial paradises, and they don't know where to put the dwelling places they bought on the installment plan! To put it bluntly, they don't know where to settle. They are travelers in transit, displaced masses currently wandering between the dystopia of The Economy and the utopia of ecology, in need of an urbanist who can design a shelter for them, show them drawings of a temporary living space. In the face of this generalized housing crisis, modesty would be treason. Anyway, philosophy does not have a tradition of being reasonable; like Gabriel Tarde, it always shouts: "*Hypotheses fingo!*"

PART 1

HOW TO MAKE AN INQUIRY
INTO THE MODES OF EXISTENCE
OF THE MODERNS POSSIBLE

Chapter 1

DEFINING
THE OBJECT OF INQUIRY

....

An investigator goes off to do fieldwork among the Moderns ⊙ without respecting domain boundaries, thanks to the notion of actor-network, ⊙ which makes it possible to distinguish networks as result from networks as process.

The inquiry defines a first mode of existence, the network [NET], through a particular "pass," or passage.

But networks [NET] have a limitation: they do not qualify values.

Law offers a point of comparison through its own particular mode of displacement.

There is thus a definition of "boundary" that does not depend on the notions of domain or network.

The mode of extension of objective knowledge can be compared with other types of passes.

Thus any situation can be defined through a grasp of the [NET] type plus a particular relation between continuities and discontinuities.

Thanks to a third type of "pass," the religious type, the investigator sees why values are difficult to detect ⊙ because of their quite particular ties to institutions, ⊙ and this will oblige her to take into account a history of values and their interferences.

L ET US IMAGINE AN ANTHROPOLOGIST WHO HAS COME UP WITH THE IDEA OF RECONSTI-TUTING THE VALUE SYSTEM OF "WESTERN societies"—a terrain whose precise boundaries matter little at this stage. Let us imagine as well that, informed by reading good recent authors, she has overcome the temptation to limit her studies among the Moderns to the aspects that superficially resemble the classical terrains of anthropology—various folklores, village festivals, ancient patrimonies, assorted archaic features. She has clearly understood that, in order to be a faithful imitator of the anthropologists who study distant societies, she has to focus on the very heart of modern institutions—science, the economy, politics, law, and so on—rather than on the margins, the vestiges, the remnants, and that she has to treat them all at the same time, as a single interconnected set.

Let us also imagine—and this is more challenging, or at least the case is less frequently encountered—that she knows how to resist OCCIDENTALISM, a form of EXOTICISM applied to what is close at hand, which consists in believing what the West says about itself, whether in praise or criticism. She has already understood that modernism's accounts of itself may have no relation to what has actually happened to it. In short, she is a true anthropologist: she knows that only a prolonged, in-depth analysis of COURSES OF ACTION can allow her to discover the real value system of the informants among whom she lives, who have agreed to welcome her, and who account for this system in terms to which she must avoid giving too much weight. This much is obvious: it is the most ordinary ethnographic method imaginable.

If the question of where to begin nevertheless strikes her as quite complicated, it is because the Moderns present themselves to her in the form of DOMAINS, interrelated, to be sure, but nevertheless distinct: Law, Science, Politics, Religion, The Economy, and so on; and these, she is told, must by no means be confused with one another. She is strongly advised, moreover, to restrict herself to a single domain "without seeking to take in everything all at once." A metaphor often used in her presence involves geographical maps, with territories circumscribed by borders and marked in contrasting colors. When one is "in Science," she is assured, one is not "in Politics," and when one is "in Politics," one is not "in Law," and so forth.

Although her informants are obviously attached to these distinctions, she comes to understand very quickly (a few weeks spent doing fieldwork, or even just reading newspapers, will have sufficed to convince her) that with these stories about domains she is being taken for a ride. She sees clearly, for example, that the so-called domain of "Science" is shot through with elements that seem to belong rather to Politics, whereas the latter domain is full of elements that come from Law, which is itself largely composed of visitors or defectors from The Economy, and so on. It quickly becomes apparent to her that not everything in Science is scientific, not everything is juridical in Law, not everything is economic in The Economy ... In short, she sees that she will not be able to orient her research according to the Moderns' domains.

How is she to find other reference points? We cannot imagine her as naïve enough to expect to find an institution wholly made up of the value in question, as if everything in Religion would be "religious," everything in Science would be "scientific," and everything in Law would pertain to "law," and so on. But we may suppose that she is intelligent enough to resist the temptation to be critical or even cynical: she is not going to waste her time being shocked that there are political "dimensions" or "aspects" in Science, or economic dimensions in Law, or legal dimensions in Religion. No, she quite calmly reaches the conclusion that the notion of distinct domains separated by homogeneous borders does not make much sense; she sees that she has to leave cartographic metaphors aside and that, if she still nurtures the hope of identifying her interlocutors' value system, she will need a very different investigative tool, one that takes into account the fact that a border indicates less a dividing line between two homogeneous sets than an intensification of crossborder traffic between foreign elements.

⊙ WITHOUT RESPECTING
DOMAIN BOUNDARIES,
THANKS TO THE NOTION
OF ACTOR-NETWORK, ⊙

Let us suppose that, by chance, she comes across the notion of NETWORK—and even, the hypothesis is not so absurd, that of ACTOR-NETWORK. Instead of wondering, for example, if Science is a domain distinct from Politics or The Economy or Religion, the investigator will be content to start with some arbitrary sequence of practices. For example, she goes into a laboratory: there she finds white lab coats, glass test tubes, microbe cultures, articles with footnotes—everything indicates that she is really "in Science." But then, with a certain obstinacy, she begins to note the origins of the successive ingredients that her informants need in order to carry out their work. Proceeding this way, she very quickly reconstitutes a list of ingredients characterized by the fact (in contradiction with the notion of domain) that they contain ever more heterogeneous elements. In a single day, she may have noted visits by a lawyer who has come to deal with patents, a pastor who has come to discuss ethical issues, a technician who has come to repair a new microscope, an elected official who has come to talk about voting on a subsidy, a "business angel" who wants to discuss the launching of a new start-up, an industrialist concerned about perfecting a new fermenting agent, and so on. Since her informants assure her that all these actors are necessary for the success of the laboratory, instead of seeking to identify domain *boundaries*, which are constantly challenged by innumerable erasures, nothing prevents her any longer from following the *connections* of a given element, it hardly matters which one, and finding out where it leads.

It must be acknowledged that the discovery of the notion of network, whose topology is so different from that of distinct domains, gives her great satisfaction, at least at first. Especially because these connections can all be followed by starting with different segments. If she chooses to use a patent as her vehicle, for example, she will go off and visit in turn a laboratory, a lawyer's office, a board of trustees, a bank, a courthouse, and so on. But a different vehicle will lead her to visit other types of practices that are just as heterogeneous, following a different order on each occasion. If she has a taste for generalizing, she may thus conclude that there is no such thing as the domain of Science, or Law, or Religion, or The Economy, but that there are indeed networks that *associate*—according to segments that are always new, and that only empirical investigation can discover— elements of practice that are borrowed from all the old domains and redistributed in a different way each time.

Whereas the notion of domain obliged her to stay in one place while watching everything else move around incomprehensibly, the notion of network gives her the same freedom of movement as those whose actions she wants to follow. To avoid

⊙ WHICH MAKES IT POSSIBLE TO DISTINGUISH NETWORKS AS RESULT FROM NETWORKS AS PROCESS.

misunderstandings, let us specify that, for this investigator, a network is not only a technological arrangement such as, for example, a network for rail transport, water supply, sewers, or cell phones. The advantage of the term, despite all the criticisms to which it has been subjected, is that it can easily be represented in material terms (we speak of sewage networks, cable networks, spy networks); that it draws attention to flows without any confusion between what is being displaced and what makes the displacement possible (an oil pipeline is no more made "of" gasoline than the Internet is made "of" e-mails); and, finally, that it establishes such a powerful constraint of continuity that a minor interruption can be enough to cause a breakdown (a leak in an oil pipeline forces the operator to shut the valves; a three-meter displacement in a WiFi zone results in a lost connection: there is no longer any "network coverage").

And yet, even if the word draws from its origins the welcome connotations of technology, materiality, and cost (without forgetting that a network must always monitor and maintain itself), the notion that interests our ethnologist is defined by a quite specific *double movement* that we must keep firmly in mind in everything that follows. The fact that information can circulate by means of a cell-phone network tells us nothing about the way the network has been put together so as to work, right now, without a hitch: when all the elements are in place and everything is working well, in the digital window of our cell phones what we can track is only the quality of a signal marked by a certain number of rising vertical bars (by convention, from one to five). The "network" in the usual sense of technological network is thus the belated result of the "network" in the sense that interests our investigator. The latter, were she to follow it, would oblige her not to verify the quality of a signal but rather to visit in turn the multitude of institutions, supervisory agencies, laboratories, mathematical models, antenna installers, standardization bureaus, protesters engaged in heated controversies over the harmfulness of the radio waves emitted: these have all *ultimately* contributed to the signal she gets on her phone. The distinction between the two senses of the word "network" would be the same if she were interested in railroads: following the tracks is not the same as investigating the French national railroad company. And it would still be the same if, taking the word more metaphorically, she wanted to investigate "networks of influence": here, too, what circulates when everything is in place cannot be confused with the setups that make circulation possible. If she still has doubts, she can rerun the video of *The Godfather*: how many crimes have to be committed before influence finally starts to circulate unchallenged? What exactly is the "offer that can't be refused"?

So under the word "network" we must be careful not to confuse what circulates *once everything is in place* with the *setups* involving the heterogeneous set of elements that allow circulation to occur. The natural gas that lets the Russians keep their empire going does circulate continuously from gas fields in the Caucasus to gas stoves in France, but it would be a big mistake to confuse the continuity of this circulation with what makes circulation possible in the first place. In other words, gas pipelines are not made "of gas" but rather of steel tubing, pumping stations, international treatises, Russian mafiosi, pylons anchored in the permafrost, frostbitten technicians, Ukrainian politicians. The first is a product; the second a real John Le Carré–style novel. Everyone notices this, moreover, when some geopolitical crisis interrupts gas deliveries. In the case of a crisis, or, more generally, in the case of a "network interruption" (we have all come to know this expression with the spread of cell phones), the two senses of the word "network" (what is in place and what puts it in place) converge. Everyone then sets out to explore *all over again* the set of elements that have to be knitted together if there is to be a "resumption of deliveries." Had you anticipated that link between the

Ukraine and cooking your risotto? No. But you are discovering it now. If this happens to you, you will perhaps notice with some surprise that for gas to get to your stove it had to *pass through* the moods of the Ukrainian president . . . Behind the concept of network, there is always that movement, and that surprise.

It is not hard to see why our ethnologist friend is interested in this single notion that can be used to cover *two distinct but complementary phenomena*: the exploratory work that makes it possible to recruit or to constitute a *discontinuous* series of heterogeneous elements on the one hand and on the other something that circulates in a *continuous* fashion, once all the elements are in place, when maintenance is assured and there is no crisis. By following the establishment of networks in the first sense, she will *also* be able to follow networks in the second sense. Just as, in physics, the resting state is an aspect of movement, a continuous, stabilized, and maintained network turns out to be a special case of a network of heterogeneous associations. It is thus indeed, as she had already suspected, the movement of association and the passage through unanticipated elements that could become her privileged tool, her Geiger counter, whose increasingly rapid clicks would signal the numerous surprises that she experiences in the discovery of the ingredients necessary to the extension of any practice whatsoever.

The notion of network can now be made a little more specific: it designates a *series of associations* revealed thanks to a *trial*—consisting in the surprises of the ethnographic investigation—that makes it possible to understand through what series of small *discontinuities* it is appropriate to *pass* in order to THE INQUIRY DEFINES A FIRST MODE OF EXISTENCE, THE NETWORK [NET], THROUGH A PARTICULAR "PASS," OR PASSAGE.

obtain a certain *continuity* of action. This principle of FREE ASSOCIATION— or, to put it more precisely, this principle of IRREDUCTION—that is found at the heart of the actor-network theory has demonstrated its fruitfulness by authorizing a number of observers to give themselves as much freedom of movement in their studies as their informants have. This is the principle that the observer-investigator counts on using at the outset.

To study the old domains designated by the Moderns, our anthropologist now has a tool, the network, defined by a particular way of *passing through*, going by way of, another element that comes as a surprise to her, at least at first. The continuity of the course of action—laboratory life, for example—would not be ensured without small interruptions, little hiatuses that the ethnographer must keep adding to her ever-growing list. Let us say that it involves a particular *pass* (as one speaks of a passing shot in basketball), which consists, for any entity whatsoever, in passing *by way of another* through the intermediary of a step, a leap, a threshold in the usual course of events.

It would be absurd to suppose that this pass would be experienced in the same way by an ethnologist who discovers the new ingredient from the outside, after the fact, as it is experienced by the laboratory director, who has discovered it earlier from the inside and in the heat of action. The surprises registered are only those of the observer: it is she, the ignorant one, who discovers as she goes along what her informants already know. All ethnologists are familiar with situations like this—and they know how indispensable such moments are to the investigation. But the notions of surprise and trial, if we shift them slightly in time, can also serve to define how the informants themselves have had to learn, in their turn, through what elements they too had to pass in order to prolong the existence of their projects. After all, the laboratory director whom our ethnologist had chosen to study at the outset had only discovered a few years earlier that he was going to have to "go through" the patent application process in order to bring his project to fruition. He "wasn't expecting that." He didn't know he would have to "pass over" that hurdle.

The notion of surprise can be understood all the more readily as common to the investigator and her informants in that they can each find themselves, in the face of the slightest crisis or controversy or breakdown, confronted with an unexpected new element that has to be added to the list, one that neither of them anticipated. For example, a disgruntled rival sues the researchers for "exceeding the patent"; they did not expect this; they have to go through lawyers or risk going under. And so the entire laboratory and its ethnologist are obliged to learn that, if they are to continue to function, a new element will have to be added to the list of things necessary for existence. Before their eyes the network is being enriched, becoming more complicated or at least more extensive.

From here on, this first mode of exploration of the entities required for the existence of another entity will be noted as [NET], for network. (Throughout this inquiry, to avoid inventing new terms, I have decided to retain the customary names of the traditional domains—Law, Religion, Science, and so on; however, when I want to give them a precise technical sense I use a three-letter code. A complete list can be found on pp. 488–489.)

Although our anthropologist is rather proud of her discovery, her enthusiasm is tempered a bit by the fact that, while following the threads of the networks, she notices that she has lost in speci- ficity what she has gained in freedom of movement. It is quite true that, thanks to the networks defined in this way, she really can wander around everywhere, using whatever vehicle she chooses, without regard to the domain boundaries that her informants want to impose on her in theory but which they cross in practice just as casually as she does. And yet, to her great confusion, as she studies segments from Law, Science, The Economy, or Religion she begins to feel that she is saying almost *the same thing* about all of them: namely, that they are "composed in a heteroge- neous fashion of unexpected elements revealed by the investigation." To be sure, she is indeed moving, like her informants, from one surprise to another, but, somewhat to her surprise, this stops being surprising, in a way, as each element becomes surprising *in the same way*.

BUT NETWORKS [NET] HAVE A LIMITATION: THEY DO NOT QUALIFY VALUES.

Now, she has a strong feeling that her informants, even when they agree to follow her in listing the truly stupefying diversity of the entities that they have to mobilize to do their work, continue in spite of every- thing (is it a matter of bad faith? false consciousness? illusion?) to assert calmly that they are indeed in the process of sometimes doing law, some- times science, sometimes religion, and so on. If the notion of domain has no meaning (she prefers not to reopen this question), everything happens as if there were indeed a boundary, a somehow *internal* limit, to the networks, one that the notion of network has not allowed her to capture, it seems. There are no borders between domains, and yet, she tells herself, there are real differences between domains.

Our friend finds herself facing an impasse here: either she retains the diversity of associations—but then she loses this second form of diversity (that of values, which "must not get mixed up"; her informants appear to hold strongly to this point)—or else she respects the diversity of values (Science isn't really the same thing as Politics; Law is not Religion; and so on), but then she has no way of collecting these contrasts except the notion of domain, and she knows perfectly well that the latter does not hold up under examination. What can she do to hold onto both forms of diversity, the first allowing her to remain attentive to the extreme heterogeneity of associations, the second allowing her, if only she has the right tool, to *determine* the type of value that seems to circulate in a particular network and to give it its specific tonality?

At first, the metaphor of the technological network continues to help her, since it allows her to differentiate the installation of a network from the result of that installation, namely, the continuous supply of a particular type of resource: a cell-phone signal, electricity, railroads, influence, gas, and so on. One could imagine, she tells herself, that the same thing holds true for the values whose system I am trying to reconstitute: to be sure, Law is no more made "of" law than a gas pipeline is made "of" gas, but still, the legal network, once it is in place (established through a multitude of nonlegal elements, she understands this now), really does ensure the supply "of law," as it were. Just as gas, electricity, influence, or telephone service can be qualified as networks without being confused with one another (even if they often share the same subterranean conduits—influence in particular!), why not use the same term to qualify "regular supplies" in science, law, religion, economics, and so on? These are networks that can be defined as series of associations of the [NET] type, and yet what circulates in them in a continuous and reliable fashion (provided that they are maintained with regularity, at great cost) does indeed supply values, services, distinct products.

With this compromise solution the anthropologist would get out of the impasse where her investigation has led her, and, still more important, she would stop uselessly shocking her informants—who have the patience to welcome her, to inform her, and to teach her their trade—by saying the same thing about all activities. She would know to doubt what she was being told—fields don't organize themselves into contiguous domains—and at the same time she could respect the diversity of the values to which her informants seem legitimately attached.

Unfortunately, it does not take her long to notice that this metaphor does not suffice to characterize the specific features of the networks she is seeking to define. If she questions gas producers, they will undoubtedly have her run through a staggering list of variables, all of which are necessary to the construction of a particular pipeline, and many of which are unforeseeable. But they will have no doubt whatsoever about the product to be transported: even if it has no smell, it is very easy to characterize by its chemical composition, its flow, and its price. More precisely, and this is what she finds most exasperating, she and her informants are capable, in any situation whatsoever, of detecting in a fraction of a second that a given phrase is "legal" whereas a different one is not, or that a certain attitude "has something scientific about it" while another one does not, or that this sentiment is "religious" and that one is impious. But when it comes to *qualifying* the nature of what is designated by these ever-so-precise judgments, her informants fall back on incoherent statements that they try to justify by inventing ideal institutions, so many castles in the air. While with the notion of network she has a tool that makes a positive empirical investigation possible, for each value her networks purport to convey she has only an ineffable "je ne sais quoi," as finely honed as it is ungraspable.

But we are dealing with a true anthropologist: she knows that she must not abandon either the empirical investigation or the certainty that those "somethings" through which values are defined are going to lead her someplace. In any case, she now has her investigation cut out for her: if the notion of domain is inadequate, so is that of network, in and of itself. So she is going to have to go a little further; she will have to begin again and again until she manages to determine the values that circulate in the networks. It is the conjunction of these two elements—she is now convinced of this—that will allow her to redefine the Moderns. However entangled the ties they establish between values, domains, institutions, and networks may be, this is where she must turn her attention. What will allow her to advance is the fortuitous realization—a real "eureka" moment for her—that, in her fieldwork, she has already encountered courses of action that have something in common with the movement of networks: they too define a PASS by introducing a discontinuity.

37

LAW OFFERS A POINT OF
COMPARISON THROUGH
ITS OWN PARTICULAR
MODE OF DISPLACEMENT.

To be sure, these are not the same passes or the
same discontinuities, but they nevertheless share
a family resemblance. The legal institution, as she
understands perfectly well, is not made up "of" or
"in" law. So be it. And yet during her investigation
our ethnologist has spotted a movement very specific to law that legal
experts designate, without attaching much importance to it, moreover,
as a MEANS. They say, every few minutes: "Is there a legal means...?"; "this
is not an adequate means"; "this means won't get us anywhere"; "this
means can take us in several different directions"; and so on. In the course
of her work, she has even followed the transformation of an ill-formed
demand made by indignant plaintiffs whose lawyer, first, and then the
judge, "extracted," as they put it, the legal "means" before passing judgment. Between the more or less inarticulate complaint, the request in
due form, the arguments of the parties, and the judgment, she is able
to trace a *trajectory* that resembles no other. To be sure, all the interconnected elements belong to different worlds, but the mode of connection,
for its part, is completely specific (we shall see this again in Chapter 13).

For any observer from outside the world of law, this movement is
discontinuous, since there is hardly any *resemblance*, at each step, between
steps n - 1, n, and n + 1, and yet the movement appears *continuous* to the legal
expert. This particular movement can even be said to define a legal expert
as someone who is capable—by dint of hard work—of grasping it in its
continuity *despite* and *owing to* the series of hiatuses that are so striking
seen from the outside. Someone who understands what the word "means"
means is a legal expert even if the word itself does not figure in specialized legal dictionaries, so obvious does it appear, precisely, "to a real legal
expert." And yet it can't be helped, the notion of *means* remains totally
obscure, marked by discontinuities whose logic completely escapes the
outside observer—and also the plaintiffs themselves.

THERE IS THUS A DEFINITION
OF "BOUNDARY" THAT
DOES NOT DEPEND
ON THE NOTIONS OF
DOMAIN OR NETWORK.

Thus there is indeed here, at least to the ethnologist's eyes, an internal *boundary* that does not trace
a border between the domain of law and what is
outside that domain (in the final ruling, the plaintiffs, the lawyers, the judges, the journalists, all
point out examples of "extralegal factors" to such
an extent that the border, if there were one, would be a real sieve) but

that nevertheless allows her to say that in the trajectory that traverses this whole medley of motifs, something specifically legal can be found. Our observer's enthusiasm is understandable: she considers that she has managed to define for law the equivalent of what a network transports without renouncing the heterogeneity, not to say the weirdness, of the elements required to maintain legal activity. No, indeed, Law is not made "of" law; but in the final analysis, when everything is in place and working well, a particular "fluid" that can be called legal circulates there, something that can be traced thanks to the term "means" but also "procedure." There is here, in fact, a *pass* particular to law; something that leaps from one step to the next in the work of procedure or in the extraction of means. In short, there is a particular type of connection, of association, that we are going to have to learn how to qualify.

If our investigator is so optimistic, it is because she soon notices that she can compare this pass, this type of transformation, with another one, just as astonishing, that she has already identified in studies bearing on the domain called "Science." It did not take her long to notice that in Science "not everything is scientific." She has even spent a fair amount of time drawing up a list, a truly dizzying one in this case, of all the ingredients required to maintain any scientific fact whatsoever (a list that nothing in the official theory of her informants allowed her to produce, moreover—here we have the contribution of the ethnography of laboratories in a nutshell). But by going into the most intimate details of knowledge production, she believes she has distinguished a trajectory characterized in its turn by a particular hiatus between elements so dissimilar that, without this trajectory, they would never have lined up in any kind of order. This trajectory, made of discontinuous leaps, is what allows a researcher to determine that, for example, between a yeast culture, a photograph, a table of figures, a diagram, an equation, a caption, a title, a summary, a paragraph, and an article, something is *maintained* despite the successive transformations, something that allows him access to a remote phenomenon, as if someone had set up, between the author and the phenomenon, a sort of bridge that others can cross in turn. This bridge is what researchers call "supplying the proof of the existence of a phenomenon."

THE MODE OF EXTENSION OF OBJECTIVE KNOWLEDGE CAN BE COMPARED WITH OTHER TYPES OF PASSES.

What really strikes our ethnographer is that, again, for someone looking at this course of action from the outside, each step in the proofs is marked by an abrupt *discontinuity*: an equation does not "resemble" a table of figures any more than the latter "resembles" the yeast cultures that were the point of departure. Even though, for an outsider, each step has "nothing to do" with the one that went before or the one that came after, for a person who is operating within this network, there is indeed *continuity*. Or rather, however strange the list of ingredients that make it possible to hold the scientific network together may be, a person who is capable of following this path by leaping from transformation to transformation in order to retain the similarity of an element that gives him a hold on another, remote until then—that person is a researcher. Had he been unable to do this, he would have proved nothing at all (we shall come back to this movement in Chapters 3 and 4). He would no more be a scientist than someone who has been unable to extract the means to proceed from a muddled dossier would be a lawyer. Two entirely different trades are nevertheless distinguished by the same capacity to grasp continuity through a series of discontinuities—and then to grasp *another* continuity by passing through *another* discontinuity. So now the ethnologist is in possession of a new pass, as discriminating in its genre as means in law, and yet totally distinct.

She is understandably excited: she believes she is capable both of defining the particular fluid that circulates within networks and of studying these networks without resorting to the notion of domain separated by borders. She believes she has discovered the philosopher's stone of the anthropology of the Moderns, a unique way to respect the values that the informants cherish above all, yet without having to believe for a moment in the distribution into domains that is supposed to justify these values.

Law is not made of "the legal," but "something legal" circulates in it nevertheless; Science is not made "of science," but "something scientific" circulates in it nevertheless. In the end, the situation is very much the same as the one that allows us to compare gas, electricity, or telephone networks, except that the definition of the values that circulate is not obvious in the least, and the theory espoused by those who work to extend these values does not permit their collection.

Our investigator now has a somewhat more robust instrument at her disposal: for any course of action whatsoever, she tries to identify the unexpected ingredients through which the actors have to pass in order to carry it out; this movement, consisting of a series of leaps (identified by the surprises encountered by the ethnologist and her informants), traces a network, noted [NET]. This heterogeneous network can in principle associate any element with any other. No border limits its extension. There is no rule for retracing its movements other than that of empirical investigation, and each case, each occasion, each moment, will be different. Every time someone confronts the observer with the existence of an impenetrable boundary, she will insist on treating the case *like a network of the* [NET] *type,* and she will define the list, specific in every instance, of the beings that will be said to have been associated, mobilized, enrolled, translated, in order to participate in the situation. There will be as many lists as there are situations.

The ESSENCE of a situation, as it were, will be, for a [NET], the list of the other beings through which it is necessary to pass so that this situation can endure, can be prolonged, maintained, or extended. *To trace a network* is thus always to reconstitute by a TRIAL (an investigation is a trial, but so is an innovation, and so is a crisis) the antecedents and the consequences, the precursors and the heirs, the ins and outs, as it were, of a being. Or, to put it more philosophically, the *others* through which one has to pass in order to become or remain the *same*—which presupposes, as we shall see later on, that no one can simply "remain the same," as it were, "without doing anything." To remain, one needs to *pass*—or at all events to "pass through"—something we shall call a TRANSLATION.

At the same time, our anthropologist has understood that another ingredient must be added to this definition of essence, one that makes it possible to go anywhere without fear: an ingredient that makes it possible to *determine,* in a given situation, the value that emanates from that situation. These trajectories have the same general form as those of a [NET]. They too are defined by leaps, discontinuities, hiatuses. But unlike networks, they create sequences that do not simply lead to heterogeneous lists of unexpected actors, but rather to a *type of continuity* specific to each instance. Our investigator has already identified at least two such types: means, in law, and proofs, in science (plus a third, networks— in the [NET] sense—through which one ultimately obtains continuity through the intermediary of discontinuities, unexpected associations, that are revealed by the course of the investigation).

The sense of a situation can thus be defined thanks to two types of data: first, the very general data of the [NET] type, which tell us nothing more than that we have to pass through surprising associations, and second, something that we have to add to these data in every case, something that will allow us to define the quality of the activity in question. The first type of data will allow our friend to explore the extraordinary diversity of the associations that define the adventure of the Moderns; the second will allow her to explore the diversity of the values they appear to cherish. The first list is indefinite, as are the entities that can be associated in a network; the second is finite, as are the values that the Moderns have learned to defend. At least we must hope that this is the case, so that the investigator will have a chance to bring her project to fruition...

THANKS TO A THIRD TYPE OF "PASS," THE RELIGIOUS TYPE, THE INVESTIGATOR SEES WHY VALUES ARE DIFFICULT TO DETECT ⊙

One more puzzle remains before she can really get started. Why is it so difficult to specify the values to which her informants seem so firmly attached? Why do the domains offer such feeble indications as to the nature of what they are thought to contain (they spill over into other domains in all directions and do not even define what they purport to cherish and protect)? In short, why is theory so far removed from practice among the Moderns? (Let us recall that our investigator has not found anything in the "theory of law" or the "theory of science" that can help her grasp these trajectories, which are so specific that it has taken years of fieldwork to make them explicit.) She cannot be unaware of this new problem, for she is not prepared to fall back on the overly simplistic idea that theory is only a veil discreetly thrown over practices. Theory must have a meaning, and the gap between theory and practice must play an important role. But what role?

Fortunately, our friend has benefited from a sound education, and she now notices (a new eureka moment) that this problem is not unrelated to an eminently classic question that she has also studied in another field, that of religion. Indeed, she recalls that the history of the Church (an institution if ever there was one!) has been traversed through and through by the question of how to be faithful to itself even as it has transformed itself from top to bottom—going all the way back to its origins.

The Church interests her all the more in that it begins by offering her *a third example* of a *pass*, but again completely distinct from the others (as we shall see in Chapter 11). Here again we find a hiatus, an agonizing one during which a priest, a bishop, a reformer, a devout practitioner, a hermit, wonders whether the innovation he believes necessary is a faithful inspiration or an impious betrayal. No institution has invested more energy (through preaching, councils, tribunals, polemics, sainthood, even crimes) than in this obstinate effort to detect the difference (never easy to formulate) between fidelity to the past—how to preserve the "treasure of faith"—and the imperious necessity of constantly innovating in order to succeed, that is, to endure and spread throughout the world.

A new pass, a new *continuity* obtained by the identification, always to be begun anew and always risky, of *discontinuities* that appear from the outside as so many non sequiturs—not to say pure inventions or, one might say, pious lies. If the legal and scientific passages gave our ignorant observer the impression of incomprehensible transformations, each in its own genre, those offered by the religious passage make her hair stand on end. And yet it is in fact this passage that the observer has to learn to compare with the others, since the transit itself, however dizzying it may be, entails a value indispensable to certain of her informants. To be faithful or unfaithful: for many of those whom she is addressing, this is a matter of life or death, of salvation or damnation.

However important this new example of a pass may be for her (it is understandable that her confidence in the success of her project has grown apace), what interests her here above all is the link between ⊙ BECAUSE OF THEIR QUITE PARTICULAR TIES TO INSTITUTIONS ⊙ this particular pass and the institution that accepts it. She is well aware that to study religion without taking this pass into account would make no sense whatsoever, since, from the preaching of a certain Ioshua of Palestine (to limit ourselves to the example of Christianity) through the

Reformation to the latest papal encyclicals, all the statements, all the rituals, all the theological elaborations bear on the touchstone that would make it possible to distinguish between fidelity and infidelity, tradition and treachery, renewal and schism. Yet at the same time it would make no sense to suppose that this shibboleth alone could explain the entire religious institution, as if Religion or even the Church consisted exclusively "in" the religious. If there is any doubt about this, our investigator has only to read a biography of Luther, a history of the papacy, or a study of the Modernist controversy (in the sense that Catholics give this late-nineteenth-century episode). Clearly, every time anyone has sought to use the fidelity/infidelity distinction as a touchstone, it has been in the midst of innumerable other considerations. All these instances of religious history would without any doubt be much better grasped by an approach of the actor-network ([NET]) type.

No, what interests our investigator about Church history is that in it the continual fluctuations in *the very relation* between these two questions—which she has still not managed to bring together—can be clearly seen. The multiple gaps between network, value, domain, and institution are not only *her* problem, as an uninformed observer, but the problem that her informants *themselves* confront constantly, explicitly, consciously. Whether it is a question of St. Paul's "invention" of Christianity, St. Francis's monastic renewal, Luther's Reform (I almost said St. Luther), each case features the relation between an aging, impotent institution and the necessary renewal that allows that institution to remain fundamentally faithful to its origins while undergoing huge transformations. And each case calls for judgment; in each case, the researcher has to make a fresh start, cast the fruitfulness of the renewal into doubt, go back to the beginning, reconsider and redistribute all the elements that had been renewed . . .

In other words, our ethnologist has a clear sense that there is here, in the history of the Church, an almost perfect model of the complexity of the relations between a value and the institution that harbors it: sometimes they coincide, sometimes not at all; sometimes everything has to be reformed, at the risk of a scandalous transformation; sometimes the reforms turn out to consist in dangerous innovations or even betrayals. And there is not a single actor who has not had to participate, during these last two millennia, in one of these judgments or another—from the secret of the confessional through the tribunals and the massacres to the scenography of the major Councils. But judgment is required on each occasion, according to a type of judgment specific to the situation.

It is entirely possible, our anthropologist tells herself, that the relation found here between value and institution is a unique case. Only in the religious domain—and perhaps only in the history of the Christian churches—would we find such a series of betrayals, inventions, reforms, new starts, elaborations, all concentrated and judged on the basis of the principal question of whether one is remaining faithful or not to the initial message. But her own idea (the origin of her eureka moment) is that the situation is perhaps the same for all the Moderns' institutions: in each case, perhaps it is necessary to imagine an original and specific relation between the history of the Moderns' values and the institutions to which these values give direction and which embrace and shelter them— and often betray them—in return.

Here is a problem that those who are busy bringing networks of gas, electricity, or cell phones and the like into cohabitation do not encounter: in each case they have a network at hand (in the sense that discontinuous associations have to be put into place). But for the case of the anthropology of the Moderns, we are going to have two types of variations to take into account: values on the one hand and the *fluctuation* of those values over time on the other. This history is all the more complex in that it will vary according to the type of values, and, to complicate things further, the history of *each value* will *interfere* with the fluctuations of all the others, somewhat the way prices do on the Stock Exchange.

⊕ AND THIS WILL OBLIGE HER TO TAKE INTO ACCOUNT A HISTORY OF VALUES AND THEIR INTERFERENCES.

What the anthropologist discovers with some anxiety is that the deployment of one value by a robust institution will modify the way all the others are going to be understood and expressed. One tiny mistake in the definition of the religious, and the sciences become incomprehensible, for example; one minuscule gap in what can be expected from law, conversely, and religion turns out to be crushed. Still, the advantage of this way of looking at things is that the investigator will be able to avoid treating the gap between theory and practice as a simple matter of "false consciousness," as a mere veil that would conceal reality and that her investigation should be content to remove. For each mode and for each epoch, and in relation to every other value and to every other institution, there will be a particular way of establishing the relation between "theory" and "practice."

Even if the task looks immense to her, our ethnographer can be rather proud of herself. She has defined her object of study; she has fleshed out her ordinary method with two additional elements specific to the modern fields: network analysis on the one hand, the detection of values on the other. Finally, she knows that she is going to have to take into account, for each subject, a fluctuating relation between the values that she will have identified and the institutions charged with harboring them. All these points are important for the way she conceives of her trade.

In fact—I should have pointed this out earlier—she is not one of those positivist ethnologists who imagine that they have to imitate the "hard sciences" and consider their object of study from a distance, as an entomologist would do with insects (the mythic ideal of research in the hard sciences, quite unfair to insects, moreover, as well as to entomologists). No, she knows that a contemporary anthropologist has to learn to talk *about* her subjects of study *to* her subjects of study. This is why she can hardly rely on the resources of critical distance. She is fairly satisfied that she knows how to describe practices through networks, even while remaining faithful to the values of her informants, yet without believing in domains and thus without believing in the reports that come from them, but also (the exercise is a balancing act, as we can see) without abandoning the idea of a possible reformulation of the link that values maintain with institutions. In other words, this is an anthropologist who is not afraid of running the risks of *diplomacy*. She knows how difficult it is to learn to *speak well to someone about something that really matters to that person.*

Chapter 2

COLLECTING DOCUMENTS FOR THE INQUIRY

··

The inquiry begins with the detection of category mistakes, ☉ not to be confused with first-degree mistakes; ☉ only second-degree mistakes matter.

A mode possesses its own particular type of veridiction, ☉ as we see by going back to the example of law.

True and false are thus expressed within a given mode and outside it ☉ provided that we first define the felicity and infelicity conditions of each mode ☉ and then the mode's interpretive key, or its preposition.

Then we shall be able to speak of each mode in its own tonality, ☉ as the etymology of "category" implies ☉ and as the contrast between the requirements of law and religion attests.

The inquiry connects understandings of the network type [NET] with understandings of the prepositional type [PRE] ☉ by defining crossings that form a Pivot Table.

A somewhat peculiar [NET · PRE] crossing, ☉ which raises a problem of compatibility with the actor-network theory.

Recapitulation of the conditions for the inquiry.

What is rational is what follows the threads of the various reasons.

I HOPE I HAVE SKETCHED IN THE OBJECT OF THIS INQUIRY WITH SUFFICIENT CLARITY: TO CONTINUE TO FOLLOW THE INDEFINITE MULTIPLICITY of networks while determining their distinctive ways of expanding. But how are we to register the documents that might allow us to give this research an empirical dimension and thereby enable the reader to distinguish the *experiences* identified in this way from the *accounts* of them that are usually offered, as well as from the accounts that will come along later in place of the first?

The source of these documents is perfectly ordinary, and the method we shall follow is quite elementary. To begin, we shall record the errors we make when we mistake one thing for another, after which our interlocutors correct us and we then have to correct—by means of sometimes painful tests—the INTERPRETIVE KEY that we shall have to apply in similar situations from that point on. It has become clear to me over the years that if we were to make ourselves capable of documenting such interpretive conflicts carefully enough, systematically enough, over long enough time periods, we would end up identifying privileged sites where contrasts between several keys are revealed. The raw material for this work is thus a vast chart in which CATEGORY MISTAKES are identified in pairs. The result is what I call a PIVOT TABLE; we shall soon see how to read its most important results.

The use of the term "category mistake" may cause some confusion in itself. The canonical example involves a foreign visitor going through the buildings of the Sorbonne, one after another; at the end of the day, he complains that he "hasn't seen the University of the Sorbonne." His request had been misunderstood: he wanted to see an institution, but he had been shown buildings . . . For he had sought in one entity an entirely different entity from what the first could show him. He should have been introduced to the rector, or the faculty assembly, or the institution's attorney. His interlocutors had misheard the key in which what he was requesting could be judged true or false, satisfactory or unsatisfactory. It is in this sense that I propose to take up the term "mistake" again—before specifying later on how "category" is to be understood.

We make mistakes so often that there would ⊙ NOT TO BE CONFUSED WITH be no value in trying to note all the ones we might FIRST-DEGREE MISTAKES; ⊙ possibly commit. The only ones of interest here are those that reveal what could be called *second-degree* mistakes and that bear on the *detection* of the causes of the mistake itself. To distinguish these clearly from the first-degree sort, let us take the exemplary case of mistakes of the senses, which have played a sometimes excessive role in the history of philosophy and in the very definition of EMPIRICISM.

Take the tower of a castle that, from a distance, looks more or less square to me. As I walk toward it, that first form shifts, becomes unstable. I then hesitate a bit: if it's worth the trouble, I change my path so as to get closer and "put my mind at rest," as it were, and I finally understand that it is *round*—that it has *proven* to be round. If I don't succeed in determining this on my own, I get my binoculars out of my bag and satisfy my curiosity through the intermediary of this modest instrument. If I cannot get close enough myself or get a sufficiently enlarged image through binoculars, I ask some better-informed local resident, I look at the geographical survey map, or I consult the guidebook I got at the tourist office. If I'm still in doubt, either about the knowledge of the local resident or the range of my binoculars or the testimony of my own eyes, I will shift my itinerary once again and go look at the cadastral survey, interrogate local experts, or consult other guidebooks.

It is easy to understand why mistakes of this type will not be of interest to us during this inquiry: they are all located, as it were, along *the same path*, that of rectified knowledge, and thus they all stem from the same interpretive key. However perplexing they may be, there is no ambiguity about the way one has to approach them in order to settle the matter little by little. As long as one is not asking for absolute and definitive certainty, one can always dispel mistakes of the senses by doing research, by perfecting instruments, by bringing together a suitable group of informants—in short, by providing time and means for the movement toward knowledge. There are no mistakes of the senses that cannot be corrected by a change in position, by recourse to some instrument, by the appeal to multiple forms of aid from other informants or some combination of these three resources. Such gropings may not succeed, but it is always in this spirit that one has to proceed if one wants to bring them to an end.

The project of listing category mistakes will not lead us, then, toward the quarrel between the epistemologists and the skeptics. Because they were trying to thwart the thoughtless demands of certain epistemologists who wanted to know with certainty but without investigation, without instrumentation, without allowing themselves the time to deviate from their path, without assembling an authorized group of reliable witnesses, the skeptics had to multiply their attacks in order to show that unfortunate human minds always found themselves caught in the net of mistakes. But if we decide not to doubt the possibility of knowledge that is rectified little by little, and if at the same time we are careful not to minimize the importance of the material and human means of which knowledge must avail itself, the skeptics' objections should not trouble us—no more than the climate skeptics' objections rattled the researcher whom I mentioned in the introduction. When Descartes wonders whether the people he sees in the streets are not robots wearing clothes, he could be asked: "But René, why don't you rely on your valet's testimony? You can't put an end to this sort of uncertainty without budging from your stove." If Descartes had to resign himself to radical doubt, it is only because he believed he had to face the monster— also hyperbolic—of a deceitful God. In the present inquiry, clearly, it is not a question either of responding to skeptics or of continuing to follow the sole path of rectified knowledge (even if we shall have to live through encounters with more than one Evil Genius).

⊕ ONLY SECOND-DEGREE MISTAKES MATTER. What will be of interest, however, are the cases in which we find ourselves confused about the very way in which the question of truth and falsity should be addressed. Not the absorption of mistakes *within* a given mode, but uncertainty about the mode itself. Not a mistake *of the senses [des sens]*, but a mistake *of direction [de sens]*. If there is no mistake of the senses that cannot be provisionally corrected by launching a research project, perfecting instruments, forming a group of reliable witnesses, in short, moving toward objectivized knowledge, there are, in contrast, mistakes of direction that often trip us up and that appear much more difficult to correct. The notebooks of our anthropologist doing fieldwork among the Moderns are full of such confusion. For, like the visitor to the Sorbonne, she often has the disturbing impression of mistaking one thing for another, making a mistake not in a given direction but *about* the very direction in which to turn her attention.

"This trial, for example, is ending at my adversary's expense, and yet I can't seem to 'get over' the wrong he did me. It's as though I can't manage to reconcile the closure of the debate (even though my lawyer assures me that there's no 'recourse' against the court's 'ruling') with the deep dissatisfaction I feel: the judge has spoken, and yet I find that, for me, nothing is really over.

"I continue to be indignant about my hierarchical superior's 'repeated lies,' without being able to spell out just what it would mean, in his case, to 'speak the truth,' since he heads a huge institution whose mechanisms are immersed, for him as well as for me, in a profound and perhaps necessary obscurity. Am I not being a bit too quick to call 'lies' what others, better informed, would call the 'arcana of power'?

"These questions bother me all the more in that the tedious moralizing discourse of this funeral ceremony strikes me as irritating, even scandalous, and in any case sickeningly sentimental; it seems to me to add a tissue of enormous lies in order to comment on texts stitched together with glosses: in the presence of this coffin holding a friend who is about to slide down a ramp into the crematorium furnace, I cannot accept the meaning they claim to give to the words 'eternal life.'

"They're still telling me that the emotion I feel, this rage that makes my heart rate go way up, lies within me and that I have to go through a lengthy analysis and plunge deep into myself, myself alone, to master it. And yet I can't keep from thinking (everyone is trying to keep me from thinking!) that I am threatened by forces that have the objectivity, the externality, the self-evidence, the power of a storm like the one that tore the tops off the trees in my garden, just a few days ago.

"For that matter, I'm not sure I really know what to make of that storm, either, for while I attribute it to the involuntary forces of nature, this magazine I'm reading calls it a consequence of global warming, which in turn is the result, or so they tell me, of recent activities of human industry—and the journalist, carried away by his own reasoning, blames the calamity that has just swept through our area on mankind's failure to act.

"I guess it's not surprising that I so often feel mistaken: what seems to me to come from the outside—my emotion just now—actually comes from inside, they say, but what I was attributing unhesitatingly to the great outdoors is now supposed to be the result of the collective will of the narrow human world. And here I fall into a new dilemma: I was getting at least a vague sense of how to control my emotions, but I haven't the slightest idea about what I could do to help overcome global warming. I have even less of a handle on collective action than on the forces of nature or the recesses of my psyche."

It is clear, in these notebook entries, that mistakes like the ones described are not at all of the same order as those of the senses. If the investigator gets lost rather easily, as we all do, it is because she fails to identify the key in which she has to grasp the direction, the trajectory, the movement of what is being asserted . . . The trajectory is the sense in which a course of action has to be grasped, the direction in which one should plunge. If we mistake a round tower for a square one, Venus for the evening star, a red dwarf for a galaxy, a simple windstorm for a tornado, a robot for a person, all these mistakes are found along the same path, which could be called "epistemological," since it concerns the path of objectivized knowledge. All these mistakes are of the same type and can be provisionally corrected by launching an investigation, perhaps a long and controversial one, more or less costly in instruments, but in any case amenable to (at least provisional) closure. As soon as knowledge is given its means, we can set aside both claims to absolute knowledge and the skeptical reactions intended to demolish such claims. But once this type of mistake has been dealt with, *all the other uncertainties* remain firmly in place. Once mistakes of the senses have been set aside, we are left with mistakes of direction.

Now, mistakes of the latter type are no longer located along the path of our inquiry like so many epistemological obstacles that would endanger the movement of knowledge alone; they do not interest either the skeptics or their adversaries. For these are mistakes that bear on *the interpretive key itself*. Every hiker knows that it is one thing to embark boldly on a well-marked path; it is quite another to decide which path to take at the outset in the face of signposts that are hard to interpret.

A MODE POSSESSES ITS OWN PARTICULAR TYPE OF VERIDICTION, ⊙

The documentary base that I have begun and that I would like to extend has to do with hesitations of this sort. The hypothesis I am adopting posits a *plurality* of sources of mistakes, each of which presents, but each time in a different order of practices, *as many obstacles* as those that one discovers at every step while pursuing the single question of objective knowledge. If the skeptics came to look something like deep thinkers, it is because they were reacting—and their reaction, although misdirected, was welcome—against the exaggeration of those who sought to make the question of knowledge the only one that mattered, because it allowed them to judge all the other modes (after it had been deprived of all means to succeed in doing so!). In other words, the skeptics, like their adversaries, unfortunately worked on *a single type of mistake*, and this is why their philosophies have so little to say about the conflicts of values in which the Moderns have found themselves entangled. As a result, by obsessively following only the obstacles to the acquisition of objective knowledge, they risk being mistaken about the very causes of the mistakes.

Still, it is not enough to recognize these moments of hesitation to know how to record them in the database. We also have to identify the principles of judgment to which each mode is going to appeal *explicitly* and *consciously* to decide what is true and what is false. This is the crucial point in the investigation, and it is probably on this subject that the endless battle between the skeptics and the rationalists has most distracted us from a descriptive task that is, however, essential. It turns out in fact that each mode defines, most often with astonishing precision, a mode of VERIDICTION that has *nothing to do* with the epistemological

definition of truth and falsity and that nevertheless warrants the quali-
fiers TRUE and FALSE.

⊙ AS WE SEE BY GOING BACK
TO THE EXAMPLE OF LAW.

This point will be easy to grasp if we return to
the example of law. When I require of a judgment
rendered in court that it also provide me with the
closure that would allow me to "get over it," as it were, I am asking the
impossible, since the type of closure provided by the legal apparatus
in no way aims—this becomes painfully clear to me—at offering reparations to my psyche; its goal is merely to connect texts with facts and
with other texts through the intermediary of opinions according to the
dizzying itinerary that is qualified, though not described, by terms we
have already encountered, MEANS or procedure. The same thing would
be true, moreover, if I expected that a ruling in favor of compensating
victims would "objectively" establish the truth of the matter. The judge,
if she is honest, will say that she has settled the "legal truth" but not the
"objective" truth in the case at hand. If she knows some Latin, as every
legal professional is prone to do, she will cite the Latin adage according to
which the judgment is *pro veritate habetur* ("taken as the truth"): neither
more nor less. If she is something of a philosopher, she will ask that
people "stop confusing" the legal requirement of truth with the scientific requirement of truth, means with proof—and even more with the
psychological requirement of intimate reparations—not to mention the
social requirement of fairness. "All that," she will say, "has to be carefully
differentiated." And she will be right.

This example, which I have deliberately chosen from the highly
distinctive world of law, proves the care with which we are capable of
identifying different orders of truth and signaling possible category
mistakes in advance: if you take the ruling of a tribunal as true, as fair,
as intimate, you are making a mistake: you are asking it for something it
will never be able to give; you are setting yourself up for wrenching disappointments; you risk being wrongly indignant if you are scandalized at a
result that the type of practices called "legal" could not fail to produce.
But your lawyer, for his part, will of course try to identify all the "legal
mistakes" that can strengthen your case in the judge's eyes. Mistakes of
this sort are committed all along the path of the law, and they are the ones
that the concerned parties, judges, attorneys, and commentators learn

to spot. So we can see, at least in this case, how to distinguish a "legal" mistake from a category mistake; the first is found along the chosen path, while the second produces hesitation about the path it would be appropriate to follow.

It is thus quite possible, at least for the case of scientific proof and that of legal means, to distinguish between two phenomena: on the one hand the detection of the difference between true and false *within* one of the two modes, and on the other the difference between *different uses* of true and false according to the mode chosen.

TRUE AND FALSE ARE THUS EXPRESSED WITHIN A GIVEN MODE AND OUTSIDE IT ⊙

The situation would be the same, moreover, if we were claiming to pass judgment on the faithfulness of a religious innovation simply by checking to see whether it follows tradition in every feature. On the contrary, it is because it does not "resemble" the tradition at all that it has a chance of being faithful, because instead of just reiterating the tradition the innovation takes it up again "in a wholly different way." This is what the audacious Jesuit fathers preached in order to modify the inflexible position of the Holy See during the Quarrel over Rites in the seventeenth century: to convert the Chinese, Church authorities would have to agree to change everything in the formulation of the rites in order to recapture the spirit of predication, doing for China what St. Paul, they said, had done for the Greeks. In their eyes, this was the only way to be faithful to their apostolic mission. What a catastrophe, if the Church were to take the letter for the spirit! Or rather, what a sin! The most serious, a sin against the spirit . . . In this new episode, the problem posed by the identification of category mistakes is clear: even if, for outsiders or uninformed people, judgments on the question of faithfulness or unfaithfulness seem to be trivial matters, proof of total indifference to the truth or, more charitably, as a "suspension" of all criteria of rational truth, there are few institutions more obsessed with the distinction between truth and falsity than the religious institution. And yet we also understand that it would be erroneous to claim to judge religious veridiction according to the entirely distinct modes of law or science.

Exactly the same thing can thus be said about the legal or religious spheres as about the epistemological sphere: just as all obstacles to knowledge have to be removed by the launching of an investigation that will make it possible to bring a provisional end to doubts while remaining throughout on a single path, so legal mistakes or religious infidelities, obstacles to the passage through law or to the expansion of the religious have to be addressed; but *within* the confines of a particular type of path that will leave, in its wake, as it were, "something juridical" or "something religious" rather than "objective knowledge." We would be making a category mistake, in this sense, if we continued to mix up the processes, networks, and trajectories that leave behind "something juridical," "something scientific," "something religious." Even if we could fill the database with judgments made all the way along a single path (involving, for example, all the methodological mistakes, all the legal mistakes, all the heresies, or all the impieties), it would still not give us any way to grasp the *plurality* of interpretive keys. If it is necessary to identify, for each type of practice, the rich vocabulary that it has managed to develop to distinguish truth from falsity in its own way, the crossings of all the modes will have to form the heart of our inquiry, for this is where the causes of mistakes are obviously the most important— and also the least well studied.

⊙ PROVIDED THAT WE FIRST DEFINE THE FELICITY AND INFELICITY CONDITIONS OF EACH MODE ⊙

To avoid getting lost, we shall need two distinct expressions to designate on the one hand the obstacles to be removed along a given path and on the other the initial choice through which we become attentive to one interpretive key rather than another.

To go back to the contrasts pointed out above, it is obvious when, in the case of law as well as in the case of knowledge or religion, either a procedure or an investigation or a predication has been launched. All three depend on a certain amount of equipment, a certain number of regroupings, expert opinions, instruments, judgments whose arrangement and use make it possible to identify in each order of truth what it means to "speak truths" and to "speak untruths." To qualify what these paths have in common—and to shift away from the metaphor of a hiking trip—I propose to use a term that is quite familiar in speech act theory, FELICITY AND INFELICITY CONDITIONS. On each path of veridiction, we will be able to ask that the conditions that must be met for someone to speak truths or untruths be specified according to its mode.

This takes care of the first term, the one that will define the armature making it possible, for example, to establish a procedure, in the case of law, or to launch a search for proofs, in the case of rectified knowledge, or an evangelizing mission, in the case of religion. But how are we to name what distinguishes one type of felicity condition from another? What I have called, in the examples above, "mistakes of law" or "mistakes of the senses" or "infidelities" will help us distinguish the veridiction proper to legal activity from the veridiction proper to the acquisition of scientific knowledge or religious piety. To designate these different trajectories, I have chosen the term PREPOSITION, using it in its most literal, grammatical sense, to mark a *position-taking* that comes *before* a proposition is stated, determining how the proposition is to be grasped and thus constituting its interpretive key.

⊙ AND THEN THE MODE'S INTERPRETIVE KEY, OR ITS PREPOSITION.

William James, from whom I am borrowing this expression, asserts that there exists in the world no domain of "with," "after," or "between" as there exists a domain of chairs, heat, microbes, doormats, or cats. And yet each of these prepositions plays a decisive role in the understanding of what is to follow, by offering the type of relation needed to grasp the experience of the world in question.

There is nothing magical about this distinction between felicity conditions and prepositions. If you find yourself in a bookstore and you browse through books identified in the front matter as "novels," "documents," "inquiries," "docufiction," "memoirs," or "essays," these notices play the role of prepositions. They don't amount to much, just one or two words compared to the thousands of words in the book that you may be about to buy, and yet they *engage* the rest of your reading in a decisive way since, on every page, you are going to *take* the words that the author puts before your eyes in a completely different tonality depending on whether you think that the book is a "made-up story," a "genuine document," an "essay," or a "report on an inquiry." Everyone can see that it would be a category mistake to read a "document" while believing all the way through that the book was a "novel," or vice versa. Like the definition of a literary genre, or like a key signature on a musical score, at the beginning an indication of this sort is nothing more than a signpost, but it will weigh on the entire course of your interpretation. To pursue the musical metaphor, if the score had not been transcribed into a different key everything would sound wrong. We can see from this example that to understand the meaning of *the proposition that is being addressed to you*, you have to have settled the initial question of its interpretive key, which will determine how you are *understand, translate, and transcribe what is to follow*.

But if the prepositions say nothing in themselves, doesn't everything depend on what follows? No, since if you take them away, you will understand nothing in the statement that is about to be made. But then, everything lies in the prepositions, and what follows is nothing but the deployment of their essence, which they would then contain "potentially"? No, this is not the case either, since prepositions engage you only in a certain way, from a certain angle, in a certain key, without saying anything, yet, about what is to come. Prepositions are neither the origin nor the source nor the principle nor the power, and yet they cannot be reduced, either, to the courses to be followed themselves. They are not the FOUNDATION of anything and yet everything depends on them. (We shall see later on how to charge the term "preposition," borrowed from linguistic theory, with a greater measure of reality.)

THEN WE SHALL BE ABLE TO SPEAK OF EACH MODE IN ITS OWN TONALITY, ☉ — The reader will have now understood the demand hidden behind that innocent claim *to speak well to someone about something that really matters to that person*: one has to seek to avoid *all* category mistakes rather than just one. The anthropologist of the Moderns is not merely trying to avoid the blunders she risks committing along the path of equipped and rectified knowledge alone; she is also trying to avoid the enormous mistake, the *mistake squared*, that would lead her to believe that there is only one way to judge truth and falsity—that of objective knowledge. She purports to be speaking while obeying all the felicity conditions of each mode, while expressing herself in as many languages as there are modes. In other words, she is hoping for another Pentecost miracle: everyone would understand in his or her own tongue and would judge truth and falsity according to his or her own felicity conditions. Fidelity to the field comes at this price.

☉ AS THE ETYMOLOGY OF "CATEGORY" IMPLIES ☉ — A project like this is not risk-free, especially since the more or less forgotten etymology of the word "category" is hardly reassuring. Let us recall that in "category" there is always the *agora* that was so essential to the Greeks. Before designating, rather banally, a type or division that the human mind, without specifying any interlocutor, carves at will out of the seamless fabric of the world's data, *kata-agorein* is first of all "how to talk about or against something or someone in public." Aristotle shifted the term away from its use in law—meaning "to accuse"—and made it a technical term, subject to endless commentary over the centuries, that was to subsume the ten ways, according to him, of predicating something about something. But let us return to the agora. Discovering the right category, speaking in the right tonality, choosing the right interpretative key, understanding properly what we are going to say, all this is to prepare ourselves to speak well about something to those concerned by that thing—in front of everyone, before a plenary assembly, and not in a single key.

Life would not be complicated if all we had to do were to avoid a single type of mistake and discriminate between speaking well and speaking badly in a single well-defined mode, or if it were enough to do this on our own, in the privacy of our own homes. The question of categories, what they are, how many there are, is thus at the outset a question of eloquence (how to speak well?), of metaphysics (how many ways of speaking are there?), and also of politics, or, better yet, of diplomacy (how are those to whom we are speaking going to react?). When we raise the big question of philosophy about categories once again—"In how many ways can one truly say something about something else?"—we are going to have to watch out for the reactions of our interlocutors. And here it is not enough to be right, to believe we are right. Anyone who claims to be speaking well about something to someone had better start quaking in his boots, because he is very likely to end up crushing one mode with another unless he also notes with extreme care, to avoid shocking his interlocutors, the relation that the various interpretive keys maintain with one another. The goal, of course, is always, as Whitehead insisted, above all, not to shock COMMON SENSE. (Common sense will always be opposed to *good sense* in what follows.)

⊙ AND AS THE CONTRAST BETWEEN THE REQUIREMENTS OF LAW AND RELIGION ATTESTS. If I have chosen the cases of law and religion as an introduction to the inquiry, it is because they are respected in contrasting ways in the public square, where the fate of categories is played out. Law benefits from an institution that is so strong, so ancient, so differentiated, that right up to our day it has resisted being confused with other forms of truth, in particular with the search for objective knowledge. It will be futile, in your rage as someone ignorant of law, to complain about its coldness, its formalism, its rulings that do not satisfy you, the "jargon" its practitioners impose on you, the "endless paperwork," the "nitpicking requirements," the signatures and the seals; the fact remains that you feel clearly, as soon as you see a judge in his robes, or an attorney behind her desk, that you are going to have to comply with an order of practices that cannot be reduced to any other, one that has a dignity of its own, an order in which the question of truth and falsity will also arise, but in a distinctive way.

Stirred up by the details of your case, you may be stupefied to discover that the very "legal means" that would have allowed you to win is lacking: for example, you have misplaced the proof of receipt that would have given you more time; the attorney, apologetic, points out that your complaint "can no longer be lodged." There is no point in becoming indignant because your case can no longer be heard owing to "such a trivial detail": you will understand very quickly that the appeal to texts or precedents, the quest for signatures and proofs, the mode of conviction, the contradictory posturing of the parties, the establishment of the dossier, all this defines a procedure capable of producing a type of truth-telling, of veridiction that, although very different from the practices of knowledge, nevertheless possesses a similar type of solidity, stability, and seriousness—and requires of you a similar respect. Law thus offers a fairly good example of an ancient and enduring institution in which a quite particular form of reasoning is preserved, capable of extending everywhere even though its criteria of veridiction are different from those of science. And yet no one will say that law is irrational. Put it to the test: you may complain all you like that law is "formal," "arbitrary," "constructed," "encumbered with mediations," but you will not weaken it in the slightest: it will always remain law, exotic perhaps, specialized, esoteric, but surely law—"*Dura lex sed lex.*"

The situation of religion is the opposite. In many circumstances, the most ignorant person around can mock religion today without risk—at least if Christianity is the target. In its most intimate mechanisms, religious life may be obsessed with the difference between truth and falsity, but to no avail: it appears too readily as the refuge of the irrational and the unjustifiable. And often—this is the strangest part—in the eyes of religious persons themselves, hastening to take shelter behind what they call a bit too quickly its "mysteries," perhaps because they too have lost the interpretive key that would allow them to speak well about what matters to them. We may be astonished at the coldness, the technical aspect of law, we may even make fun of it, but we do not dismiss it with scorn. It seems that people have full latitude to scorn religion, as if the difference between salvation and damnation no longer mattered, no longer traced a perilous path of exhortations that no one dare neglect without risking his life—or at least his salvation.

It will be clear why I spoke earlier of the importance of noting carefully the fluctuations in the relations that the different modes have maintained among themselves over the course of history. The institution of law has visibly better resisted the test of modernism—to the point where our investigator can take it, unlike religion, as an example of veridiction that has maintained a dignity equal, inter pares, to that of the search for objective knowledge. It is impossible to do the same thing with religion. We are going to have to try to understand why there are so few types of veridiction in the history of the Moderns that have managed not to clash with others.

To define the project of this inquiry, all we have to do now is link the results we have just obtained to those of the previous chapter. Every **COURSE OF ACTION**, let us say, every situation, can be grasped, as we have seen, as a network (noted **[NET]**), as soon as we have recorded the list of unexpected THE INQUIRY CONNECTS UNDERSTANDINGS OF THE NETWORK TYPE **[NET]** WITH UNDERSTANDINGS OF THE PREPOSITIONAL TYPE **[PRE]** ⊙

beings that have had to be enrolled, mobilized, shifted, translated in order to ensure its subsistence. The term "network" reminds us that no displacement is possible without the establishment of a whole costly and fragile *set of connections* that has value only provided that it is regularly maintained and that will never be stronger than its weakest link. This is why it still appears indispensable, if we are to keep from getting lost, to raise the following question at the outset: "In what network do we find ourselves?" The great advantage of this mode of understanding is that it allows the analyst as much freedom as that of the actors in the weaving of their worlds; it frees the field entirely from its organization into domains. Especially when we learn to liberate ourselves from some of the supposedly uncrossable borders—which the Moderns constantly cross, however—between nature and culture, for example, or power and reason, the human and the nonhuman, the abstract and the concrete.

But as we have just seen, the same situations can be grasped in an entirely different way, one that we can readily identify as soon as we begin to compare the different ways of making a series of associations, using *prepositions* that provide the key in which what follows is to be interpreted. This second mode will be noted [PRE] (for preposition). The interest of this second understanding is that it allows us to *compare* the types of discontinuities and consequently the trajectories that these discontinuities trace, one pair at a time. In this new type of understanding, comparisons are made by characterizing as precisely as possible the discontinuities and hiatuses through which the continuities are obtained. We have already checked off three of these: legal means, scientific proof, and religious predication (all obviously provisional terms). To these we must now add [NET].

We shall thus say of any situation that it can be grasped first of all in the [NET] mode—we shall unfold its network of associations as far as necessary—and then in the [PRE] mode—we shall try to qualify the type of connections that allow its extension. The first makes it possible to capture the multiplicity of associations, the second the plurality of the modes identified during the course of the Moderns' complicated history. In order to exist, a being must not only pass *by way of another* [NET] but also *in another manner* [PRE], by exploring other ways, as it were, of ALTERING itself. By proceeding in this way, I hope to remedy the principal weakness of every theory that takes the form of an association network (it is a weakness of all monisms in general, moreover): the ethnographer will be able to retain the freedom of maneuver proper to network analysis, while respecting the various values to which her informants seem to cling so strongly.

With the [NET · PRE] link, we meet the first instance of what I call a CROSSING, the meticulous recording of which makes up the raw material of the ⊙ BY DEFINING CROSSINGS THAT FORM A PIVOT TABLE.

Pivot Table (by convention, the order of the modes will always be that of the table on pp. 488-489). It is in fact by situating ourselves at these cross-ings that we can grasp the irreducible character of their viewpoints: this is where we shall be able to see why the conclusion of a trial bears no resemblance to that of a scientific proof and why one cannot judge the quality of a predication either through law or through science. A crossing makes it possible to compare two modes, two branchings, two types of felicity conditions, by revealing, through a series of trials, the CONTRASTS that allow us to define what is specific about them, as well as the often tortuous history of their relations. We must expect to treat each crossing, each contrast, as a separate subject that will require its own elaboration in each case. It is becoming clear why the inquiry is going to take time and will expand quickly!

The [NET · PRE] crossing is rather special, since it is the one that authorizes the entire inquiry. From the standpoint of descriptions of the [NET] type, all the A SOMEWHAT PECULIAR [NET · PRE] CROSSING, ⊙

networks resemble one another (this is even what allows our investigator to go around freely, having extricated herself from the notion of DOMAIN), but in this case the PREPOSITIONS remain totally invisible except in the form of mild remorse (the investigator has a general feeling that her descriptions fail to capture something that seems essential in the eyes of her informants). Conversely, in an exploration of the [PRE] type, networks [NET] are now only one type of trajectory among others, while the modes have become incompatible, even though their felicity conditions can be compared for each pair, but *only from the standpoint* of [PRE].

Readers of a sociological bent will not have failed to notice that the [NET · PRE] crossing raises a problem of "software compatibility," as the computer scientists would say, between the ACTOR-NETWORK ⊙ WHICH RAISES A PROBLEM OF COMPATIBILITY WITH THE ACTOR-NETWORK THEORY.

THEORY [ANT] and what we have just learned to call [PRE]. Clearly, in order to be able to continue her investigation, the anthropologist of the

Moderns must now get over her exclusive penchant for an argument that had nevertheless freed her from the notion of distinct domains.

This theory played a critical role in dissolving overly narrow notions of institution, in making it possible to follow the liaisons between humans and nonhumans, and especially in transforming the notion of "the social" and SOCIETY into a general principle of free association, rather than being an ingredient distinct from the others. Thanks to this theory, society is no longer made of a particular material, the social—as opposed, for example, to the organic, the material, the economic, or the psychological; rather, it consists in a movement of connections that are ever more extensive and surprising in each case.

And yet, we understand this now, this method has retained some of the limitations of critical thought: the vocabulary it offers is liberating, but too limited to distinguish the values to which the informants cling so doggedly. It is thus not entirely without justification that this theory is accused of being Machiavellian: everything can be associated with everything, without any way to know how to define what may succeed and what may fail. A tool in the war against the distinction between force and reason, it risked succumbing in turn to the unification of all associations under the sole reign of the number of links established by those who have, as it were, "succeeded." In this new inquiry, the principle of free association no longer offers the same metalanguage for all situations; it has to become just one of the forms through which we can grasp any course of action whatsoever. The freest, to be sure, but not the most precise.

RECAPITULATION OF THE CONDITIONS FOR THE INQUIRY.

I can now recapitulate the object of this research. By linking the two modes [NET] and [PRE], the inquiry claims to be teaching the art of *speaking well* to one's interlocutors about what they are doing—what they are *going through*, what they are—and what they *care about*. The expression "speak well," which hints at an ancient eloquence, sums up several complementary requirements. The speaker who speaks well has to be able

· TO DESCRIBE networks in the [NET] mode, at the risk of shocking practitioners who are not at all accustomed, in modernism, to speaking of themselves in this way;

· To VERIFY with these same practitioners that everything one is saying about them is indeed exactly what they know about themselves, but only in practice;

· To EXPLORE the reasons for the gap between what the description reveals and the account provided by the actors, using the concepts of network and preposition;

· Finally, and this is the riskiest requirement, TO PROPOSE a different formulation of the link between practice and theory that would make it possible to close the gap between them and to redesign institutions that could harbor all the values to which the Moderns hold, without crushing any one of them to the benefit of another.

The project is immense, but at least it is clearly defined, all the more so in that each of its elements is the object of a specific test:

· THE FIRST is factual and empirical: have we been faithful to the field by supplying proofs of our claims?

· THE SECOND requires an already more complicated negotiation, something called the *restitution* at the end of investigations: have we succeeded in making ourselves understood by those whom we may have shocked, without giving up our formulations?

· THE THIRD is both historical and speculative: have we accounted for the historical fluctuations between value and network?

· THE FOURTH presupposes the talents of an architect, an urbanist, a designer, as well as those of a diplomat: in the plan proposed for a habitat, are the future inhabitants more comfortable than they were before?

The reader may object that I have already failed to measure up to this admirable program by failing to take up the adjective "rational" on my own behalf. And yet it would be so convenient to be able to reuse

WHAT IS RATIONAL IS WHAT FOLLOWS THE THREADS OF THE VARIOUS REASONS.

the venerable term REASON. After all, if we don't want to fool ourselves, it is because we too want to find the reason for things, to be right, to resist settling down complacently in error, to live as rational beings, and so on. There may be other vocations under the sun, other cultures, even other civilizations, but the form of life into which we were born, the one that the Moderns would really like to inherit, the one about which they passionately wish not to be mistaken, must indeed have something to do, in spite of everything, with a *history of Reason* with a capital R.

The problem is that we have been presented with a form of reason that is not reasonable enough and above all not demanding enough, since it has always been divorced from the *networks* we have just identified, and since it has only interrogated truth and error in *a single* key. Now, if the term "rational" can be given a precise meaning, if it can designate the veridiction within a network that is proper to that network, for example to law, or knowledge, or religion, it no longer has any meaning once it has been deprived of its conditions of exercise. Reason without its networks is like an electric wire without its cable, gas without a pipeline, a telephone conversation without a connection to a telephone company, a hiker without a trail system, a plaintiff without legal means. If it is true that "the heart has its reasons that Reason does not know," we have to acknowledge that each mode has its networks that Reason does not know.

But, someone will say, this is exactly where you've taken the wrong path: there are not several ways to distinguish truth from falsity; "Reason" can't be declined in the plural but only in the singular; or else words don't mean anything any longer and you are asking us to sink along with you into the irrational. Now, it is in this assimilation of the rational with the singularity of a particular type of trajectory that the most dangerous and least noticed source of mistakes, mistakes squared, seems to lie. If it is true that the notion of category allows us to multiply the ways of speaking well of a given thing, we may be astonished that this is always to reply to the single question of equipped and rectified knowledge. It is precisely because we want to take up the adventures of reason anew that we have to make the notion of category mistakes capable of following a plurality of reasons.

This is why I would like us to be granted the right, in what follows, to use the adjective RATIONAL to designate from now on the step-by-step, thread-by-thread tracing of the various *networks*, to which we shall add the various *trajectories* of veridiction or malediction, each defined by a separate *preposition*. To understand rationally any situation whatsoever is at once to unfold its network and define its preposition, the interpretive key in which it has to be grasped ([NET · PRE]). After all, isn't wanting to speak well about something to someone, standing in the agora, a fairly good approximation of what the Greeks called *logos*?

Let us note, to conclude, that here we are encountering a first example of the "diplomatic representations" that we shall have to carry out all along the way: can we reassure the rationalists as to the solidity of their values, even as we refine what they cherish in a way that makes it unrecognizable, at first glance? Can we really convince them that their values, thus represented and redefined, will turn out to be *better grounded* than in the past? I do not imagine a reader impatient enough, cruel enough to demand that I succeed in such a negotiation after only a few pages, but he has the right to ask for a reckoning at the end of the exercise, for this is indeed the direction in which I am obliged to set out by the definition of reason I have just proposed.

Chapter 3

A PERILOUS CHANGE OF CORRESPONDENCE

....

To begin with what is most difficult, the question of Science ⊙ by applying principles of method that entail identifying passes, ⊙ which allow us to disamalgamate two distinct modes of existence.

Description of an unremarkable itinerary: the example of a hike up Mont Aiguille ⊙ will serve to define chains of reference and immutable mobiles ⊙ by showing that reference is attached neither to the knowing subject nor to the known object.

The notion of Subject/Object correspondence conflates two passes ⊙ since it is clear that existents do not pass through immutable mobiles in order to persist in being.

Although there is no limit to the extension of chains of reference [REF] ⊙ there are indeed two modes of existence that co-respond to each other.

We must therefore register new felicity conditions ⊙ that will authorize a different distribution between language and existence.

This becomes particularly clear in the prime example of the laboratory.

Hence the salience of a new mode of existence, [REP], for reproduction ⊙ and of a crossing [REP · REF] that is hard to keep in sight ⊙ especially when we have to resist the interference of Double Click.

I F DEBATES OVER THE DEFINITION OF THE RA-
TIONAL AND THE IRRATIONAL ARE SO VIGOROUS,
IF THE PROSPECT OF NEGOTIATING THE FORM OF
institutions finally cut out for the work of reason
seems so remote, it is because of a major problem in the anthropology
of the Moderns: the enigma posed for them by the irruption of the sciences, starting in the seventeenth century and continuing today. This
enigma has been made insoluble by the immense abyss that developed,
in the course of the Moderns' history, between the theory of Science and
the practice of the sciences, an abyss further deepened with the emergence of ecology, which obliges us to take into account what is called the
"known and inhabited world" in an entirely new way.

The anthropologist studying the Moderns can only be struck by the
importance her informants attribute to the themes of Reason, rational
explanation, the struggle against beliefs and against irrationality, and, at
the same time, by the lack of realism that characterizes the descriptions
the same informants provide for the advance of rationality. If we were
to believe what they say, officially, about Reason—and Science is almost
always the highest example of Reason, in their eyes—this Reason could
never have obtained the material and human means for its spread. Since,
to hear them tell it, capital S Science in theory needs only purely theoretical methods, the small s sciences would have found themselves long since
with no funding, no laboratories, no staff, no offices: in short, reduced to
the bare minimum. Fortunately, and this accounts for the discreet charm

of expeditions among the Moderns, their right hand is not fully aware of what their left hand is doing. The sciences turn out to be equipped, in the end, but until very recently no one has felt a need to provide a more or less credible description of the process. To keep the reader with us here, it would be useful to summarize all the work that has been done on the institution of science, and we should follow the networks that allow us to outline its astonishing practices. However, I have chosen to focus not on an anthropology of scientific institutions but rather on just one of its ingredients, one whose specific tonality warrants special emphasis, in my view: the assurance that scientific results do not depend on the humans who nevertheless produce these results at great cost.

Everything hinges on the question of the CORRESPONDENCE between the world and statements about the world. Some will say that if there is any subject that ethnology ought to avoid like the plague, it is this famous *adequatio rei et intellectus*, at best good enough to serve as a crutch for an elementary philosophy exam. Unfortunately, we cannot sidestep this question; it has to be faced at the start. Everything else depends on it: what we can expect of the world and what we can anticipate from language. We need it in order to define the means of expression as well as the type of realism that this inquiry has to have at its disposal. By way of this apparently insoluble question, nothing less is at stake than the division between reality and truth. The opacity peculiar to the Moderns comes from the inability we all manifest—analysts, critics, practitioners, researchers in all disciplines—to reach agreement on the condition of that correspondence. We shall never be able to define the other modes if we give up on this one at the outset.

But on what? What to call it? This is precisely the problem: in what way is what has been designated up to now by the adjective "scientific" a particular mode of veridiction? Indeed, whenever we talk about correspondence between the world and statements about the world, we don't know exactly *what* we're talking about, whether we're dealing with the world or with Science. As if the two, through the fuzzy notion of correspondence, had actually amalgamated to the point of being indistinguishable. On the one hand, we are told that they are one and the same thing; on the other, that they have nothing to do with each other and that they relate as a thing relates to a mind. As if the world had become knowable—but through what transformation? As if words conveyed reality—but through what intermediary? This notion of correspondence is a real muddle. When someone wants to define legal veridiction, and, even more to the point, religious veridiction, everyone has a clear sense that this person's felicity conditions must be very precisely defined, and that these conditions are specific to the mode that leaves something legal or something religious, as it were, in its wake. But if we assert the same thing "about" what is scientific, people will no longer know what we're talking about: are we supporting the critical position that assimilates science to one mode of "representation" among others, following the legal or even (horrors!) the religious model? Or is it that we are amalgamating in a single definition a statement and the world whose mere existence validates what is said about it? If we accept the first version, *all modes* of veridiction, science included, turn into simple manners of speaking without access to reality; if we accept the second version, this mode, by what miracle we do not know, would (alone?) be capable of fusing truth and reality. While we might be able to agree to attribute the births of Venus or the Virgin Mary to miracles, it is a little embarrassing, for the Moderns' simple self-respect, to attribute the birth of Reason to the operation of the Holy Spirit. We still have to seek to understand. We still have to trust reason "for real."

If we must begin by facing this difficulty, it is because, were we to leave it behind unexamined, it would poison all the diagnoses we might want to posit about the other modes. These modes could never benefit from their *own way* of linking truth and reality. In fact, as strange as it may seem, it is this so ill-composed notion of correspondence, despite—or perhaps because of—the obscurity it projects, that has served to judge the quality of all the other modes! After having absorbed all of reality, it has left to the other modes only the secondary role of "language games." Through a paradox whose most unanticipated consequences we shall never stop assessing, it is the deformed offspring of a category mistake that has ended up in the position of supreme judge over the detection of *all* the other category mistakes! By undoing this amalgam at the beginning of the inquiry, I hope to remove one of the chief obstacles to the anthropology of the Moderns. There will always be time later to come back to a description of scientific networks in the manner of science studies: practice will always stay in the foreground rather than disappearing mysteriously along the way.

As one might imagine, it is impossible to approach the question head on. Happily, we learned in the previous chapters that our inquiry bore on the identification of a type of TRAJECTORY whose seeming

⊙ BY APPLYING PRINCIPLES
OF METHOD THAT ENTAIL
IDENTIFYING PASSES, ⊙

continuity was actually obtained by a particular way of leaping over discontinuities that were different in each case. We have already identified four of these PASSES: legal means, scientific proof, religious predication, and an all-terrain mode, networks of associations; we have also learned that, to resolve the contradiction between continuities and discontinuities, each pass or each mode had defined its own forms of veridiction that allowed it to define the conditions for the success or failure of such a leap; finally, we have understood that modes can be compared in pairs when they intersect in CROSSINGS, occasions revealed most often by a test of category mistakes bearing on one of the felicity conditions.

It should be fairly easy to recognize that the proliferation of propositions about the traps and impasses of the correspondence between the world and statements about the world reveals such crossings, at least symptomatically. To judge by the scale of the anxieties that this branching generates, something essential must have been knotted up here. In the absence of agreement on the description of this correspondence, we can at least rely on the passions aroused by the very idea that one might wish to describe the sciences as a practice grasped according to the mode of networks of the [NET] type. The reader will undoubtedly acknowledge that science studies are rather well endowed with passions of this type.

In this section, we are going to try to identify the crossing between two types of trajectories that the argument of a correspondence between things and the mind, the famous *adequatio rei et intellectus*, reveals (as a symptom) and conceals (as a theory).

⊙ WHICH ALLOW US
TO DISAMALGAMATE
TWO DISTINCT MODES
OF EXISTENCE.

We shall try to *insert a wedge* between two modes that have been amalgamated with each other so as to respect two distinct passes and register the effects of this category mistake on which, one thing leading to another, all the others depend. Observing that *reproduction* must not be confused with *reference*, we shall give them distinct names here that will be defined later on. Let us acknowledge that it is somewhat counterintuitive to define as a crossing between two modes something that the informants claim precisely is *not* a crossing, or worse, something that they interpret as the equivalence between a knowing mind and a known thing. But the reader is now prepared; a similar elaboration will accompany every branching.

DESCRIPTION OF AN
UNREMARKABLE ITINERARY:
THE EXAMPLE OF A HIKE
UP MONT AIGUILLE ⊙

In order to break apart the connection that we would risk missing if we had misunderstood the activity of equipped and rectified knowledge, let us take a very simple case that does not depend on a discovery (to focus on a discovery would be excessively simple, owing to the novelty of the objects that it would imply), but that relies, on the contrary, on some ancient knowledge that is the object of well-anchored habits in harmony with the hiking metaphor. At the end of the road we shall come back to the idea of correspondence, to find that it has undergone a small but decisive modification. It will no longer be a question of relating a mind and a thing but of bringing into correspondence two entirely distinct modes of verification while respecting the *break in continuity* that must always distinguish them.

Since the argument is frightfully difficult, let us begin with what will not be a simple little stroll for our health. Let us go back to the hiking trails, and—why not?—to the French geological survey map 3237 OT "Glandasse Col de la Croix Haute," which I made sure to buy before setting out on the Vercors trails. As I was having trouble finding the starting point for the path leading to the Pas de l'Aiguille, I unfolded the map and, by looking from the plasticized paper to the valley, located a series of switchbacks that gave me my bearings despite the clouds, the confusion of my senses, and the unfamiliarity of the site. I was helped by the yellow markers that punctuated the route, and by the fact that the tourist office was kind enough to associate those markers with the map so carefully that one can go back and forth and find the same words, the same distances and times, and the same turns on both the map and the landscape—although not always.

The map, the markers, the layout of the path are, of course, different, but once they are *aligned* with one another they establish a certain *continuity.* Moreover, in cases of uncertainty, the steps of countless hikers who had gone before or the little piles of fresh donkey manure would add a welcome confirmation of the circuit I was to follow. As a result, although I was unquestionably enjoying the privilege of being "outdoors," "in fresh air," "in the bosom of nature," "on vacation," I was definitely *inside* a network whose walls were so close together that I chose to lean on them every ten minutes or so, verifying whether the map, the markers, and the approximate direction taken by other hikers were indeed in *correspondence,* forming a sort of coherent conduit that would lead me up to the Pas de l'Aiguille. Not a superfluous verification, since the High Plateau of the Vercors (as the topographic guide I bought as an extra precaution had just warned me in rather frightening terms) is known for its fogs, its crevasses, and its deserts, and it is not marked by any signals or signposts, only by cairns standing here and there. If you doubt that I needed to stay within a network ("Don't leave the marked trails!"), you're welcome to go get lost up there in my wake, some foggy day when you can't see the tips of your shoes.

However, I have to admit that I was dealing with a particular conduit, whose walls, although materialized (otherwise I would not have checked my path with such anxiety), are not made of a material as continuous as, for example, the walls of a labyrinth or those of a gallery inside a mine: the two-dimensional paper map, the wooden signposts painted yellow, the trail marked by trampled grass and blackened leaves, the landmarks spotted (cairns or just piled-up stones? I hesitated at every turn), none of these elements resembles any of the others in its matter. And yet they did *maintain* an overall coherence that allowed me to "know where I was." The discontinuity of the landmarks ended up producing the continuity of indisputable access. For they formed the quite particular type of pass we encountered in Chapter 1, when I spoke of the movement of proofs.

The particularity of such linkages is that they establish a connection that maximizes two apparently incompatible elements: *mobility* on the one hand, *immutability* on the other. Map 3237 OT folds up to fit very easily in the pocket of my backpack; I can carry it the whole way and unfold it at any point to see, for example, whether the expression "Refuge du Chaumailloux" corresponds to the particular hut I see, unless it is that other one, a few steps farther on, in which case the unmarked path would begin there on that slope and not here in this valley . . . (I haven't yet acquired a GPS, which would end up making me so thoroughly surrounded that I wouldn't even have to look at the landscape "outside" to know where I am; it would be enough to keep my eye fixed on the screen, like a totally blind yet perfectly oriented termite.)

But this transportable, mobile, foldable, tear-proof, waterproof map establishes relations with the signposts and with the peaks and valleys, plains and cliffs (and indeed with the remarkable signals established for triangulation purposes by the old topographers, later with aerial photos, today with superimposed, artificially colored satellite images whose slight discrepancies allow us to determine the relief), relations, as I was saying, that *maintain intact* a certain number of geometrical liaisons, appropriately called *constants*. If I have not left my compass at home, I can verify that the angle formed by the edge of the southern escarpment of Mont Aiguille and the cross on the monument to Resistance victims is actually the "same"—except for mistakes in the viewing angle and the coded declinations—as the one on my 3237 OT map. So it is possible to establish an itinerary (I am not forgetting that this has required three centuries of geographers, explorations, typographical inventions, local development of tourism, and assorted equipment) thanks to which one can maximize both the total dissimilarity—nothing looks less like Mont Aiguille than the map of Mont Aiguille—and the total resemblance— the angle that I am targeting with my compass is indeed the same as the one printed on the map. I can *refer* to the map to locate Mont Aiguille; I can refer to Mont Aiguille to understand what the map means; if everything is in place—if there is no fog, if some goofball hasn't turned around the signposts or kicked over the cairns, if my senses do not deceive me—I can move along the path with complete safety, because at the same time I can *go back and forth* along a continuous road paved with *documents*, even though none of these has any *mimetic* resemblance to the one that precedes or the one that follows. What is more, it is precisely *because* the map does *not* resemble the signposts, which do not in any respect resemble the prominent features, which in no way resemble the cliff of Mont Aiguille, but because all of them *refer* to the previous and subsequent items by remaining constant across the abyss of the material dissimilarities, that I benefit from the comfort of this network: I am not

lost, I know where I am, I am not making a mistake. This comfort is relative, however, because no matter how secure my own displacement may
be, I still have to sweat my way up the steep slope!

To mark the originality of these networks clearly, ⊙ WILL SERVE TO DEFINE
let us agree to designate their trajectory by the expres- CHAINS OF REFERENCE AND
sion CHAINS OF REFERENCE, and let us say that what char- IMMUTABLE MOBILES ⊙
acterizes these highly original chains is that they are
tiled and covered over with what I have come to call IMMUTABLE MOBILES.

The oxymoron is intentional. The expression can indeed be taken in
two opposite ways. It can either be understood as seeking to emphasize
through high tension two and a half millennia of original inventions on
the part of countless learned disciples working to solve the key question
of reference by maximizing the two opposing requirements of maximum
mobility and maximum immutability; or, conversely, as *presupposing the
problem solved* by acknowledging as self-evident that a displacement can
be achieved without any transformation, through a simple glide from
one entity through another identical to it and on to another... In the first
sense, the expression "immutable mobiles" sums up the efforts of the
history and sociology of the sciences to document the development of
the technologies of visualization and INSCRIPTION that are at the heart of
scientific life, from the timid origins of Greek geometry—without trigonometry, no topographical maps—up to its impressive extension today
(think GPS); in the second sense, the same expression designates the *final
result* of a correspondence that takes place *without* any discernible discontinuity. Quite clearly, the two meanings are *both true at the same time*, since
the effect of the discontinuous series of markers has as its final product
the continuous itinerary of the sighting that makes it possible to reach
remote beings without a hitch—but only when everything is in place.
This is what I said earlier about the two meanings of the word "network":
once everything is working without a hitch, we can say about correspondence what we would say about natural gas, or WiFi: "Reference on
every floor."

The important point for now is to note that the ⊙ BY SHOWING THAT
itinerary of these chains of reference in which immu- REFERENCE IS ATTACHED
table mobiles circulate in both directions would not NEITHER TO THE KNOWING
be clarified in the least if we introduced into their SUBJECT NOR TO THE
midst the presence of a "human mind." We gain KNOWN OBJECT.

access to the emotion elicited by the High Plateau of the Vercors only if we do not stray an inch from the composite network formed by the roads, paths, maps, tourist offices, hotel chains, hiking boots, backpacks, and the walkers' habits introduced by Jean-Jacques Rousseau, along with the clichés developed during the nineteenth century expressing admiration for the heights. Without mediation, no access. But this itinerary would not be clarified, either—symmetry has its importance—if we introduced the notion of a "thing known." Borges has clearly warned us against the dream of a map at full scale, since any knowledge that "covered" the world would be as profoundly obscure as the world itself. The *gain* in knowledge allowed by immutable mobiles stems precisely from the fact that the map *in no way* resembles the territory, even as it maintains through a continuous chain of transformations—a continuity constantly interrupted by the differences between the embedded materials—a very small number of constants. It is through *loss* of resemblance that the formidable effectiveness of chains of reference is won.

In other words, the network manages to spread precisely because it does not establish any type of relation between the *res* and the *intellectus*, but it never stops erecting bridges between one inscription and the next. This accounts for all the weirdness of this business of knowing, and it is why James, with his customary humor, introduced his "*deambulatory theory of truth*": instead of a "mortal leap" between words and things, he said we always find ourselves in practice facing a form of crawling that is at once very ordinary and very special, proceeding from one document to another until a solid, secure grasp has been achieved, without ever passing through the two obligatory stages of Object and Subject.

If this point has been well understood, it will be obvious that chains of reference trace in the territory a particular type of network that maintains constants provided that it breaks at every step with the temptation of resemblance, to obtain at last a displacement that seems (here's the crux of the matter) to proceed from same to same despite the abyss of differences. If we do not observe closely how the documents line up one after the other in both directions, we have the impression that these immutable mobiles are almost miraculous! It is certainly true that at the outset we have before our eyes, as soon as I unfold my map and *relate* it to the landscape—never "directly," of course, but through the intermediary of signposts and all the rest—a form of transsubstantiation: the signs inscribed on the waterproof paper are gradually *charged*—as I keep on going back and forth—with certain properties of Mont Aiguille and allow me to come closer to it. Not all its properties (I shall come back to this): not its weight, not its odor, not its color, not its geological composition, not its full-scale dimensions; and this is a good thing, for otherwise I should be crushed under its weight, as in Borges's fable. Conversely, the map manages to *extract* from Mont Aiguille a certain number of remarkable features, an extraction facilitated moreover by the magnificent sheer drop of its cliffs, making it eminently "recognizable," as if it were already a sort of seamark cut out in advance to be included in a guide.

It is clear, then, that to capture the originality of a chain of reference we can never limit ourselves to two extreme points, the map and Mont Aiguille, the sign and the thing, which are only provisional stopping points: we would immediately lose all the benefits of the "network setup." No, it is the whole series of points along the way that make it possible to verify the quality of our knowledge, and this is why I call it a chain or a linkage. To get a good sense of its expansion, we need rather to imagine a strange means of transportation whose continuous back-and-forth movement along a fragile cable—all the more continuous in that it will be discontinuous, leaping from one medium to another!—gradually charges the map with a minuscule portion of the territory and extracts from the territory a full charge of signs. (We find this warning written on the map, moreover: "If the route indicated on the map differs from the signage on the ground, it is advisable to follow the latter"! And, further on: "Users of this map are urged to let the IGN [National Geography Institute] know of any mistakes or omissions they have observed.") No question: to refer, as etymology tells us, is thus always to *report*, to *bring back*.

Even if there are no miracles here, it is nonetheless fitting to admire an operation that sums up, brings together, draws aside, and compresses hundreds of person-years and some of the most innovative, audacious, stubborn, and also costly human endeavors. To be convinced of this, just think of the price that had to be paid in terms of bureaucracy, atomic clocks, satellite launches, and standardizations in order to obtain, through cross-checking, the little "click-click-click" of a GPS seeking its "cover" of three satellites. (My venerable topographical map adds proudly on its cover, moreover, that it is "GPS compatible," a new indication offering ample evidence that, through immense effort and at great cost, we have succeeded in adding a new layer of normalization to increase still further the "safety" of hiking trails.) Historians of science have spent a lot of time following the invention, installation, extension, maintenance, and dissemination of these sorts of "cables," which make the comings and goings, the reports, and the work of reference possible. Even the splendid view that one embraces from the Vercors plateau fascinates me less, in the end, than the humble effectiveness of map 3237 OT.

We have to be careful, however, not to transform this straightforward progression, this fascinating stroll—empirically attributable and describable from one end to the other—into an unfathomable mystery that would threaten to deprive reason of the only chance it has to be reasonable. There is in fact nothing difficult, in principle, about doing justice to the itineraries of reference, as long as one agrees to take reason not, as it were, stark naked, but on the contrary clothed, that is, instrumented and equipped. After the previous chapter, we now know that we can consider reason with or without its networks; apart from networks, as we have understood, it remains unattributable: it has no more meaning than map 3237 OT tucked away in the depths of a library, or a painted wooden signpost in the storeroom of the Isère Department's tourist office, before it has been planted in the ground. But of course once it has been reincorporated, reinserted, reaccompanied, rearticulated into the networks that give it its direction, reason in the sense of reference immediately points both to a series of discontinuities, hiatuses, steps, leaps, each of which separates one stage from the next, and to the result of a continuity that allows access.

The lines traced by these chains will now allow us to unsettle the ordinary notion of correspondence. In fact, what are usually called the "knowing mind" and the "known object" are not the two extremes to which the chain would be attached; rather, they are *both* products arising from the lengthening and strengthening of the chain. A knowing mind and a known thing are not at all what would be linked through a mysterious viaduct by the activity of knowledge; they are the progressive result of the extension of chains of reference. In fact, if we are so readily inclined to speak of "correspondence" between the two, it is because they both indeed arise from the *same operation* as the two sides of the same coin. It is as though one could "collect" the scientific mind and the thing known at every point throughout networks of equipped knowledge—somewhat— but the metaphor is too prosaic—as one would do with the rubber that seeps from a rubber tree. Paradoxically, either one concentrates on the extremes (a known thing and a knowing subject) and sees nothing of the chain, which can no longer be extended; or else one concentrates on the chain: the known thing and the knowing subject both disappear, but the chain itself can be extended. There is nothing astonishing about this, as every mountain guide knows: the person who equips a major hiking trail, who carves a path into a cliff, produces by the *same action* both a mountain that is at last accessible and also a hiker or climber *capable* of attacking it. Chains of reference are not rope bridges strung between the mind on one side and reality on the other, but snakes—don't we associate snakes with knowledge?!—whose heads and tails grow further and further apart as their bodies grow longer and stouter.

Careful, here: we've unmistakably come to a branching point that we mustn't miss, a bifurcation on the path of our inquiry where we'll do well to set down our packs and spread out the map. Is it possible to speak of the world to which reference gives access other than as that *res* that would serve as a counterpart to *intellectus*? What happens to it that is peculiar to it? How can we describe that through which a territory *passes*, when it does not *pass* through the map? Here is where we need to slow down, because we risk failing to do justice to the "world" if we treat it as a "known thing"—just as we would fail to do justice to scientists in treating them as "knowing minds." Is there a mode of description that will allow us to consider existents and the map *at one and the same time*? The map in the world, or rather on the world, no, let us say as an add-on, an incision, a precision, a fold of worlds? In other words, can we bring to the surface at once the world and the map of the world, without amalgamating them too hastily through the notion of correspondence? If the principle of our inquiry is valid, we know that this question must be raised in the following form: is it possible to identify a HIATUS, a step, a leap, a pass that will allow us to define existents *also* as a particular manner of establishing continuity through discontinuities? If we were capable of

THE NOTION OF SUBJECT/ OBJECT CORRESPONDENCE CONFLATES TWO PASSES ⊙

differentiating them, we would then have *two distinct modes* that would then indeed enter into *correspondence with each other*—common sense was right—but only *after* having clearly *distinguished* and *without confusing* this crossing with the equivalence between a known thing and a knowing mind—good sense was wrong.

⊙ SINCE IT IS CLEAR THAT
EXISTENTS DO NOT PASS
THROUGH IMMUTABLE
MOBILES IN ORDER TO
PERSIST IN BEING.

At first glance, we can hardly make out the branching point, so thoroughly has it been obscured by the notion of correspondence between minds and things. And yet the crossing becomes easier and easier to detect as we equip and clothe and materialize the reference-producing networks and their own trajectories, which seem perfectly orthogonal to the famous Subject/Object relation. If I stress this point, it is to recall why this anthropology of the Moderns had to begin with the ethnography of laboratories and had to depend, more generally, on the development of science studies.

In fact, the more we foreground the "wandering" of reference, the more unlikely it seems that we are dealing here with the *very mode* through which the old "known thing" had to pass in order to maintain itself in existence. For, after all, I don't yet know where Mont Aiguille is headed, ultimately, but if it is to maintain itself in existence, if it is to remain the same yesterday, today, and tomorrow, one thing is certain: it does not leap from one immutable mobile to another in order to discern, through the discontinuity of materials, the maintenance of a geometric constant compatible with the inscription on a map. It does *maintain* itself, since it exists and endures and imposes itself on my steps as it does on the instruments of geomorphologists, but in any case, and this is hard to doubt, *what* is maintained in it and through it does not have the *same* properties of inscription, documentation, or information as the properties that come and go along chains of reference. One can call this "essence," "permanence," "subsistence" (we shall soon give it a precise name), but it is certain that it is not in the same sense and within the same type of network as the constants that make it possible to produce rectified knowledge—by simultaneously creating, as *derived products*, an objective mind and a thing objectively known. As a result, if we do not want to make such a sweeping category mistake, we must no longer confuse the displacement of immutable mobiles along the cascades of reference with the displacement of Mont Aiguille along its path of existence. We are dealing here with two clearly distinct trajectories.

The confusion was possible only owing to a description of knowledge so nonmaterial that it could be detached without difficulty from its networks and attach itself mysteriously to what it knew, to the point of merging with it (we shall see how later on); whence this impression of resemblance between the two, this mystery of equivalence; and also this uneasiness before such a conjuring trick, as if there were something deceitful in such an amalgamation—something that skeptics have sensed without being able to pinpoint the cause of their dissatisfaction. Once again, one can revisit the Moderns' single-minded obsession with the enigma of correspondence only by materializing the work of knowledge; as soon this work is idealized, the problem seems to go away, since the crossing between the two modes has simply disappeared. This is what explains the lasting incomprehension that has marked science studies. We must not attribute any ill intentions to their adversaries, here: for them, the problem simply does not exist, since knowledge costs nothing.

But as soon as we begin to make chains of reference visible and perceptible, their extraordinary originality also becomes apparent, and, as a result, so does the implausibility of requiring that existents themselves pass through such trajectories. As soon as the description is made more realistic, more material, the least sophisticated observers sense that it is as implausible to make the objects of the world transit through these chains as it is to make an elephant jump through the hoop of a lion tamer—or to make a camel pass through the eye of a needle. It is only when we have brought chains of reference into view that the metaphysical question can take on its full relief: what happens to existents themselves? How do they pass? And this new question— here is the essential point—is not at all an insult, an offense to equipped, instrumented, and rectified knowledge, but only a specification as to how it can be localized.

As the entire inquiry depends on avoiding this primary mistake, it is important to understand clearly that knowledge is not limited in its extension—and all the less limited by the subjective frameworks of the human mind! For years historians and

ALTHOUGH THERE IS NO LIMIT TO THE EXTENSION OF CHAINS OF REFERENCE [REF] ⊙

sociologists of science have been studying the extension of networks of reference, and they have always found them capable of extending their grip—provided, of course, that these scholars pay the price, allow enough time, invent adapted equipment, assemble an appropriate group of specialists—without failing to procure financing for all these *impedimenta*. As for the limits of the mind, one never finds them, since, if you confine a scholar to a "limited point of view," to a given "standpoint," he will immediately find you a dozen arrangements capable of *displacing* the viewpoint through the invention of an instrument, a mission, a research project, a collection, a well-designed experimental test. *Displacing the viewpoint* is something at which chains of reference excel: the theory of relativity allows a cosmologist to circulate among the galaxies without leaving her little office in the Paris Observatory, as surely as I know where I am in the Vercors thanks to the topographic map. In this sense, scientific knowledge is indeed *limitless*.

And yet there is a limitation that follows this knowledge wherever it extends, albeit one that is in a sense *internal* to its expansion. Once again, the trace of its trajectories provides a much better identification of this internal frontier: however far they go, however well equipped they are, however fine the mesh, however complete their "coverage," however competent their operators, chains of reference can never be *substituted* in any way for what they know. Not at all because the known "eludes" knowledge in principle and resides in a world "of its own," forever inaccessible, but quite simply because existents *themselves are also going* somewhere, but *elsewhere*, at *a different pace*, with a different *rhythm* and an entirely different *demeanor*. Things are not "THINGS IN THEMSELVES," they belong "to themselves"—a different matter altogether. Still, none of this deprives knowledge of access. On the contrary, it accedes marvelously well to whatever network, whatever reason, it has to grasp. There is thus, properly speaking, no *beyond* of knowledge: either knowledge is truly beyond us—along a trajectory *different* from that of chains of reference—and then we are not dealing with equipped and rectified knowledge—or else there is access—by a new method, a new instrument, a new calculus—and we remain in fact *within* the limits of knowledge, not at all beyond. Those who seek to humiliate the sciences with "higher" or "more intimate" knowledge behave like people who want to reach something without establishing access.

The ethnologist finds something almost comic in the endless complaint invented by CRITIQUE: "Since we accede to known things by way of a path, this means that these things are inaccessible and unknowable in themselves." She would like to answer back: "But what are you complaining about, since you have access to them?" "Yes," they keep on whining, "but that means that we don't grasp them 'in themselves'; we don't see them as they would be 'without us.'" "Well, but since you want to approach them, if you want them to be as they are 'without you,' then why not simply stop trying to reach them?" More whining: "Because then we'd have no hope of knowing them." An exasperated sigh from the ethnologist: "It's almost as though you were congratulating yourselves that there is a path to Mont Aiguille, but then complaining that it has allowed you to climb up there . . . " Critique behaves like blasé tourists who would like to reach the most virgin territories without difficulty, but only if they don't come across any other tourists.

On reflection, our ethnologist understands that this inconsistency on the part of Critique is symptomatic of an entirely different phenomenon: the notion of "known thing" does not in fact exhaust what can be said about the world. Not at all because scientists are "limited" in their knowledge of things that would remain unknowable, since they accede to them quite well and know them admirably, but because the expression "objective knowledge" (provided that it is materialized) designates a progression, an access route, a movement that will cross paths with other types of movements to which it cannot be reduced and that it cannot reduce, either. This impression that there is always something *more* than what is known in the thing known does not refer at all to the unknowable (the complaint of Critique is in no way justified) but to *the presence of other modes* whose equal dignity EPISTEMOLOGY, despite all its efforts, has never allowed to be recognized. Knowledge can grasp everything, go everywhere, but in its own mode. It is not a DOMAIN, whose expansion has to be limited or authorized. It is a network that traces its own particular trajectory, alongside other, differently qualified trajectories, which it never ceases to crisscross.

⊙ THERE ARE INDEED TWO MODES OF EXISTENCE THAT CO-RESPOND TO EACH OTHER.

And now, finally, we can talk about *correspondence* again, but this "co-response" is no longer the one between the "human mind" and the "world." No, we now have a tense, difficult, rhythmic correspondence, full of surprises and suspense, between the risk taken by existents in order to repeat themselves throughout the series of their transformations on the one hand and the risk taken by the constants in order to maintain themselves throughout another no less dizzying series of transformations on the other. Do the two series sometimes respond to each other? Yes. Do they always do so? No. If it is true that it takes two to tango, it is equally true that it is meaningless to speak of *co-responding* unless there are two movements in the first place, each of which will respond to the other—often multiplying their missteps. What the canonical idea of objective knowledge never takes into account are the countless failures of this choreography.

WE MUST THEREFORE REGISTER NEW FELICITY CONDITIONS ⊙

But what about the felicity conditions that would allow us, as I said, to define a mode? Can we refer without being ridiculous to the "veridiction proper to Mont Aiguille"? Of course we can, since it is a question of recognizing steps and passes. Maintaining oneself in existence, being rather than not being, is without question one of the components—and perhaps the most important one—of what we usually call "true" or "false." Consequently, instead of having on the one hand a language that would say what is true and what is false—but without being able to follow the reference networks—and on the other hand "things" enunciated that would be content to verify the utterances by their simple presence or absence, it is more fruitful to give up both notions, "word" and "thing," completely, and to speak from now on only of modes of existence, all real and all capable of truth and falsity—but each according to a different type of veridiction.

Here is where we are going to begin to understand why our inquiry bears on modes of *existence*. At first glance, the idea of attributing the term "existence" to the two trajectories that cross paths can be surprising, because the tradition passed along to us asserts, rather, that there are "existents" on one side—Mont Aiguille, for example—and knowledge on the other, knowledge that states, when it is well conducted, the truth or falsity of these existents. Now it is precisely this *division of tasks* whose relevance we shall have to challenge. The distribution is awkward on both sides: it gives at once too much and too little to equipped knowledge; too much and too little to the known. Too much to the first: knowledge moves around everywhere without our knowing how; too little, because knowledge no longer has the means to establish its access routes. Too little to the second, which no longer has anything to do but be stupidly there, waiting to be known; too much, since the known alone validates what is said of it without knowledge being involved at all. To avoid such category mistakes, we shall have to propose another sort of transaction, perhaps the most difficult of the diplomatic representations to come: we need to harmonize the notion of mode of existence with the work of reference, and, conversely, recognize in existents the capacity to be true or false, or at least, as we shall see, to be ARTICULATED in their own way.

I could be completely mistaken, but it seems to me that the example of the Pas de l'Aiguille is going to allow us to cross a col that overlooks the entire Plateau. In fact, I have attempted to invert the inversion and to redescribe the landscape, including in it from here on at least *two types* of distinct displacements: the one through which the mountain goes its own way and the one, just as venerable, just as interesting, of equal dignity, but quite different, through which we know the mountain. The world is articulated. Knowledge as well. The two respond to each other sometimes—but not always. Is there something here that should frighten common sense? Is it really asking the impossible, doing violence to ordinary intuitions? I am asking no more than that we stop confusing the territory with the map, the equipment of a road with the cliff that it makes accessible.

And yet our task has been complicated by the example of a mountain. The distinction would have appeared simpler with a living being, for example, a cat. When an analytic philosopher asserts that one must establish a "correspondence" between the statement "the cat is on the mat" and the presence of said cat on said mat in order to be able to validate the statement's truth value, he is surely right (although one can hardly do a good job describing the peregrinations of a chain of reference with only two terms). But the philosopher forgets to speak of the *other correspondence*, an equally important one *for the cat itself*: the one that allows it to exist in time t + 1 after having existed in time t. Now there is a truth value to which the cat holds—in every sense of the word "hold"! There is a hiatus that the cat has to cross, and that every living being has to pass through with fear and trembling. In addition to this quite particular leap on the part of an utterance verified by a state of things, there is thus always that *other pass*, also dizzying, also worthy of attention, made by the state of things that remains similar to itself through the test of subsistence.

⊙ THAT WILL AUTHORIZE A DIFFERENT DISTRIBUTION BETWEEN LANGUAGE AND EXISTENCE.

Why is the analytical philosopher interested only in the abyss that he has to cross in order to "give up" his quest for answers, and not in the abyss that his cat has to cross to remain on its mat? (To be a little more thorough, here, we would also have to be interested in the harder-to-dramatize pass that allows the mat to keep on existing!) Yes, there is indeed *correspondence*, but this fine word has to designate the relation maintained between the two risky passages, not just the first.

At this stage, I am not asking the reader to be convinced but simply to accept the project of an inquiry into the modes of *existence*, an inquiry that will proceed step-by-step and in step with other crossings, as with the enigma of knowledge that was obliging us to separate, mistakenly, "truth conditions" on the one hand and "existence" on the other. If the reader finds it frankly bizarre to trace in the imagination the narrow pathway along which reference circulates, and still more bizarre to speak of the "network" in which a mountain makes its way in order to maintain itself in existence, it is because he has not correctly measured the profound obscurity in which we find ourselves plunged whenever we maintain the fiction of a pair of china dogs glaring at each other: a language that would speak of things. By accompanying objective knowledge in the chains of reference, by *granting* it the ontological dignity of being a mode of existence, but while *refusing* to allow it to substitute itself, through an overly tempting interpolation, for what it succeeds in knowing, it must be possible to sketch out a different landscape. In any case, we want to place ourselves in a position that will allow us to celebrate Mont Aiguille and the map of Mont Aiguille simultaneously, without having to forget either one, and without having to reduce one to the other.

If we have to grant such importance to labora- ⊙ THIS BECOMES
tory studies, it is because they let us see even more PARTICULARLY CLEAR
clearly how rare and complicated it is to establish a IN THE PRIME EXAMPLE
correspondence between the two modes, something OF THE LABORATORY.
that the idea of an *adequatio rei et intellectus* completely
concealed. In fact, the demonstration would have been simpler if instead
of taking an example from cartography—a science so ancient, so established, so instituted that it is almost impossible to bring out the way its
network was set up—we had chosen, as is customary in the history of
science, a discovery *in the process of being made.*

Let us take the example of a laboratory studying yeasts. It is impossible to limit oneself to two segments along the chain, as with the cat
and its proverbial mat. Beer yeasts were in no way prepared to become
the experimental material through which the "yeastists" in Bordeaux
made them capable of making themselves known. These yeasts had been
making grapes ferment as long as there have been grapes, and producing
grape must as long as there have been farmers, but they had never before
caused brains to ferment, or contributed to the writing of blog posts
and articles. They had never before undergone the astonishing transformation that consists in being altered to the point where their profile
now stands out vividly against the white bottom of petri jars. They had
never been fixed, along the path of their existence, through the effect of
controlled and calibrated freezing thanks to a vast refrigerator whose
opening and closing organized the life of the entire laboratory. Now,
after a few years, despite all the problems that the artificiality of their
new conditions threatened to cause, they are becoming quite adept at
producing documents, at training their yeastists to recognize them, at
providing INFORMATION about themselves, and each one finds itself—is
there any other way to put it?—"embarked" on the back-and-forth movements of which we spoke earlier, since in many of their stages they have
maintained their two faces, "tails"—the document—and "heads"—the
experimental material; this double aspect allows them to participate in
the journey of reference, at a pace increasing by the day. Each yeast has,
in part, become one of the numerous stages in the race to instrumented
knowledge (only in part, because they also continue to follow their own

paths). How can we doubt that here is a decisive branching in the path of their existence? An *event* for them as well as for the yeastists?

Here in a laboratory's grasp of things that it has chosen to engage in the destiny of objectivity we have an example of what it means for two modes of existence to interact to some extent, to correspond to one another gradually; and this reality is specific, sui generis. Let us not be too quick to say that this grasp necessarily mobilizes either things or words or some application of words to things. We would lose all we have gained in our exploration, and we would forget that it is one of the effects of reference to engender both a type of known object and a type of knowing subject at each of its extremities; object and subject are then no longer the causes but only the *consequences* of the extension of such chains and, in a way, their products. The more these chains lengthen, thicken, and become more instrumented, the more "there is" objectivity and the more "there is" objective knowledge that circulates in the world, available to speakers who want to plug into it or subscribe to it.

It is easy to see from examples of this type that, as long as the event of discovery lasts, no researcher is unaware of the potential dangers of establishing a correspondence between the dynamics of things and the work of reference. They all know that they are transformed by the event, they themselves and the things on which, finally, after so many failures, they have a grip—provided that they contain these things firmly all along the path of experimentation, modelization, re-creation, and calculation. The danger of "missing the connection" is what keeps researchers on edge at work. The following argument ought to be advanced with more diplomacy than we are capable of for the moment, but the category mistake would be to believe that the world *before* the invention of knowledge was *already* made of "objective knowledge." This does not keep us from saying (on the contrary, it is what allows us to say) that *after* chains of reference have been set up and gradually charged with reality, yes, undeniably, there is objective reality and there are scientist subjects capable of thinking it.

Wasn't it the most famous scientist of them all who used to say that "the most incomprehensible thing in the world is that the world is comprehensible"? The second part of the aphorism is true, unquestionably: the world is comprehensible. But Einstein was mistaken is saying that it is incomprehensible that this should be so. There is no mystery, no miracle: there has been a series of risky events in which at each point we can see the emergence of a double discontinuity, in the reproduction of the world and in the extension of reference along with the pas de deux through which the encounter with "thought collectives"—to borrow Ludwig Fleck's lovely expression—adapts. It is on the basis of such collective events that we must understand the surprise of knowledge that marks the scientist transformed by her discovery just as much as it marks the object grasped by the scientist.

Before we fall pell-mell into the dispute between realism and CONSTRUCTIVISM, those who are about to do a nosedive should ask themselves if they are really underwriting the guarantees of knowledge by *depriving* it of the possibility of being taken as a mode of existence that is *complete in its kind*. Is it really praising something to speak of it by denying in the first place that it is a matter of a reality sui generis? Denying that the only reality grasped by knowledge is exactly the same as it was before it was known? That nothing has changed? What? All that work for *nothing*— for nothing, ontologically speaking? It is not certain that it is useful for the defense of the scientific institution to render invisible the admirable contrast extracted in the world by the invention of objective knowledge.

But it is too early to formulate peace proposals in what inevitably becomes a subject of dispute with those who hold knowledge to be the supreme value (and they are right to hold to it, as we saw in the introduction, even if they are surely wrong about the shape of the institution charged with protecting it).

We have not solved the problem of knowledge? No, of course not; but we have begun to unblock the intersection that the notion of equivalence had covered over with a thorny thicket. Or at least we have opened up a wedge between two modes, and thus have to redefine the notion of correspondence, this time in positive terms.

HENCE THE SALIENCE OF A NEW MODE OF EXISTENCE, [REP], FOR REPRODUCTION ⊙

As we have done up to now, we are again going to plant our own little signposts along these major trails to mark the branching point whose importance we have just measured. Let us thus use [REP], for REPRODUC-TION (stressing the "re" of re-production), as the name for the mode of existence through which any entity whatsoever crosses through the hiatus of its repetition, thus defining from stage to stage a particular trajectory, with the whole obeying particularly demanding felicity conditions: to be or no longer to be! Next—no surprise—let us note [REF], for REFERENCE, the establishment of chains defined by the hiatus between two forms of different natures and whose felicity condition consists in the discovery of a constant that is maintained across these successive abysses, tracing a different form of trajectory that makes it possible to make remote beings accessible by paving the trajectory with the two-way movement of immutable mobiles.

Readers will better appreciate the full difficulty of doing the anthropology of the Moderns if they now compare the [REP · REF] crossing we have just identified with what virtually all official representations of the question of knowledge have designated as the relation between a "knowing mind" and a "known thing." There is in fact *no sort of resemblance* between the strange Subject/Object amalgam and what can be expected from the risky ties between reproduction [REP] and reference [REF]. (When I have to pin down ambiguous terms that belong to several modes, I shall specify them by placing their prepositions immediately before or after.) And yet what vast investments have been made in this relationship! How much anguish at the idea that the bridge might end up falling down! Can a subject know an object? Yes; no; not always; never; never completely; asymptotically, perhaps; as in a mirror; only through the bars of the prison-house of language. Now, those who have succeeded in identifying the movement of these two modes will have noticed that the figure of the "subject" is completely absent (in what respect is [REF] a "knowing subject"?—it is a network of instruments and formalisms that produces, at its opposite ends, knowledge and knowers); and the object is even more absent (in what respect does [REP] resemble an "object to be known"?—it has something entirely different to do!). In the final analysis, there are neither Subjects nor Objects, but a knowing subject and a known object, the twin results of the extension of proven knowledge; it is not surprising that they resemble each other and correspond to each other, since they are the same entities counted twice! The gap between the Moderns' theorizations and their practices is greater here than with any other crossing. This gap would not pose any problems (after all, we can survive perfectly well without explaining what we are doing) if it had not cast everything else into deep darkness through a sort of cascade effect, which we shall have to trace to its source in the next chapter.

If we are astonished that the Moderns have not maintained more carefully, through very elabo- rate institutions, a crossing that seems so essential to their sustainability, it is because it would take ⊙ AND OF A CROSSING [REP · REF] THAT IS HARD TO KEEP IN SIGHT ⊙

almost nothing to make the crossing vanish, owing to its very success. We understand now that there are paths of reference that resemble gas pipelines or mobile phone networks: once they are in place, no one (except someone responsible for maintaining them) is interested in the other meaning of the word "network" (the one involving hetero- geneous associations that were necessary for putting the functioning networks in place). As soon as someone "subscribes" to chains of refer- ence, gets used to them, their thickness, their materiality, their equip- ment disappear, and all the discontinuities required in order to follow them fade away. Once all the intermediate steps have vanished, only the two extremities remain to be considered: the mind and the world. As if there were no longer any need for transformations, passes, discon- tinuities. And to make matters worse, this is true only once the network is well established and only if it is continually maintained. Then yes, in such cases, the subject *has* something of the object, just as every floor in the building *has* some gas for the cook.

It is at this very moment that a sort of Evil Genius comes into play, having waited for the chains of reference to be deployed and stabilized before it intervened. In an allusion to the digital mouse, we ⊙ ESPECIALLY WHEN WE HAVE TO RESIST THE INTERFERENCE OF DOUBLE CLICK.

are going to call this devil DOUBLE CLICK (and note it [DC]). Based on a real enough experiment—reference permits access—this Evil Genius is going to whisper in your ear that it would surely be preferable to benefit from free, indisputable, and immediate access to pure, *untrans- formed* information. Now, if by bad luck this ideal of total freedom from costs served as the standard for judging between truth and falsity, then *everything would become untruthful, including* the sciences. This is hardly surprising, since we would be demanding the impossible: a displace- ment *without transformations of any sort*—beyond mere DISPLACEMENT. If you make the absence of any mediation, leap, or hiatus pass the one and only test of truth, then everyone, scientists, engineers, priests, sages, artists, businessmen, cooks, not to mention politicians, judges, or moralists, you all become manipulators and cheaters, because your hands are dirtied by the operations you have carried out to maintain in working order the networks that give direction to your practices. You will always be accused of passing through heterogeneity to obtain homogeneity, of introducing scandalous discontinuities in what ought to be smooth and continuous. You will be caught with your hand in the till; you will have lied about it.

By a dangerous inversion of the two senses of the word NETWORK, Double Click has begun to propagate everywhere an accusation of irrationality about everything that needs, if we are to be able to tell what is true from what is false, a certain number of operations of transformation or DISPLACEMENT—operations that are, however, as we have seen, a matter of reason itself. As if the accuser had in front of him the recipe for obtaining directly, without any mediation whatsoever, a displacement capable of going from one identical entity through another to another. Even worse, through a perversity whose origins we shall come to understand later on, this devil (for he is truly diabolical!) has begun to stigmatize, under the expression "RELATIVISTS," those who want reason to pay for the means of its extension in networks. Without seeing that the inverse position, the one that claims that displacements without transformation exist, deserve no label but "absolutism." We really don't want to deceive ourselves, we want to be able to say that one thing is rational and another irrational, this thing true and that other thing false, but we especially don't want to deceive ourselves about deception itself to the point of embracing absolutism! By claiming to give a unique and inaccessible model—displacement without transformations, reason without networks—to all forms of veridiction, this Evil Genius would by contrast make all other distinctions between truth and falsity irrational and arbitrary.

Our ethnologist must thus teach the Moderns to protect themselves against Double Click. It is the struggle against relativism that threatens, if they are not careful, to efface, to obliterate one by one the types of veridiction necessary to the exercise of their civilized life—and, to begin with, a paradox within the paradox, scientific activity itself, which will have become unattributable. We must learn to find in relativism, or, better, in RELATIONISM, that is, in the establishment of *networks of relations*, the fragile help that will allow us to advance in the inquiry, feeling our way without going too far astray. If the history of modernism is defined, in a highly canonical fashion, as "the appearance and extension of the reign of Reason," it is clear that the direction of this history will not be the same depending on whether we call "Reason" the extension of double-click information or the jealous maintenance of distinct sources of truth. In the first case, the more modernist we are the more likely we are to dry up all the sources but one—which does not exist. In the second case, the more we envisage becoming "resolutely modern" at last, the less we shall confuse the sources of veridiction. These are the two alternative histories whose threads we shall have to learn to untangle. If there is one source of mistake that has to be brought to an end, it is the one that claims to be putting an end to mistakes by rendering all practices irrational and arbitrary—and first of all those of the sciences!

LEARNING
TO MAKE ROOM

....

To give the various modes enough room ⊙ we must first try to grasp existents according to the mode of reproduction [REP] ⊙ by making this mode one trajectory among others ⊙ in order to avoid the strange notion of an invasive material space.

If those who have occupied all the space nevertheless lack room ⊙ it is because they have been unable to disamalgamate the notion of matter ⊙ by the proper use of the [REP · REF] crossing.

Now, as soon as we begin to distinguish two senses of the word "form," ⊙ the form that maintains constants and the form that reduces the hiatus of reference, ⊙ we begin to obtain a nonformalist description of formalism, ⊙ which turns out, unfortunately, to be wiped out by a third sense of the word "form."

At this point we risk being mistaken about the course followed by the beings of reproduction ⊙ in that we risk confusing two distinct courses in the idea of matter.

A formalist description of the outing on Mont Aiguille ⊙ generates a double image through a demonstration per absurdum ⊙ that would lead to a division into primary and secondary qualities.

But once the origin of this Bifurcation into primary and secondary qualities has been accurately identified ⊙ it becomes a hypothesis too contrary to experience ⊙ and the magic of rationalism vanishes ⊙ since we can no longer confuse existents with matter, ⊙ a matter that would no more do justice to the world than to "lived experience."

WE ARE GOING TO NOTICE THAT IF THE MODERNS HAVE NEVER BEEN ABLE TO TAKE THE EXPERIENCE OF THE VARIOUS modes as a guide, it is for want of enough room to shelter them all, in particular the trajectories whose autonomy we have just recognized, and among these especially the one called reproduction [REP]. For reasons we shall try to sort out in this chapter and to some extent in the next, the Moderns have chosen to institute not a mode but an amalgam between two modes [REP · REF] that everything should have encouraged them to distinguish carefully. The most common name for this amalgam is "material world," or, more simply, "matter." The idealism of this materialism—to use outdated terms—is the main feature of their anthropology and the first result of this inquiry, the one that governs all the others.

For a clear understanding of what follows, the reader must be prepared to stop considering this "matter" as a province of reality, but rather as an extremely bizarre institution, one that has had the rather unfortunate consequence, moreover, of creating, by contrast, a "knowing subject" and even a "mind" capable of extracting itself from "matter" by projecting an "external world" "outside" itself, a world whose existence has become uncertain, furthermore. It is this strange series of inventions that has made the Moderns opaque to themselves, and, what is more serious, it has left them unable to grasp the "other cultures," which had been getting along perfectly well without either the "material world" or "subjects."

Indeed, this is why anthropology has never been able to encounter the others except precisely as "CULTURES." To get back to the thread of experience, to become capable of learning from those who have worked out their relations with existents quite differently, and to understand, finally, why the verb ECOLOGIZE is going to serve as an alternative to the verb MODERNIZE, we shall have to highlight the mode of reproduction and then make it clear through what operation it has been confused with that of reference so as to engender "matter."

Our ethnographer is at first glance quite power-less to define the mode of reproduction, since among the Moderns no institution is available to help her locate it. Every time she defines it, she risks appealing to what she "knows" about it according to the mode of reference [REF] alone, and thus, thanks to the positive sciences, she too hastily obliterates the correspondence whose strange pas de deux we have just reconstituted. She is then in danger of settling for the standard versions of scientific cosmology, deploying the series of atoms, quantas, planets, genes, cells, living organisms, that would always land her on some Master Narrative leading from the Big Bang to human evolution, from Lucy in the Great Rift Valley to the gangs in suburban Los Angeles. Or, worse still, she might rely on the countless efforts, as old as the scientific revolution itself, to grasp the world "outside," "alongside," or "beyond" Science; if she were to do this, she would be settling for a more "immediate," more "naïve," more "sensitive," more "sensual," more "alive," perhaps more "romantic" grasp—in any case a less well equipped one; but then she would find herself brought back to simple human subjec-tivity and thus as far as possible from the originality proper to this mode, which is as distant from Subjects as from Objects. As Whitehead indi-cates so vividly, no question about such a trajectory can be clarified by adding the presence of a human mind contemplating it.

The strangeness of reproduction would be better captured by a sort of *negative* metaphysics: no, reproduction is surely not "NATURE," a premature unification of all existents, probably political in origin; nor is it the cosmos—too nice a setup, aesthetic in origin; nor is it the spec-tacle of sublime landscapes suited to elevate the soul by imitating moral law; nor is it the world indifferent to human feelings, since the world of

⊙ WE MUST FIRST TRY TO GRASP EXISTENTS ACCORDING TO THE MODE OF REPRODUCTION [REP] ⊙

reproduction is swarming with differences, and the fact that it in no way targets persons is not even attributable to indifference toward them. It is hardly probable that this world obeys laws, for there is not yet any law and still less any obedience; it would be useless to supplement it with mind, with anthropomorphism, humanity and souls; and, of course, the world of reproduction is not objective, either, since OBJECTIVITY comes to it only through a crossing with reference; to say that this world is "before" everything, like a "background," does not advance us any further, for the world is as much tomorrow's as it is today's, as remote as it is close, and it applies to all sorts of EXISTENTS. And if, despairing before this apophantic metaphysics, the ethnographer resigns herself to saying that there is nothing specific about this world, that perhaps it simply doesn't exist, all the existents that can be grasped according to the mode of reproduction press forward and insist stubbornly on being recognized for themselves and in their own names. If they demand to be thought for themselves, it is because they do not want to be mistaken for mere supporting players or accomplices of knowledge.

⊙ BY MAKING THIS MODE ONE TRAJECTORY AMONG OTHERS ⊙   Fortunately, the anthropologist of the Moderns is now equipped with a questionnaire that allows her to determine TRAJECTORIES fairly precisely without having to involve them in the major issue of OBJECTS and SUBJECTS (from here on always in capital letters as a reminder that we are steadily distancing ourselves from them). Every instance of continuity is achieved through a discontinuity, a HIATUS; every leap across a discontinuity represents a risk taken that may succeed or fail; there are thus FELICITY and INFELICITY CONDITIONS proper to each mode; the result of this passage, of this more or less successful leap, is a flow, a network, a movement, a wake left behind that will make it possible to define a particular form of existence, and, consequently, particular BEINGS.

When we use this questionnaire with beings of reproduction, we understand why it would be very unsatisfactory to qualify them by saying that they form a simple "material world" or that they are "PRELIN-GUISTIC." On the contrary, they express themselves, they predicate themselves, they enunciate themselves, they articulate themselves admirably. To be sure, they reproduce themselves almost identically, but that is no reason to believe that they do not have to pay for maintaining themselves in existence by passing through other beings, thus by a particular PASS. Indeed, this is probably what qualifies them best: they insist on existing *without any possibility of return*. The risk they take in order to continue in existence can never be taken a second time; if they fail, they disappear for good. No mode is more demanding in terms of the difference between success and failure.

We can recognize them first in two forms, as LINES OF FORCE and as LINEAGES, two distinct ways of defining the minuscule or massive hiatus that separates their antecedents from their consequents. The difference between these two types of alignments is well marked by Whitehead when he points out humorously that museums of natural science keep crystals in glass cases, but they have to keep living creatures in zoos and feed them!

The insistence proper to lines of force—these entities called, too disparagingly, "inert beings"—has repetition and quantity as its consequences; they are numerous, no, they are countless, *because* they repeat themselves and insist. The very notion of FORCE, which will be such a useful handhold when physics and then chemistry are born, is the consequence of this repeated insistence and of this proliferation. But if these entities form *lines*, alignments, it is because, despite the hiatus, despite the leap from one instant to the next (a leap impossible for human eyes to discern), each occasion inherits something that allows it to sketch out, as Whitehead says (he was their mentor and, as it were, their protector!), "historic routes." The notion of a "material world" would be very ill suited to capturing their originality, their activity, and especially their diffusion, for it would transform into a full, homogeneous domain what has to remain a deployment within a network of lines of force.

But it is with *lineages* that the distortion would be greatest, if someone were still stubbornly insisting on talking about a "material world." Here the existents in question are much less numerous than the lines of force, much more complex and sensitive to all sorts of influences and opportunities; in order to endure, they must not only insist by repeating themselves, they must first of all succeed in enduring, and then in reproducing themselves—in the usual sense of the term—by running the truly frightening risk of disappearing entirely if they fail to pass something along—but what?—to the next generation. And all this with no possibility of returning to the past; no second chances. Living beings—for these are at issue here—sketch out more regional entanglements, to be sure, but also more folded, more heterogeneous, more inventive ones as well. Thanks to DARWINISM, we have been familiar for a century and a half with the risk taken by the entities that thrust themselves into subsistence through the intermediary of reproduction. We experience without difficulty the richness of the "almost," since we are, literally, its descendants. We have finally understood that there was no Idea of a Horse to guide the proliferation of horses. Here ends, on this point at least, the quarrel of the Universals. If each mode of existence defines a form of ALTERATION through which one must pass to subsist, then lineages continue to have much to teach us about the alterations and detours necessary to their subsistence.

But the grasp of existents according to the mode of reproduction is not limited to lines of force and lineages; it concerns everything that maintains itself: languages, bodies, ideas, and of course institutions. The price to pay for the discovery of such a hiatus is not as great as it appears, if we are willing to consider the alternative: we would have to posit a substance lying behind or beneath them to explain their subsistence. We would certainly not gain in intelligibility, since the enigma would simply be pushed one step further: we would have to find out what lies beneath that substance itself and, from one aporia to another, through an infinite regression that is well known in the history of philosophy, we would end up in Substance alone, in short, the exact opposite of the place we had wanted to reach. It is more economical, more rational, more logical, simpler, more elegant—if less obvious in the early phases owing to our (bad) habits of thought—to say that subsistence always pays for itself in alteration, precisely for want of the possibility of being backed up by a substance. The landscape discovered in this way seems surprising at first glance, but it has the immense advantage of being freed from any ultraworld—substance—without loss of continuity in being—subsistence. There is nothing beneath, nothing behind or above. No TRANSCENDENCE but the hiatus of reproduction. This newly acquired freedom of movement (in the world and in the language of the world alike) will count for a lot when we have to become authentic "materialists" and when we redefine, in Chapter 10, what must be understood by IMMANENCE.

Ethnography is obviously not in a position to sketch out the institution that would shelter these beings—we would need a whole new diplomacy, whose lineaments we shall discover only later. What

⊙ IN ORDER TO AVOID THE STRANGE NOTION OF AN INVASIVE MATERIAL SPACE.

is important to us here is simply to situate the trajectories of reproduction outside the stifling clutch of a "material world," or, worse, of an "external world," by recognizing that they have a capacity for ARTICULA-TION, and thus for expression, which makes them *comparable* to the other modes we have already recognized, since they are able to respond to the same questionnaire. Here an approach that is not so much positive as *defensive* will suffice to keep us from smoothing over all the leaps with the notion, though it is a very widespread one—we are about to find out why—of an "external" or "objective world" subjected solely to the reign of "laws of causality." This is because, if we are to get across all the modes, including reproduction, we need room, and the institution of a "material world" does not have enough to give us.

We know, thanks to Malinowski, that every anthropologist has his moments of weakness (is it the heat? exhaustion? homesickness? the mosquitoes?) when in spite of himself he gives in to exoticism: "These people are really too weird; their customs are absolutely atrocious; I want to go home." He gets over it, of course, but still, he occasionally succumbs. Our ethnologist, too, gives in from time to time, out of weakness, to Occidentalism. Especially when she hears some popularizer explain to her in tremulous tones that the quantum world "is not restricted to the three dimensions of common sense." What she finds really bizarre is not the quantum world, it is that predicator's idea of the common world. What! It has only three dimensions?! She turns and looks in every direction, but to no avail: she does not understand where the famous "Euclidean space" might be, a space that is supposed to be equally suited to all the world's objects and that would stand in such striking contrast to the breathtaking proliferation of the quantum worlds. She is no more convinced when someone adds to the ordinary world, in a sort of concession, the "fourth dimension" of time. She cannot keep from wondering: "How can these people believe for a second that they are living in a world of 3 + 1 dimensions? They are really too absurd. I want to go straight home.

Let's leave the Moderns to their weirdness." Only there's the rub: she has no other home to go to!

IF THOSE WHO HAVE OCCUPIED ALL THE SPACE NEVERTHELESS LACK ROOM ⊙
And yet the problem remains: through what cascade of category mistakes have the Moderns managed to start thinking that they live in a four-dimensional world when nothing in their experience, nothing at all, validates this astonishing reduction? If we do not succeed in understanding this dizzying gap between experience and its representation, we shall never understand the sort of frenzy for which they need treatment. This is moreover the only way to define the term MODERN, which we have been using from the start rather too casually, though we shall really be able to account for its meaning only if the inquiry succeeds.

We could give a more precise definition even now by saying that a Modern is someone who thinks he lives in a world of 3 + 1 dimensions. Provided that we add: and who then wonders with increasing anxiety where he is going to be able to localize the set of values to which he holds. In other words, a Modern is someone who, believing himself to be submerged in a world of 3 + 1 dimensions, is distressed to see that it literally no longer has any room, anywhere, for him to deploy his values. He considers the importance of law, morality, fiction, politics, the economy, organizations, perhaps religion, even psyches, collective actions, seeking to anchor them somewhere, all in vain: there is no longer any place to put them. He is groping in the dark. "The Son of Man has no place to lay his head."

This frenzy that has struck all observers since the adjective "modern" came into use stems less from a utopian dream than from the sort of wandering explained by the brutal expulsion not from an earthly paradise but from the entire habitable Earth. Modern man has been seeking to settle down for centuries, yet he has voluntarily chosen displacement, exile, in a terra incognita. As if the Whites, wherever they landed, left blank (white!) spots on the map. Because they believe they are living in a 3 + 1 dimensional world, precisely. Were they chased away from their homes? No, they expelled themselves! In thought, at least, for, in practice, on the contrary, they have settled in everywhere . . . they have conquered the world and yet they still lack room! These internal exiles are still fighting for their "living space," their "breathing room." We have to admit that, for our inquiring ethnologist, the paradox warrants a closer look, and we shall readily excuse her brief lapse into Occidentalism.

All the more so because it is precisely from here that, from the beginning of the modern era on (in the historians' sense, this time), all the poisons and perfumes of exoticism are going to emerge. The more the Moderns expel themselves from all habitable lands, the more they believe that they have discovered among the "others" peoples that are, unlike themselves, solidly attached, anchored, rooted, yes, "autochthones," as we say or, better yet, "natives." Oh, how they are going to start envying those noble savages! "If only we had been able to remain like them!" And there will never be a shortage of reactionaries to fuse these two forms of exoticism, the distant and the near, by starting to dream of a utopian utopia, utopia squared: "If only we could become like them again!" Once again rooted, once again native, once again autochthonous, once again "really at home." A recipe for creating the most dreadful barbarities. Inevitably, since the Moderns, to begin with, have never left home! Have never been modern! How could they have survived for a moment if they had really lived in this 3 + 1 dimensional world? A strange adventure, believing they are a people wandering in the desert searching for a promised land when they haven't even gotten out of Egypt! We told you that the Moderns warranted an in-depth anthropological study, that they too are really interesting . . . that they need us to approach their wounds with caution. That they are worth comforting; we might even contemplate caring for them.

Once again, these questions are much too vast to be confronted head on, and yet it is indispensable to do the genealogy, however sketchily, of this idea of a SPACE so invasive that it would stifle all modes of existence. As if the Flood had devastated everything, and there were only a few rafts floating on the waters, vessels on which the exiled Moderns had piled up in haste the few values they wanted to save.

It would not be wrong to define the Moderns as those who believe they are materialists and are driven to despair by this belief. To reassure them, it would not make much sense to turn toward the mind, that is, toward all the efforts they have deployed as a ⊙ IT IS BECAUSE THEY HAVE BEEN UNABLE TO DISAMALGAMATE THE NOTION OF MATTER ⊙

last resort, all the lost causes (and causes are indeed at issue here!) in order to situate their values in "other dimensions," as they say—dimensions other than that of "strict materialism" since matter, as we are beginning to understand, is the most idealist of the products of the mind. The operation we must undertake leads us in exactly the opposite direction: we have to de-idealize matter in order to arrive at immanence and find the means, at last, to follow experience. When everything is submerged in matter, there is no raw material, no accessible reality, no experience to guide us. The reconquest of the "living space" necessary for the deployment of the full set of modern values comes at this price.

⊙ BY THE PROPER USE OF THE [REP · REF] CROSSING.

Even though this is an extremely complicated issue, we are not completely helpless, since we have identified the crossing noted [REP · REF]. We have already understood that matter is a COMPOSITE arrangement that amalgamates, to the point of indistinguishability, the requirements of knowledge—a transfer of constants, or, to use the technical term, of IMMUTABLE MOBILES—and the requirements of SUBSISTENCE—maintenance in existence through the leap of reproduction. It is as if the mode of displacement necessary to reference had been mistaken for the mode of displacement of the beings of reproduction to which reference accedes. In other words, the notion of matter is going to come in and hide the [REP · REF] crossing by making it undergo this minuscule and nevertheless decisive modification that will make it impossible to tell the two hiatuses apart even though they are radically distinct.

NOW, AS SOON AS WE BEGIN TO DISTINGUISH TWO SENSES OF THE WORD "FORM" ⊙

The operator that is going to allow this slight displacement in the idea of displacement itself is the possibility of producing a description of formalism that is not itself formalist. The development that follows will seem somewhat brusque to the reader, but it is indispensable to the comprehension of the whole. If we do not succeed in deploying this crossing, all the rest of the inquiry may go up in smoke. (It might be more expedient, moreover, to skip over the rest of this chapter and come back to it after finishing the book, after verifying whether it is true or not that there is indeed now room to accommodate other modes of existence rather than simply multiple representations of a single world.)

It all depends on the possibility of redescribing the notion of FORM as a practice. The work of reference, as we now know, relies on the establishment of a series of transformations that ensure the discovery and the maintenance of constants: continuity of access depends on discontinuities. This is the only means—but a means whose practical discovery is always perilous and fragile—to ensure the back-and-forth movement, the coming and going owing to which one can start from a given point (a laboratory, an institute, a computer center) and reach another, more or less remote. Think about the hundreds of successive operations required by an electron microscope through which a researcher ensures access to the division of a cell that cannot be seen by the naked eye. Think about the strings of calculations needed for the spectrum analysis owing to which an astronomer ensures access to a galaxy, also invisible to the naked eye. Two infinities that should not scare us, since biologists and astronomers both have access to them without the slightest vertigo, from their laboratories (provided that they have gone through the tollbooths of their "access providers," a term from information technology that would provide a pretty good basic definition of the sciences). The thing known comes closer as the steps taken to reach it multiply!

There is no difficulty here: a *form* is what is maintained through a series of *trans*formations. Suspend the alignment of the transformations and the form vanishes at once. Using the metaphor of hurdle jumping or relay races, form, in this first sense, thus occupies the position of the *runner.* No matter how good an athlete one is, even during training, at every hurdle, at every passing of the baton, one always feels the little flutter that raises the heartbeat of the champion, of course, but also of the coach and the spectators. Why? Well, because he might fall, knock over the hurdle, drop the baton. It *can go wrong.* Form is what must be called a dangerous sport.

Let us look now at the makeup of the successive stages along the risky course of reference: it is composed of forms, this time in the very concrete sense of the term (it is interesting to note that in French *forme* and *fromage* have the same root). A form or shape, in this second sense, is always an object (an instrument, a document, an image, an equation) that allows *putting into form,* or *shaping,*

⊙ THE FORM THAT MAINTAINS CONSTANTS AND THE FORM THAT REDUCES THE HIATUS OF REFERENCE ⊙

because it ensures the *transition* between the "tails" side of the coin, closer to the original raw materials, and the "heads" side, which brings us closer to the stage of putting into words or calculations.

Here, too, this is very ordinary business: delicately placing a specimen brought back from an archaeological dig in a drawer lined with cotton is "putting into form," since the drawer is marked by a label with a number that will make it possible to categorize the specimen, and the white cotton lining makes the specimen's shape more visible (it was hard to make out when it was only a brown spot on brown soil). The drawer has its "tails" side—it takes in the fossil—and its "heads" side—the fossil receives a label and reveals its outlines more readily. Something like an *ideography*. A minuscule transition, to be sure, but indispensable in the long series of transformations that permit, in the end, perhaps, if the paleontologist is lucky, the reinterpretation of the fossil.

Through the centuries, every discipline has developed thousands of these arrangements for putting into form, from the humblest, like the drawer, to the most audacious equations. (But let's not belittle the file cabinets, the ring binders, the card files, the cupboards: you would be surprised at the number of sciences that depend on them!)

⊙ WE BEGIN TO OBTAIN A NONFORMALIST DESCRIPTION OF FORMALISM, ⊙ The key point is that each of these shapings, these putting-into-form events, has meaning *only through the stage n - 1 that precedes it and the stage n + 1 that follows*; only the set of successive embeddings allows this highly paradoxical back-and-forth movement that obtains continuity of reference (the runner) through the discontinuity of the stages (the hurdles, the passing of the baton). When we speak of form in the concrete sense of the term, then, we are designating the framework and the chain of all reference networks. There are *longitudinal* forms, as it were, that replay the constants through the transformations, that thus achieve immutability through mobility, and there are *lateral* forms that authorize the passage of the longitudinal ones by multiplying the transitions, gradually paving over the distance that separates one place from another. A bit like a ladder, which needs both vertical rails and horizontal rungs to be used to reach something.

The metaphors of hurdle-jumping, relay-racing, and ladders obviously have their limitations, for here it is by multiplying hurdles or relays or rungs that we speed up the course! The more numerous the stages to cross, in fact, the more the forms are separated by tiny hiatuses, the faster reference will move to catch what it is to *bring back*. It is as though by multiplying the transitions we can ensure coverage of great distances. In the end, when everything works, when the network is in place, access is indeed obtained; you put your finger on a map, a document, a screen, and you have in your hand for real, incontestably, a crater of the Moon, a cancerous cell deep within a liver, a model of the origin of the universe. You really do have the world at your fingertips. There is no limit to knowledge. To describe, as do the history and sociology of the sciences, the circulation of these veins and arteries of objective knowledge, from Greek geometry to CERN's huge detectors, is to appreciate the enthusiasm that these have generated. It is also to measure their fragility. Not only because any little thing may interrupt them but also because if they work too well they risk disappearing from view. This disappearance of the risky character of form—in the first sense—and of the concrete character of forms—in the second sense—is going to lead to the invention of a wholly parasitical sense of this same word. It is through the very success of reference that things begin to go badly.

Let us suppose that, through a mix of enthusiasm for the results obtained and a sudden outburst of laziness, which will be mingled, as we shall see in the following chapters, with powerful political and even religious motives, we were to start to take the word "form" in a *third sense*. This time, we are going to pay no attention at all to the back-and-forth movement of reference, and we are going to select *only some* of the stages covering the chain, without taking into account all the movement and all the apparatus necessary to the work of reference.

⊙ WHICH TURNS OUT, UNFORTUNATELY, TO BE WIPED OUT BY A THIRD SENSE OF THE WORD "FORM."

What forms are we going to choose? Certainly not those found at the beginning of the transitions (like the storage drawer!), for these are too material, too humble, too unworthy of respect to play the role we want them to play. No, we are going to isolate instead those *at the end*, those that have the consistency of a number, or, better, of mathematical signs. We are going to start saying that what really counts in reference (there is no more chain, no more linkage) is form in the new sense of a suspended *notation*, a *document* whose movement has been interrupted, a *freeze-frame*. And here is where the whole danger of the operation comes in: we are going to find ourselves tempted to believe that the true basis for knowledge lies there, and there alone. With this third sense of the word "form," we introduce a *formalist* definition of FORMALISM that is going to exploit parasitically the *nonformalist description* of forms in the first two senses. An isolated document, shifted 90°, is now going to be taken for the entirety of the risky transfer of immutable mobiles that the network as a whole made possible. It is as though we were to take the fascinated contemplation of a single hurdle, the last one, for the whole hurdle course, or if we were to take the baton dropped on the ground for the whole relay race, or the top rung for the whole ladder!

AT THIS POINT WE RISK BEING MISTAKEN ABOUT THE COURSE FOLLOWED BY THE BEINGS OF REPRODUCTION ⊙

As we saw in the previous chapter, the work of abstraction is a concrete job: it is the labor of a whole chain of proof workers, from those whose hands are black with dirt to those whose hands are white with chalk. Why, our ethnologist wonders, would one give this an abstract description that would result only in *interrupting* the movement of knowledge? Here is indeed a category mistake that we run no risk of making. And yet this confusion seems to be central to the definition of modernity. Why? Because an accident of history has come along to *combine* it with another suspension, another freeze-frame, that is going to be practiced now on the *other side* of the crossing, on the movement of reproduction [REP]. And it is through this double category mistake that the notion of "matter" emerges, sometime in the seventeenth century—let us say, as a reference point, at the moment when the RES EXTENSA is being invented, around Descartes. The Moderns—this is what will define them—begin to believe that the thought of matter describes real things, whereas it is only the way the RES COGITANS—itself dreamed up—is going to start *imagining matter*.

We have seen earlier that a mountain, a cat, a yeast, in short any line of force or any lineage at all, necessarily had to pass through a series of discontinuities [REP] to achieve continuity. To obtain being, otherness is required. Sameness is purchased, as it were, at the price of ALTERATIONS. These discontinuities are totally different from those of forms in the sense that I have just defined, but they compose the passes, the passage, the past thanks to which this particular type of insistence and persistence is achieved. This is what allows the mountain to remain the same, and the cat, even if it grows old, to prolong its meditation on its proverbial mat without being interrupted by the meditation of the no less proverbial philosopher drinking his white wine fermented by yeasts. All of these (mat, cat, mountain, yeasts, and even the philosopher) move along surprising trajectories, yes, networks, composed, as we have just seen, of *their antecedents and their consequents* separated by a slight gap, a little leap.

It just may be tempting, however, to erase these discontinuities, these filiations, these risks, in their turn (we shall soon see where this temptation comes from). Especially because they are not always visible, or because one may choose not to emphasize them. Now, as soon as we fail to note the hiatus of persistence in being, we are surreptitiously introducing a SUBSTANCE *underneath* subsistence. We are thus starting to imagine that there would be, "underneath" the beings of reproduction, a support, a subversive agent, a console, a seat that would be *more durable than they are* and that would ensure their continuity without having to take the trouble, themselves, to leap over the discontinuities required for existence. We would be starting from a *passage* of the same through the other, and then slipping unawares toward a *maintenance* of the same on the same.

At first glance, making the beings of the world go through this abrupt interruption that would isolate them from their antecedents and their consequents may seem to be no more meaningful than interrupting the course of reference by isolating one form from those that precede it and those that follow. In the first case, one would be interrupting the movement of reproduction; in the second, that of knowledge and access. So this second supposition appears as improbable as the first. One can thus rely on common sense to make sure that these two hypotheses are never entertained.

Except, precisely, if you *cross* the two suppositions, the two interruptions, and you make the *form taken on the side of reference* the thing that would ensure *substance on the side of reproduction.* Then you   ⊕ IN THAT WE RISK CONFUSING TWO DISTINCT COURSES IN THE IDEA OF MATTER.
eliminate all the risks, all the movements, all the leaps. You explain in a single stroke the famous correspondence between the world and

knowledge. You obtain simple DISPLACEMENTS, whereas up to this point you had to concern yourself with translations. In the place of the crossing, only a simple *transport of indisputable necessities* appears (we shall see in the next chapter what the adjective "indisputable" is doing in this muddle).

It's a bit as though the Evil Genius we met earlier, Double Click, had managed to wipe out everything that made both modes of existence risky, as if he had succeeded in erasing the two series of discontinuities that had made subsistence possible for the one, access to remote beings for the other. Everything is seemingly still in place, and yet everything is profoundly different, since on both sides the motor that made it possible to achieve displacements is missing. The race is always already won in advance—without any need to budge. We are now going to act as though there were cost-free displacements of constants both in the world—*res extensa*—and in the mind—*res cogitans*. And the two are going to become inextricable: the world is knowable; thought grasps the world. From here on it will be as though, from the fact that knowledge is possible, we had drawn the conclusion that the world was itself *made of* "knowability"! Matter becomes this ideal world that might be called *res extensa-cogitans*.

A FORMALIST DESCRIPTION OF THE OUTING ON MONT AIGUILLE ⊙ To flesh out this operation that seems so implausible, let us go back to the example of Mont Aiguille, and try to explain how the map works, setting aside everything that we learned in Chapter 3 as well as the first two definitions of the word "form."

In this demonstration *per absurdum*, we are going to try to account for the mystery of reference. To begin with, we shall get rid of the whole jumble of networks, geometricians, pack mules carrying geodesic reference points; we shall do without the whole slow accumulation of mediations, cartographers, national geographic institutes, and tourist offices; we shall thus force ourselves to ignore everything about the path of existence that Mont Aiguille has to follow in order to continue to exist. We shall skip over both modes of existence [REP · REF] at once. This time, there will be nothing painful about our hike; it's really just a stroll for our health. The explanation is self-evident. Necessarily, because it has no more obstacles to overcome! We shall say that the map and Mont

Aiguille "are alike" because they both *share the same form* (in the third sense of the word).

That the map is made up of forms (in the second sense) is unquestionable. It has meaning only because it *inscribes*, little by little, linked angles—obtained earlier at great cost by the perilous missions of geometricians equipped with geodesic targeting equipment and moving from triangle to triangle, starting from a base measured with precision (and a lot of hassles) by a surveyor's chain. But it is precisely all that rich practical experience, all that labor, that we have decided to erase. On this base map, the geographers then learned to respect the various constraints of two-dimensional projection, to draw elevation curves, to add shading, and then they taught us how to respect the typographical conventions as well as the color codes. Let's forget that too. But this still doesn't explain the effectiveness of my map, for Mont Aiguille itself isn't two-dimensional; it still doesn't fold up to fit in my pocket, it still doesn't seem marked by any elevation curves, and today, moreover, as it disappears into the clouds, it completely lacks the aspect of the little pile of calibrated scribbles marked in oblique letters in fifteen-point type, "Mont Aiguille," that appears on my map. How am I to superimpose the map and the territory?

All I have to do is act as though Mont Aiguille *itself, basically*, in its deepest nature, were *also* made of *geometric forms*. This is where the Evil Genius, the serpent of knowledge (though not that of good and evil) becomes truly dangerous. Then here everything is indeed explained all at once: the map resembles the territory because the territory *is* basically *already* a map! Map and territory are the same thing, or rather have the same form, because *things* are basically *forms*. I then obtain a term-for-term superimposition that gives the notion of correspondence an indisputable validity. The operation is painless, the passage surreptitious, the temptation immense. And it is true that at first glance, such an explanation appears so enlightening that it would explain the Enlightenment itself. It's Columbus's egg, the one that opens the way, not to the Indies but to the continent still more mythical than the one Columbus hoped to discover: the immense terra incognita of KNOWLEDGE, that continent formed by equating a mind (which *thinks* form) and things (which *are* forms). The idea is so impressive that the divine Plato himself draws from it the very idea of the *Idea*. The Object finds a Subject worthy of itself since both are made of thought.

⊕ GENERATES A DOUBLE
IMAGE THROUGH A
DEMONSTRATION
PER ABSURDUM ⊙

In this enlightening explanation there is obviously one tiny detail that seems not to fit in very well at the start and that makes the supposition absurd: Mont Aiguille, which I am going around on my hike, stubbornly continues not to resemble in the slightest the map that I unfold from time to time. It continues to bear down with its full weight, to veil itself in the scattered mist, to gleam intermittently with colors that the map does not register, and, especially, it continues to exist at scale 1: there is no way I can fold it up or make it change scale.

Now it is at just this point, in order to respond to a common-sense objection, an objection truly as massive as Mont Aiguille itself, that one may let oneself be tempted by a second supposition, a consequence of the first. To respond to this common-sense objection, let us agree that Mont Aiguille *has a double*. We are going to pretend that what, in the mountain, resembles the map, is its form, in the third sense of the word, and we are going to make this its real *basis*, its *true substance*, while setting aside all the rest, claiming that it is unimportant, that it is in fact *insubstantial*. Even if we are obliged to acknowledge (how could we do otherwise?) that this form remains *invisible* (except, precisely, through the intermediary of the map!) and that it appears solely to the universe of thought—still, this form is what remains real, objective, and even—here is where the amalgam is produced—*material*. Descartes would not have hesitated to subject Mont Aiguille to the same treatment as his famous ball of wax: subjected to erosion, everything would disappear, except extension. And the best proof of this fundamental and ineradicable objectivity is that it is indeed this formal half of Mont Aiguille that resembles the map, which is made, it is true, we have acknowledged this, of geometrical forms (to which have been added some typographical conventions that can just as well be ignored). The reasoning is logical from start to finish even if its consequence is not very rational, since it has lost the thread of the reasoning, that is to say, let us not forget, of the trajectories and the networks.

And this is not all. What are we to do, indeed, with the rest? For, after all, we cannot just wipe away with a stroke of the pen all that cumbersome accumulation of *dissimilarities*: the mountain remains in

⊙ THAT WOULD LEAD TO A DIVISION INTO PRIMARY AND SECONDARY QUALITIES.

its irreducible mismatch, which—a major drawback—cannot be missed by any observer, from a four-year-old child to the most seasoned climber: you freeze your fingers when you reach the summit of Annapurna, not when you unfold the *map* of Annapurna. To get out of the jam without abandoning the foregoing reasoning, there's no other solution, in the face of the indignant protests of the most widespread experience, except to take one more step in this sort of coherent madness: we shall now suppose that all these properties, these dissimilarities, are in fact superfluous, since they do not touch the formal essence, the rational objectivity of Mont Aiguille but belong rather to the "subjective" impression that the mountain inspires among the mere mortals whose minds remain, alas, "too limited" to grasp the thing "in its essence," that is, "in its form," by "thought alone." This is where the rest (which is almost everything!) is going to become and remain from now on a heap of peripheral attributes, devoid of reality, with respect to the unique real substance whose existence can be proved, moreover, in case of doubt, by the map (as long as the network that gives the map its meaning is left out, a network whose outline, by making the equipment visible, would instantly annul the so-called proof!).

In the seventeenth century, to designate this real, invisible, thinkable, objective, substantial, and formal Mont Aiguille, grasped by the cartography whose practice had been obliterated, people fell into the habit of speaking of its PRIMARY QUALITIES—the ones that most resembled the map. To designate the

BUT ONCE THE ORIGIN OF THIS BIFURCATION INTO PRIMARY AND SECONDARY QUALITIES HAS BEEN ACCURATELY IDENTIFIED ⊙

rest (almost everything, let us recall), they spoke of SECONDARY QUALITIES: these are subjective, experienced, visible, perceptible, in short, *secondary*, because they have the serious defect of being unthinkable, unreal, and not part of the substance, the basis, that is, the form of things.

At this stage of reasoning, Mont Aiguille indeed has a double. As Whitehead would put it, the world has begun to *bifurcate*. On the one hand there is an invisible but formal reality—which explains the effectiveness of the map since, at bottom, the map and the territory are each reflected in the other; and on the other hand we're left with a whole set of features, accessible to the senses, to be sure, but unreal, or in any case devoid of substance; the map, indeed, can get away with neglecting them (the dissimilarity is thus well "explained"!) since they refer simply to the perceptual requirements of human hikers. And this division of labor will recur every time a discipline—geology, agriculture, meteorology—approaches Mont Aiguille without foregrounding the instruments of its knowledge and its access. This multiplied BIFURCATION is going to make the reconciliation of modern philosophy with common sense infinitely difficult; its genesis is what will allow us to explain in large part the opposition between theory and practice that is so characteristic of the Moderns.

Because of this Bifurcation, or, better, these multiple Bifurcations, we see the emergence of that strange artifact of matter, RES EXTENSA-COGITANS, this world of displacements without transformation, of strict linkages of causes and effects, of transports of indisputable necessities. The fact that this world is impossible and so opposed to experience will not be held against it; on the contrary, that it is contrary to experience *proves* its reality. In the grip of such a contradiction, Reason herself cannot help but cry out: "*Credo quia absurdum!*" "I believe because it is absurd."

⊙ IT BECOMES A HYPOTHESIS
TOO CONTRARY TO
EXPERIENCE ⊙

Is such an operation feasible? Can such a cascade of implausible consequences be reasonably—let us not say rationally—sustained to the end? Even though the answer is obviously "No, of course not!" from the standpoint of common sense, it turns out that this operation has been sustained and extended to everything and everyone by the Moderns: it has reached the point of defining solid, serious, brute materiality. This is one of the knots of our entire history. No one believes himself to be a realist—not among the Moderns, at least—if he is not a "materialist" in the sense in which we have just defined the term, if he does not believe that everything visible that exists is forever carved out of the unique fabric of "knowability"—a composite and toxic product that does not even have the advantage of ensuring knowledge for us, since it is as far removed from the networks of reference as it is from the paths of existence that allow beings to continue existing. As if all the objects of the world had been transfused and turned into something like zombies.

The reader unfamiliar with the anthropology of the Moderns will object that putting Mont Aiguille through such an indignity by this doubling, imposing on common sense such a flagrant contradiction and, more strangely still, rendering the very establishment of chains of reference—the only guarantors of both the production and circulation of equipped and rectified knowledge—unthinkable, incomprehensible, while depriving the world's beings themselves of any path of existence: performing such a series of operations has no chance to succeed. One cannot found Reason by rendering the world insubstantial, experience vain, science itself unattributable. This whole matter of matter has to have remained just a simple mind game.

This reader could reassure himself moreover by telling himself that practicing scientists, those who work proofs, in short, all those whose direct interests require the establishment and continual maintenance of knowledge networks, will rise up to prevent anyone from giving a version of his work that would so manifestly interrupt its course. How could scientists allow notions to be developed that would no longer make it possible to equip the paths of knowledge with all their heavy apparatus of forms and instruments? How could they not be the first to make sure that the conditions have all come together to capture the rare events called discoveries? It isn't done by corresponding with zombies, this much they know perfectly well. Moreover, the more the sciences develop and insinuate themselves everywhere, the more the continual pulsation of these networks becomes visible, and the less one will risk confusing their mode of displacement with the others. This vascularization of the sciences is as visible as the veins and arteries on the inside of the wrist. Not a single scientist would let himself be taken in by this childish example in which the form of Mont Aiguille on a map is naïvely mistaken for its fundamental reality.

Yet the reader should not count too much on the resources of good sense. This would mean forgetting the immense gap between Science and the ⊙ AND THE MAGIC OF RATIONALISM VANISHES ⊙ sciences, a gap that divides all the practitioners themselves—as we have seen from the introduction on—and that explains the decision to begin

the inquiry with this question, so off-putting in appearance. In the genealogy I have just sketched (one that would take volumes to complete),
matter does not emerge as a part of the world, a demarcated domain of
a much broader ontology (alongside or underneath another domain
that would be "thought" or "mind"!): it is an institution, an organization, a distribution of the powers of thought (and of politics, as we shall
soon see). Although it may seem strange, we can speak appropriately
from now on of the *institution of matter* at the heart of the history of the
Moderns. This is what has brought about the quite peculiar designation
of the *res extensa-cogitans* and what has over time produced the strange
scenography of a being that believes itself to be a "Subject" in the face of
what it believes to be an "Object."

How can such an institution have been established despite the
continual denials of common experience as well as of scientific experimentation? We have to remember that there may be situations so
perilous that people will prefer the *irrealism of description* to the *power of
the effects achieved*. There might exist such powerful motives that they
will sweep away all objections and give such a thought experiment some
rather solid support. Such are the motives at the heart of the RATIONALISM
to which Moderns believe they must cling as to the apple of their eye.

We shall have an initial idea of these motives if we recall the point
of departure: erasing the two types of hiatus, of breaks in continuity,
of mediations that make the detection of constants allowing access
to remote beings [REF] as risky as the discovery of discontinuities that
permit subsistence [REP]. With one stroke of the magic wand—and it
is really a question of magic here, except that magic is the source of the
idea of Reason!—we mow down all the difficulties, we eliminate all the
risks, we forget all the failures, we have no more need for any costly, local,
material conditions. Necessities (that no one has produced) are trans-
ferred (untransformed) without conduits, without networks, without
cost, throughout a world at once real and knowable, composed entirely
of forms that are the only substances. The construction may appear mad,
but the gains are enormous.

⊙ SINCE WE CAN NO LONGER
CONFUSE EXISTENTS
WITH MATTER ⊙

How can we manage to put a stop to this
Flood that is drowning existents under the waters
of matter—of the *thought* of matter? The difficulty
ceases, the waters begin to recede, as soon as we

notice that this RES RATIOCINANS is never *of* or *in* space. If it gives the impression of being "everywhere," this is because it is literally *nowhere*, since it does not pay for its displacements by setting up networks. If it is capable of invading everything (in thought), this is because it never controls the budget for its extension, because it wipes out all the gaps, short-circuits all the passes, and acts as though there were only undistortable displacements of necessities, concatenations of causes and effects in which even the little leap, the little break in continuity, the hiatus between cause and effect, had disappeared—necessarily, because the relay, as we are about to see, has been surreptitiously taken over by an argument whose goal is to put an end to a debate.

If the popularizer I was mocking earlier, dumbfounded by the multiplicity of quantum worlds, could believe that the "common world" unfolded in a "space of only three dimensions," it is because it cost him nothing to believe that the microphone into which he was speaking, the rostrum from which he was pontificating, his own body, his genes, the walls of the room, the audience that he was carrying along in his frenzy, all that too was bathed in a Euclidean space. He has to believe that, since he submerges in thought all the dimensions that compose them, bathing them in that *res extensa* that renders unattributable any operation of measuring and any element of proof. Experience will never contradict him, because he has lost the thread of experience.

If he had begun to *take the measure* of what he was saying in a somewhat serious way, for example, if he wanted to have a carpenter make him a copy of the rostrum that had brought him such success, he would have had to take a woodworker's tape measure, a square, a piece of paper, and a pencil out of his pocket; and all this wouldn't have been enough, for he would have had to draw the piece of furniture in perspective or as a projection before he sought out a color specialist to choose the tint and put together another set of samples so he could decide on the quality of the wood; and even so, a rostrum is easy to draw, it bears enough resemblance to a thinkable object in Euclidean space. How could we imagine the work he would have had to undertake to capture the dimensions, spatial relations, temporal relations, and rhythms of the set of beings gathered together to listen to him? One hopes, for him, that, after interrupting his talk for several minutes in order to ponder all the dimensions of the problem, he would have modified his conclusion and admitted that the quantum world is child's play in comparison with the multiplicity and complexity of the dimensions that are simultaneously accessible to the most minimal experience of common sense.

Unless, and this would be cleverer, he were to conclude that, all things considered, once rid of its transfusion of *res ratiocinans*, the common-sense world, with all those leaps, discontinuities, and unexpected branchings, closely *resembles* the quantum worlds. Except for one detail: it has been *infinitely less explored* than the other! What a fine paradox: we have gotten so much in the habit of thinking that we believe in the *res ratiocinans*, thinking that we live, as it were, "submerged," that the world of common sense has become less thinkable, less calculable, less describable than that of the infinitely small . . .

⊙ A MATTER THAT WOULD NO MORE DO JUSTICE TO THE WORLD THAN TO "LIVED EXPERIENCE."

We see now why we have been able to define the Moderns as a people who believe themselves to be materialist and despair at the thought. For the Moderns are never entirely comfortable with this position. Who would want to live indefinitely flooded, under water, without access to dry land, to terra firma, lacking even raw materials? This is in fact the strangest consequence of the extension of the *res ratiocinans*: it does not even let us do justice to the values of the Moderns, to the values they themselves most obviously hold dear, and, to begin with, the sciences themselves.

In everything I have said so far, I have not claimed that materialism, unfortunately, missed the subjective, the intimate, "lived experience." In the example chosen, I did not try to make my reader resonate with the warmth of my *feeling* for Mont Aiguille, a feeling that "will never be captured by the frozen knowledge of geologists or mapmakers." Quite to the contrary: the establishment of chains of reference, the history of cartography, geology, trigonometry, all this was just as warm, just as respectable, as worthy of attention as my pale expressions of admiration, as my emotions as an amateur hiker and as the shiver I feel when the wind comes up and chills the sweat running down my chest. By splitting Mont Aiguille into primary and secondary qualities, making it bifurcate into two irreconcilable modes, what is neglected is not only subjectivity, "lived experience," the "human," it is especially Mont Aiguille itself, in its own way of persisting, and, *equally*, the various sciences that have striven to know it and that depend on its durability to be able to deploy their chains of reference. In this matter, it is not only humans who lack room, it is first of all Mont Aiguille itself, and second, the various sciences that allow us access to it! If the splitting had caused only the neglect of human feelings, would the loss be so great? The danger is that this loss threatens to deprive us of both the map and the territory, both science and the world.

Our investigator understands perfectly well that in criticizing materialism she risks getting mixed up in a defense of "spiritualism"— which would lead her straight to the nineteenth century—or in a struggle against "reductionism"—and we would then be right back in the twentieth. But she understands now that these two battles have ceased for want of combatants, or, more precisely, *for want of matter*. There is no matter at all. The *res ratiocinans*, that strange composite of *res extensa* and *res cogitans*, is not the basis for the world. We don't have to struggle "against" it. We can just do without it, as physicists have learned to do without ether. It is a badly conceived institution, in fact, the effect of a badly written Constitution intended to establish an awkward compromise between entirely contrary constraints, the result of a *conflict of values* [REP · REF] that has had as an unintended consequence the digging of an abyss between theory and practice, the relegation of experience to the inexpressible; and it has ended up hiding the very materiality of materials under its profound ignorance.

As we shall observe in subsequent sections of this book, even technology, even the economy, these triumphs of modern "materialism," are not made more comprehensible than the sciences if one confuses their raw materials with matter. Imagine this: a people for whom common sense is less familiar than the quantum world and that cannot account for its own greatest exploits, technology, the economy, objective knowledge, three of its principal sources of pride! We should not be astonished that the Moderns have been rather surprised to see the specter of Gaia suddenly fall upon them.

To exit from matter and allow comparison with the "other cultures," we must not look up, for example, toward the mind, but rather *down*, toward the solid ground whose damp, rich, and fertile forms are beginning to reveal themselves. If matter does not exist, then the waters have already receded. The institution of matter distributes the competencies of beings as poorly as possible, and ensures no protection whatsoever for the deployment of modes of existence. In the face of this institution, anthropology (conjugated with militancy, and almost half diplomacy in any case) finds itself rather like General de Gaulle champing at the bit with impatience before the Constitution of the Fifth Republic during his time away from power. Anthropology knows that nothing good will come of it; it has in mind a different distribution that would share powers appropriately and would liberate the energies that are continually hampered by the current arrangements; it is waiting for grave events to overturn the old procedures. Are we not also waiting, we too, for grave events to overturn the outworn institution of matter and are we not also, we too, expecting an entirely different Constitution? Moreover, haven't these grave events already taken place?

REMOVING SOME SPEECH IMPEDIMENTS

....

If we had to begin with the hardest part ⊙ it was because of an insistence on "straight talk" that connects formalism with closing off discussion.

Although this straight talk cannot rely on the requirements of reference [REF], ⊙ it leads to the disqualification of all the other modes ⊙ by creating a dangerous amalgam between knowledge and politics [REF · POL], ⊙ which makes it necessary to abandon the thread of experience in order to put an end to debates.

Fortunately, the method that allows us to recognize a crossing ⊙ will succeed in identifying a veridiction proper to politics [POL], ⊙ which has to do with the continual renewal of a Circle ⊙ that the course of reference cannot judge properly.

Thus we have to acknowledge that there is more than one type of veridiction ⊙ to foil the strange amalgam of "indisputable facts" ⊙ and thus to restore to natural language its expressive capacities.

The most difficult task remains: going back to the division between words and things ⊙ while liberating ourselves from matter, that is, from the res ratiocinans ⊙ and giving ourselves new capacities of analysis and discernment ⊙ in order to speak of values without bracketing reality.

Language is well articulated, like the world with which it is charged, ⊙ provided that we treat the notion of sign with skepticism.

Modes of existence are indeed at stake, and there are more than two of these, ⊙ a fact that obliges us to take the history of intermodal interferences into account.

I OWE THE READER AN APOLOGY FOR HAVING
HAD TO BEGIN WITH THE MOST CHALLENGING
ASPECT OF OUR INQUIRY. THERE WAS NO OTHER
way to get started, since all the questions that make the anthropology
of the Moderns so obscure have played out over time around the place
to be given to equipped and rectified knowledge. If we had not first un-
raveled the confusion between the modes of reproduction [REP] and reference [REF], in order to bring out their correspondence, we would have
been caught in the Subject/Object vise, and we would have had to believe in the existence of an "external material world known by the human mind." In particular, we would have been obliged to consider the
Moderns as "NATURALISTS" and thus forced to share the prejudice with
which they view all the other cultures: unlike "us," it seems, these others have "confused" matter with its "symbolic dimension." Now, if there
is one civilization that has dreamed matter, imagined it, blended it with
all sorts of "symbolic and moral dimensions," it is certainly that of the
Moderns: the idealism of the Moderns' materialism is what strikes our
ethnographer, on the contrary. We shall see later on that, if we accept this
viewpoint, new possibilities will open up for comparing COLLECTIVES and
for entering into relations with existents. For the time being, we are still
far from having recovered the freedom of maneuver necessary for our inquiry. We still have to remove some *speech impediments.*

First, because we have not yet discovered the reason why this
amalgam of the two modes [REP · REF], despite its implausibility, ended

up triumphing over experience, even though experience is so contrary to it. In fact, the Moderns do not live, any more than other peoples do, in a world that has really split into primary and secondary qualities. They would not have survived. They may *believe* they are modern, but they cannot actually be so. Thus powerful reasons were needed, as I indicated earlier, for the implausibility of Bifurcation to be preferred over the warning signals of common sense; the analyst now has to try to understand these reasons.

We take a giant step in the anthropology of the Moderns when we discover the curious link they have set up between FORMALISM (in the third sense of the word "form") and a certain type of interlocutory situations—situations we shall have to call political. There is an event here, a knot, a conjunction, a convergence of circumstances that determined for the Moderns, over a very long period of time, all the ways of *speaking well* about something to someone.

⊙ IT WAS BECAUSE OF AN INSISTENCE ON "STRAIGHT TALK" THAT CONNECTS FORMALISM WITH CLOSING OFF DISCUSSION.

Our investigator noted some time ago that the Evil Genius, Double Click [DC], claimed to engage in *straight talk*. From the start, the requirement had always struck her as absurd. What? One would go from proof to proof, transporting what one meant to say without chicanery, without exoticism, without eloquence, without provocation, without rhetorical embellishment, without fanfare? There would be no break in reasoning, no hiatus in expression, no turns of phrase or circumlocutions, no impromptu shifts of ground, no metaphors, for sure, no tropes, either (both tropes and metaphors being forms of drifting, impulsivity, deviation, seduction)? In short, people would speak *literally*? They could maintain from one paragraph to the next a path that would pass from necessity to necessity, through simple DISPLACEMENT, without ever jumping through any operation of TRANSLATION? They would achieve total coincidence between words and meaning and thereby succeed in stating what is, since what is—that, too—advances from necessity to necessity by mere displacement? They would speak straightforwardly, *with no detours*, about what is patently obvious, under the heading of "common sense"?

Here is a requirement that would seem untenable in any other civilization. Unless it had been decided that such a way of speaking would make it possible to cut through endless disputes, to humiliate phrase-makers, to restore dignity to the weak, to bring down the powerful, and to take a fast road through the woods, the sunken roads, the swamps, and the booby traps of ordinary ways of speaking and believing . . . This would have implied a radical opposition between "knowledge" on the one hand and opinion, that wretched *doxa*, on the other. Here is a feature that needs to be analyzed in detail: the claim of straight talk that it obeys the movement of knowledge, but for polemical reasons, battlefield reasons.

ALTHOUGH THIS STRAIGHT TALK CANNOT RELY ON THE REQUIREMENTS OF REFERENCE [REF], ⊙

Following what has just been said, it will be clear that political speech cannot be assessed by contrasting it with Science, since the paths of reference do not run straight either. To be sure, they are not exactly crooked: in the end, when everything is in place, they do ensure the *direct* access that may in effect be taken for a straight road. Still, as we have just seen, their ultimate rectitude is the *result* of the establishment of chains of reference. Consequently, Double Click can in no case allege familiarity with the paths of knowledge as a pretext for casting doubt on the quality of political speech.

The history of the sciences never tires of showing us through what blind gropings, by what twists and turns scientific collectives reach this correspondence, this adjustment with the beings of the world. Access to some remote object depends, as we have seen, on the paving constituted by multiple intermediaries and on the rapid but always precarious shifting of a constant. Not counting the fact that the setup and linking of such a string of forms (in the first two senses of the world recognized above) require a long apprenticeship that is always achieved by trial and error. The movement of these forms looks more like the agitation of an anthill than the passage of a high-speed train. Scientists can obviously decide to limit themselves to the final results alone, but then they will have dried up the resource that would allow them to gain access to *new* beings and to trace *new* paths—and they will quickly lose access, because they won't know how to maintain the roads and the means that have already been established. Neither straight lines nor crooked ones, networks of reference move along in their own way, and if they have to be compared to something, it should be to the everyday work of a Department of Roads and Bridges, with its corps of structural engineers and the back-and-forth movements of bulldozers on the construction site of a public works project. Wanting to use straight talk thus does not mean, despite the claim often put forward, that one is going to speak "as the sciences do" or "the way a scientist talks," but that one wants to *imitate the results* without having to encumber oneself with *imitating the burdensome process as well.*

Someone will argue that, at least in some of the more formalized disciplines, it is possible to maintain necessities throughout an entire discourse, and that it is not clear why, by following the scientists' model, Double Click would not succeed in linking his demonstrations and finally produce straight talk. Isn't this what is called writing *more geometrico?* But this would mean forgetting everything we have learned over the last thirty years of science studies about the *mores* and the *manners* of geometricians, and about the ever-so-concrete work by which formalisms are produced. If the sciences are indeed capable of carrying out transfers of necessities, it is only on condition that these necessities be made to traverse an always stupefying series of transformations every time; it never happens through the maintenance of an identity (which is impossible in any case). What is striking, rather, in the establishment of chains of reference, is the continual invention of modes of writing, types of visualizations, convocations of experts, setups of instruments, new notations that permit the cascades of transformations we have noted above. IMMUTABLE MOBILES do end up traversing the universe, but it is because they pay for each transport with a transformation. Without these series of innovations, there is no objectivity, no necessity, not even any *apodictic* proof.

What strikes every ethnographer of the Moderns is not that straight talk has the slightest plausibility (how can one speak without tropes, without figures, without metaphors, without drifting, without virtuosity, without scrambling sideways), but that merely evoking it suffices to discredit all other forms of speech.

⊙ IT LEADS TO THE DISQUALIFICATION OF ALL THE OTHER MODES ⊙

Strangely, it is the very implausibility of straight talk that makes it astoundingly effective for disqualifying all the other modes. Once one makes the mere *supposition* of a transfer of indisputable necessities, of information without any transformation, of displacement without any translation, all the other ways of speaking are suddenly subject to deep suspicion. And, even more strangely, the other modes start to doubt themselves and their own capacity to distinguish between truth and falsity.

In comparison with this unattainable ideal, the advocates of straight talk set about using terms like "figurative," "ordinary," or "incomplete" to label all the displacements, manipulations, operations, all the discontinuities, all the understandings that have become more or less illegitimate but that are used by those who do not engage in straight talk to express themselves more or less awkwardly: poets, rhetoricians, common people, tradesmen, soothsayers, priests, doctors, wise men, in short, everyone—and of course scientists, whose ways and means will cruelly *disappoint* the Double Click sectarians who would so much like to imitate their effects [REF · DC]. Once the standard of *straight* talk has been invented, everyone else suddenly begins to engage in *crooked talk*; they become double-dealers, liars, manipulators. And then begins the crushing labor of the rationalists: they start seeking to *rectify* everyone through a sort of generalized speech therapy. If the domain of what they call the "irrational" is so vast, it is because the rationalists adopt a definition of "rational" that is far too unreasonable and far too polemical. If we, too, give in to the temptation of straight talk, a whole series of modes of veridiction, so decisive for common life, risk falling into oblivion, henceforth unable to have their own criteria of truth and falsity, or at least incapable of achieving their full measure of realism, their ontological dignity. The danger is greater still, since EXPERIENCE itself may well stop being expressible. If the only shibboleth becomes that of Double Click information and straight talk, then *all* experience starts to *ring false*. It will never be expressible because it will never be formalizable (in the third sense of the word "form"). Experience will have been lost from sight, and with it, of course, any possibility that the Moderns may be EMPIRICAL, that is, may draw lessons from their experiences.

⊙ BY CREATING A DANGEROUS AMALGAM BETWEEN KNOWLEDGE AND POLITICS [REF · POL], ⊙   The analyst finds herself here before one of the knots in the inquiry that must be patiently untangled: the principal impediment to speech among the Moderns comes precisely from their strange idea of a speech that nothing would impede any longer! Compounding the strangeness, they have borrowed this model of obstacle-free speech from mathematical demonstrations, whereas mathematicians, on the contrary, know very well, and from firsthand experience, what obstacles have to be removed one at a time if they are to succeed in transporting a necessity from one point in the reasoning to another. And yet Double Click uses this manifest counterexample— after clearing away all the mediations—to criticize all the other modes. How could such an operation stand a chance? This is the third strange thing about it: *the reasons are political.* We shall understand very little about this curious crossing between knowledge and reproduction if we fail to approach a different crossing, this time the one between knowledge and the conditions of public life [**REF · POL**]. Our ethnologist managed to jot down this observation in her notebook: "The Moderns are those who have kidnapped Science to solve a problem of closure in public debates."

One doesn't have to be a genius to imagine situations in which the adventures of knowledge can be put on the wrong track in this way. We discover such situations as soon as we turn back to the agora where, as we have seen, the fate of **CATEGORIES** is decided— ⊙ WHICH MAKES IT NECESSARY TO ABANDON THE THREAD OF EXPERIENCE IN ORDER TO PUT AN END TO DEBATES.
amid the tumult, the controversies, the endless quibbles, the polemics always at risk of slipping into violence and rampage. Let's suppose that there arises, here, a way of "saying something to someone," a life form, a literary genre, that procures the unexpected advantage of giving the impression to anyone who uses these tropes that they *put an end* to debate, *bring closure* to controversies, by mobilizing, through various conduits, various ploys (which will never appear as ploys), these transfers of necessities, transfers that have now become *indisputable*. What appeared so improbable, in the previous chapter, in the amalgam between ways of knowing and the peregrinations of existents now takes on a formidable efficacity: through a subtle *bypass* operation, a seemingly metaphysical question (*of what* is the world made?) is linked to a question of argumentation (*how* can we put an end to the endless squabbling?).

The scene is very familiar; it has been brilliantly studied by excellent authors: it reproduces, this time through arguments, the doubling that we have just seen emerge in the institution of matter *with no possibility of distinguishing between the doubles*. The most decisive effect of such a bypass is that from then on we shall never know, when someone is speaking of any subject at all, whether the speaker is seeking to close off the discussion or to transport a series of causes and effects. Here lies the specific genius—the deviltry, rather—of this amalgam. When you turn toward the world, it already has the aspect of an argument that someone is going to make against you; when you turn toward the interlocutory situation, you find the train of the world itself heading toward you full steam ahead with all its boxcars, without giving you time to react. All the distinctions between what the world is made of, how one can know this, and how one can talk about it vanish. It is no longer a question of *res extensa*; it is not even a question of what I have called *res extensa-cogitans*; it is another creature entirely, another ether, another ideal, which we shall finally have to call RES RATIOCINANS. A determining invention for the idea that the Moderns come to have about themselves. It is indeed a *res*, but now it is also *ratiocinans*.

Before very long our investigator begins to suspect that it is not only to correct the bad manners of tinsmiths, wet nurses, or helmsmen, or even to do justice to expert geometricians, that Double Click got the idea of rectifying all menacing discourses. The slightest knowledge of the history of philosophy is enough to let us situate the origin of this invention of straight talk, which is nevertheless so remote from the practice of objectivity. The ones whose divagations needed to be straightened out were obviously the most dangerous of men: the Sophists, the Rhetoricians, the Politicians. These are the Philosophers' only real rivals on the agora, those who are capable, and culpable, through the agitation of their sharp tongues alone, of drowning all proofs under sarcasm, of stirring up or calming down the crowd, of getting the assembly to vote either for Helen of Troy, the holy martyr, or against Helen of Troy, the filthy slut; in short, they are constantly stirring up the witches' cauldron that is called public life. Those people, it is said, are indifferent to the search for truth. They lie shamelessly. They manipulate brazenly. Their speech is crooked and twisted—no, *they* are crooked and twisted. Such at least is the Master Narrative in which the beginnings and the necessity of Philosophy are located: against the dangers of Politics, Reason must be able to serve as counterweight; otherwise, they threaten, all hell will break loose.

The ancient version of this battle is renewed from one century to the next through formulas, different every time, that always revive the same split: on one side the indisputable demonstration based on facts that are themselves indisputable; on the other, eloquence, rhetoric, propaganda, communication. There is an unbroken thread, a basso continuo, from Socrates's arguments—"You don't know your geometry, Callicles!"—to attacks on the purported relativism of science studies. There are speakers capable of serving as conduits for transfers of indisputable necessities who are the absolute opposites of smooth speakers, fine speakers: the latter's speech, even if it has effects, succeeds in transmitting only questionable approximations. On one side, *apodeixis*, demonstration; on the other, *epideixis*, rhetorical flourishes. As if, throughout history, the people most passionate about argumentation had endlessly rediscovered the betting strategy that would put an end to argumentation!

We shall never manage to keep the CONTRAST identified above between reproduction [REP] and reference [REF] open very long if we don't succeed in bringing out this other crossing between reference and political speech. The historical question is still under dispute, but it seems as though the Moderns may have actually been the only people in the world to have imagined such a setup. In any case, the two are too closely linked by history for us not to treat them in sequence, and, of course, each is defined by the other: the crooked line has been imagined in terms of the straight line and vice versa. People have told themselves that they had to talk *straight* to have a chance to straighten up those who talked *crooked*; as a result, *two modes* have been made to deviate from well-formed speech instead of just one.

Just as we had to undo the composite notion of MATTER to extract from it the crossing of which it is only the AMALGAM, we now have to untangle another badly composed notion that makes the opposi- FORTUNATELY, THE METHOD THAT ALLOWS US TO RECOGNIZE A CROSSING ⊙ tion between Reason and Politics the preferred Gigantomachy of the Moderns. But we now know how to proceed: frontal opposition doesn't impress us any more than factitious confusion does. We too, from here on, have our reasons that in no way resemble what good sense tells us about Reason. Consequently, nothing prevents us from substituting the

CROSSING between two modes of veridiction for the great battle between demonstration and rhetoric.

The investigator now knows the three criteria by which one can recognize a mode of existence. First, thanks to a category mistake: she feels vaguely, in the beginning, and then more and more precisely, that she is missing something, that she isn't getting what is said in the right tonality, that she hasn't preceded it with the right PREPOSITION. Alerted by this feeling that she has blundered, she understands that she must look, second, to see if there is some type of discontinuity, some HIATUS that would account for a particular type of continuity and that would thereby trace a TRAJECTORY, its own particular PASS. Finally, she knows that she has to find out whether there are FELICITY AND INFELICITY CONDITIONS that would make it possible to say of a mode of existence in its own idiom under what conditions it is truthful or deceitful.

⊙ WILL SUCCEED IN IDENTIFYING A VERIDICTION PROPER TO POLITICS [POL], ⊙

Someone will object that this is precisely what is impossible in the case of political speech (let us use the word "speech" for now while waiting to invest it, in Chapter 12, with its full charge of reality), since this speech is in the first place an art of manipulation and lies. Yes, but the ethnologist is now on the alert: the accusation of lying is brought by a way of speaking that claims for its part, *not to pass through any pass*. This smacks of category mistake and then some. For straight talk—that serpent's language of the demon Double Click—cannot serve as judge, since it never allows a category to be heard in its own language. Beware: everything it suggests calls for skepticism.

What happens if we try to define political speech *in its own language*? The category mistake that straight talk seeks to impose on it is glaringly obvious: political speech cannot seek to transport *indisputable* necessities because it was born *in* and *through* discussion, at the heart of controversies, amid squabbling and often in the grip of extreme violence. To those who fight so that their own desires and intentions will be debated, indisputable necessities are not of much use. Asking political speech to engage in straight talk is as absurd as asking a florist to send the azaleas you promised your mother-in-law through the telephone line you have just used to order them: a manifest branching mistake; the wrong choice of conduit.

Must we ask political speech to establish chains of reference to accede to remote states of affairs? The conduit mistake would be almost as flagrant. It is not that objectivity would bother it or be burdensome, it is that political speech has nothing to do with the form of objectivity: for the former, as Callicles puts it so well, in the heart of the agora, with urgency, in the middle of the crowd, it is always a matter of responding, on the fly, without full knowledge of the ins and outs of the issue, to a whole series of questions in which the life and death of the collective are at stake. For political speech, resolving the problems of access to remote states of affairs one by one would be of no use at all for resolving its own problems, which have to do with urgency, with multitudes, and—especially—with turmoil.

If it is in the grip neither of straight talk nor of the quest for access to remote beings, does political speech lead anywhere? It seems as though all politicians, militants, activists, citizens, endlessly reply "yes." Their most ordinary experience, their deepest passions, their most stirring emotions, involve bringing to the surface a group that has unity, a goal, a will, and the ability to act in an *autonomous*, that is, free, fashion. Every word spoken, every gesture made, every intervention, every situation finds itself traversed by a trajectory whose trace seems quite specific, since it produces temporarily associated wills. It is not very risky to identify POLITICS as a *movement*. Can we recognize the hiatus and the pass that would be responsible for this movement? Probably, and it is even against the hiatus or the pass that straight talk constantly wags a finger: there is nothing more fragmented, interrupted, repetitive, conventional, and contradictory than political speech. It never stops breaking off, starting over, harping, betraying its promises (from the standpoint of the straight path), getting mixed up, coming and going, blotting itself out by maneuvers whose thread no one seems to be able to find anymore.

But how does political speech *itself* judge these hiatuses? How does it evaluate the forward or backward thrust of its own movement? Through its capacity to obtain unity from a multitude, a unified will from a sum of recriminations; and then through another capacity, its ability to pass, by just as dizzying a series of discontinuities, from

⊙ WHICH HAS TO DO WITH THE CONTINUAL RENEWAL OF A CIRCLE ⊙

provisional unity to the implementation of decisions, to the obedience of those who had been uttering recriminations, despite the continual transformation that this multitude imposed on the injunctions while resisting through every possible means. What was united disperses like a flock of sparrows. And everything must be begun again: this is the price of *autonomy*. This continual renewal or REPRISE of a movement that cannot rely definitively on anything is probably the most characteristic feature of political speech; the obligation to start everything all over again gives political speech perhaps the most demanding of all felicity conditions, and it explains the choice of the adjective "crooked." To sketch it out, we shall henceforth speak about the CIRCLE, since it is indeed a matter of ceaselessly retracing one's steps in a movement of *envelopment* that always has to be begun again in order to sketch the moving form of a group endowed with its own will and capable of simultaneous freedom and obedience—something that the word AUTONOMY captures perfectly.

⊙ THAT THE COURSE OF REFERENCE CANNOT JUDGE PROPERLY.

We shall have to return at greater length to this mode—unsurprisingly noted [POL]. Our immediate concern is to show to what extent the veridiction of this mode cannot *be properly judged* either by straight talk or by reference. Yes, in politics there is indeed a sort of lie, but lying, for politics, does not mean refusing to talk straight, it means *interrupting* the movement of envelopment, suspending its reprise, no longer being able to obtain continuity, the continuation of the curve through the multitude of discontinuities: screams, betrayals, deviations, panic, disobedience, untanglings, manipulations, emergencies, and so on. And this truth is quite as demanding as all the others. Let us recall the difficulties reference has to overcome in order to win the hurdle race: a clever ploy is needed to make a constant leap from one form to another in order to obtain access to remote beings; but who can measure the courage of the men and women capable of making the political juices flow across these other breaks, these other hiatuses, and capable of winning that other race, the race for autonomy? And on the pretext of this ploy, some would belittle their courage?

The only result that matters to us here is showing that political speech is by no means *indifferent* to truth and falsity: it defines them in its own terms. What it is capable of generating in its own wake will be given it by no other form of veridiction; without political speech there is no autonomy, no freedom, no grouping. A capital discovery. An essential contrast. A supreme value. Here is surely a mode of veridiction that the Moderns, who are so proud of it, would not want to crush by entrusting to it, in addition, the impossible task of transferring referential truths that it is in no way made to transport [REF · POL]; or, what is worse, by exaggerating its falsity, requiring it to *abandon* any requirement of truth, as if it were condemned to do nothing but lie for all time. In this mode of existence there is something sui generis, in the literal sense of "self-engendering," something the Greeks called AUTOPHUOS, and that we are going to have to learn to treat with as much respect and skill as we grant to chains of reference.

Now, here is the full strangeness of the situation imposed by the mere suggestion of straight talk: how can such requirements possibly be judged by Double Click, who already understands nothing about science, since he is content to imitate its final results by a simple ersatz demonstration? If Double Click managed to discipline political speech, to rectify it, he would only produce a much worse monster, since one could no longer obtain the proper curvature that outlines the collectives in which we learn to protest, to be represented, to obey, to disobey—and to start over [POL · DC]. Add some transparency, some truth (still in the sense of Double Click), and you still get only dissolution, stampede, the dispersal of that very agora in which the fate of all categories is judged.

How could we say that he is speaking in the right category, someone who needs to empty the public square in advance of all those beings, too numerous, too ignorant, too agitated, with whom he must converse? Woe to anyone who claims to speak well and who begins by emptying the auditorium of those by whom he was supposed to be understood! Didn't Socrates boast of demonstrating a theorem to a single handsome young man while the infamous orators carried on in the public square with the whole crowd in a rage? But who speaks better, who is more sensitive to the requirements of this veridiction? The one who learns to speak "crooked" in an angry crowd, looking for what it wants, or the one who claims to speak straight, perhaps, but leaves the crowd to its disorderly agitation? In the agora, at least, the answer is clear. And yet isn't it strange that we continue to abhor the Sophists and heap praise on the hemlock drinker?

The reader will understand without difficulty that if there is a confusion to be avoided, it is that of mixing the requirements of objective knowledge with the movements necessary to political expression (profuse apologies to the Socrates of the *Gorgias*). If we do this, we inevitably lose on both counts. If there is a crossing that the Moderns ought to have protected like the apple of their eye against amalgams, this is the one. Unfortunately, it is also the one that the matter of links between "Science and Politics" has most thoroughly muddled, and one that will require a great deal of patience if we want to disentangle it.

THUS WE HAVE TO ACKNOWLEDGE THAT THERE IS MORE THAN ONE TYPE OF VERIDICTION ⊙

We see how advantageous it would be to do entirely without even the idea of straight talk to reject the temptations of Double Click in full. It would in fact become possible to replace a false opposition (the one between Reason and Rhetoric) by recognizing a crossing between two forms of veridiction, each of which misunderstands the other by translating it into its own terms. At that point nothing would keep us from liberating ourselves little by little from the speech impediments created by the impossible requirement that we speak straight to avoid having to speak "crooked."

Moreover, the distinctions we have just introduced will be still more enlightening if we remember that, in Chapter 1, we brought out two other modes, both of them also totally original: religious speech and legal speech, henceforth noted [REL] and [LAW] (we shall encounter them again in Chapters 11 and 13, respectively). They too passed through discontinuities in order to obtain continuities (religious predication, legal means), and they too would not have been judged properly either by straight talk or reference or the quest for political autonomy—nor would they have been able to judge these others properly in turn. If I stress this aspect of the inquiry now, it is to remind the uneasy reader that the rectifications and clarifications the inquiry introduces, although they have nothing to do with the ones that could be expected (that were expected) from straight speech, are nevertheless truly rectifications and clarifications. It is a matter of reason, error, and truth—but with the new constraint of not mixing up the different ways of speaking truth. The philosophy called

"ANALYTIC" was right to want to *use analysis* to clear up the sources of confusion introduced into thought. It was limited by its belief that to do so, it had to start from language alone and entrust all hope of clarity to the most obscure of enlightenments: to Double Click himself!

Someone may object that there is no reason not to attribute to straight talk, too, the status of mode of existence. (In Chapter 10 I shall account for its emergence while treating it more charitably.) But we see ⊙ TO FOIL THE STRANGE AMALGAM OF "INDISPUTABLE FACTS" ⊙ clearly that it cannot be a mode, since Double Click completely denies that information needs to *pass through* any hiatus, any discontinuity, any translation whatsoever—the counterexample also enlightens us as to its method. Its felicity conditions are thus unattributable.

Moreover, straight talk cannot even justify its own existence since, in order to speak, it has to *contradict itself* by going from figures to metaphors, from metaphors to tropes, from tropes to formalisms, from formalisms to figures, and so on. Straight talk is a literary genre that only *imitates*, by its seriousness, by its authoritative tone, sometimes simply by inducing boredom, certain truth conditions whose pertinence it is forever unable to manifest. This is even what makes it so dangerous, for it costs straight talk nothing, as we have seen in the previous chapter, to insinuate itself everywhere and to disqualify all propositions that would seek to acknowledge, quite humbly, the series of transitions—mediations—through which they have to pass to reach truth.

But what makes straight talk even more contradictory is the superimposition of a way of speaking straight on the support that that impossible speech claims to receive from another speech device so bizarre that even in the most seasoned ethnologists it triggers a slight regression back into Occidentalism: "FACTS THAT SPEAK FOR THEMSELVES." This device is all the more astonishing in that it is going to be used to cast doubt on the cultures of all the other peoples whose way of life seems to have deprived them of the support of these mute chatterers! The poor creatures: they have been deprived of knowledge of the facts; until the Whites landed among them, they had to stick to beliefs and mere practice . . .

The phenomenon has become entirely commonplace; nonetheless, the investigator is still struck by it. How could she fail to be astonished at hearing talk of "indisputable facts"? Does this not consist precisely in mixing of two distinct domains?—or else the word "indisputable" no longer has any meaning. And the operation is all the more astounding in that, after having thus mixed speech and reality in a single trope, the same people will explain to you, with the same authority, the same good faith, the same self-confidence, that "ontological questions and epistemological questions must *be absolutely distinguished*" . . . They claim that the very same concept that made it impossible to attribute any realistic, reasonable, rational distinction between the world's pathways and reference networks must be protected from any inquiry, even from any objection, by making the distinction between ontology and epistemology an absolute!

There is an extremely crucial *prohibition on speech* here that explains in large part the gap, among the Moderns, between theory and practice: how could people who agree—first stage—to fuse speech and reality by making the two inseparable, and who then—second stage—require that the world and words be separated by an insurmountable barrier, be *comfortable speaking?* If one could reconstitute in all its twists and turns the relation between the Moderns and that astonishing speech device the "voice of facts" discussed by "straight talk," we would be able to give a much more precise description of them—we would assemble them in quite a different way. (And, to begin with, we could talk to them about themselves without instantly infuriating them: will we ever be capable of this?)

⊕ AND THUS TO RESTORE TO NATURAL LANGUAGE ITS EXPRESSIVE CAPACITIES.

It will be said that, by depriving herself of straight talk and indisputable facts, our ethnologist may have carried out a useful task of clarification, but that she now finds herself totally without means to speak. In this view, she finds herself in the same impasse as the analytic philosophers, who, by dint of cleaning, scrubbing, disinfecting philosophy's dirty dishes, have forgotten to fill the plates at its banquet... Might we have lost everything by losing the dream of straight talk? This is what the Moderns seem to believe. How strange they are: they had language, and they did not hesitate to deprive themselves of it, to take up a different one, impossible, unpronounceable! By seeking to invent a language that would be absolutely true, outside of any context, apart from any translation, by a rigorous stringing-together of identical entities, they heaped scorn on *natural language*, the only one available to us. Now, if this natural language remains incapable of carrying out displacements without transformations—no demonstration has ever managed this—it remains admirably well adapted to follow, in their smallest sinuosities, the very movements of displacement and translation. On this point at least, no nullifying flaw will keep it from advancing as far as we like. Speech flows, it descends, it advances, it turns against itself—in short, it reproduces exactly the movement of what it is talking about and what it is seeking to capture by following its course.

Directly? No, but provided that it is self-correcting, as they say, it "tries and tries again." To say something is to say differently, in other words, it is to comment, transform, transport, distort, interpret, restate, translate, transpose, *that is to say* metamorphose, change form, yes, if you insist, "metaphorize." To speak literally one would either have to keep totally silent or else settle for stammering uh, uh, ah, ah, uh . . . Opposed to the impossible dream of "straight talk" is what we already have at hand: speech *that takes itself in hand*. Can we reassure the Moderns on this point? Assure them that this speech will offer them all they need to follow what counts above all else for them: the possibility of tracing the *displacements* proper to the different modes of veridiction. To tell the truth—to tell *the truths*—natural language lacks for nothing. This is to draw the *positive* conclusion where Wittgenstein, obsessed by his critique of rationalism, drew only the negative. One can perfectly well ask language to speak "with rigor," but without asking it for all that to pretend to progress from necessity to necessity by laying down a string of identical entities. This would be to ask it either to run in place or to lie—in short, once again, to be irrational, since it would have lost the thread of reasons. No, as we well know, the only rigor that matters to us is learning to speak in the right tonality, to speak well—shorthand for "speaking well in the agora to someone about something that concerns him."

Let us recognize that our ethnologist is advancing—slowly, to be sure, but she is moving ahead. She has unmasked the fantasy of straight talk; she has been able to avoid getting caught up in the Gigantomachy of Reason versus Politics; she has begun to grasp the plurality of modes of veridiction; she has

THE MOST DIFFICULT TASK REMAINS: GOING BACK TO THE DIVISION BETWEEN WORDS AND THINGS ⊙

regained confidence in her natural language and in her mother tongues—
provided that she doesn't hesitate to correct herself. The hardest part still
lies ahead: how could she claim to be speaking *well* about something to
someone if her speech were not *also* engaging the reality of what is said?

Now, as it happens, the misunderstandings created by the impossible requirement of straight talk have had the regrettable consequence
of distinguishing two questions for separate consideration within the
same problem: reality and LANGUAGE. If the ethnologist of the Moderns
accepts this distinction, and she is well aware of this, she runs the risk
of letting the scientific, legal, religious, or political modes that she has
begun to distinguish clearly start out with a terrible handicap: they
would completely lack all reality except for the linguistic version. We
can be pretty sure that the solution won't come from a clarification of
language. It will come from something that analytic philosophy has,
however, refused to do, although doing so would truly justify its name:
analyzing why the question of what language allows is always asked as
a *different* question from the question of what reality allows. Except in
the sole case of knowledge, where it was necessary to introduce a wedge
between the two in order not to fall into the weirdness of "indisputable
facts that speak for themselves"!

It is clear that we cannot let this distinction go simply by tossing the
words "pluralism" and "speaking well" into the air. The ethnologist does
not have access to the resources of critical thought. She cannot show up
in the agora and start by saying that, "quite obviously," politics, law, religion, and even science are only "ways of speaking," "fictions" that do not
engage reality in any way. She would be thrown out! And the authors of
fiction (to whom we shall return in Chapter 9) won't be the last to protest,
right after the guardians of the sciences. It isn't easy to see how anyone
could claim to be respecting her interlocutors if she started by denying
the reality of what they are talking about, or claimed that mute things
speak on their own *without her.*

⊙ WHILE LIBERATING
OURSELVES FROM MATTER,
THAT IS, FROM THE RES
RATIOCINANS ⊙

Fortunately, to give more realism, that is, more
reality, to the verb "speak"—and to bring ourselves
closer to the ancient meaning of *logos*—we are not
completely without resources, since in the previous
chapter we managed to shake loose what had crushed

the expression of reality under the amalgam of the RES RATIOCINANS. The operation of returning to reality in and through language can now take place in two phases: by finding space again, and thus room to breathe. Because in order to speak, we have to have air in our lungs.

Let us note first of all that with the notion and even the connotations of the word "network" we have gained *room* and *space* where we can collect values without merely mouthing the words. If we get into the habit of speaking of trajectories and passes that are limited and specific to each occasion for the paths of persistence [REP], chains of reference [REF], law [LAW], or political autonomy [POL], the landscape spread out before the observer is already entirely different from the one that obliged him to believe himself surrounded on all sides by an "external material world" that would have invaded the entire space and that would have forced all the other values to retreat little by little. But to go where? Into the mind? Into the brain? Into language? Into symbolics? No one knew. It was a black hole. Stifling. Suffocating.

From this point on, observers no longer find themselves facing a world that is full, continuous, without interstices, accessible to disinterested knowledge endowed with the mysterious capacity to go "everywhere" through thought. By taking apart the amalgam of *res ratiocinans*, we have become able to discern the narrow conduits of the production of equipped and rectified knowledge as so many slender veins that are *added* to other conduits and conducts along which, for example, existents can run the risk of existing. These networks are more numerous than those of references, but they are no less localizable, narrow, limited in their kind, and, too, a sketch of their features—this is the essential point—reveals as many empty places as peaks and troughs. The stubborn determination of things to keep on existing does not saturate this landscape any more than knowledge could.

Once we have accompanied knowledge [REF] in its networks— finally giving it the means to go as far as it wishes, but always provided that it pays the price of its installation and its extension—and above all, once we have accompanied the beings of the world in the conduits where they find the consecution of their antecedents and their consequents [REP], neither knowledge nor beings can fill in the landscape with empty padding. They can no longer overflow, dribble out. They have clearly marked edges—yet this feature in no way prevents their extension; indeed, that very demarcation is what allows the extension. The networks have in fact relocalized the DOMAINS. When a subscriber to a mobile telephone service exclaims "I've lost the network," she doesn't mean to imply that the "coverage" of a supplier could truly "cover" the entire surface, but that its "network" (the word is well chosen) resembles a lacy fabric full of holes. The ethnologist finds herself henceforth before a sort of macramé thanks to which the passage of law, for example, can finally slide through with ease: she adds a bit in a new color, which reinforces the solidity of the whole, but which is not capable either of "covering" reality or of "padding" it. This thread, this channel, this network, too, like the others, on the same basis as the others (knowledge [REF], persistence [REP], predication [REL], autonomy [POL]), leaves space, room, emptiness, still more emptiness, around each of its segments. In any case, room for the others.

Clearly, what matters in all these somewhat awkward metaphors is the attention they allow us to pay to materiality rather than to words, and to the empty spaces rather than the full ones. And they allow us in particular to feel that the unfettered circulation of one value no longer has the ability to make another one completely *disappear* by disqualifying it from the outset on the pretext that there is "no place" for it to go. The one can no longer *derealize* the other a priori. They can all start circulating *side by side*.

⊙ AND GIVING OURSELVES NEW CAPACITIES OF ANALYSIS AND DISCERNMENT ⊙

Side by side: this is the key. To move forward in this inquiry, we need an ontological pluralism that was scarcely possible before, since the only permissible pluralism had to be sought perhaps in language, in culture, in representations, but certainly not in things, which were entirely caught up in that strange concern for forming the external world on the basis of an essentially argumentative matter, the *res ratiocinans*. At the very heart of the notion of matter, there was a polemic intent that conceived of it as made entirely of transfers of "indisputable" necessities. By disamalgamating it, we are going to be able to restore to *discussion* the task of bearing, for each case, its *reality test*. Yes, there are things to discuss. Yes, there are beings that do not deserve to exist. Yes, some constructions are badly made. Yes, we have to judge and decide. But we shall no longer be able, a priori, without any test whatsoever, to discredit entire classes of beings on the pretext that they have no "material existence," since it is matter itself, as we have understood, that is terribly lacking in material existence! It is in the public square and before those who are primarily concerned by it that we have to run the risk of saying: "This exists, that does not exist."

Our method thus does not imply asserting that "everything is true," "that everything is equal to everything else," that all the versions of existence, the bad as well as the good, the factitious along with the true, ought to cohabit without our worrying any longer about sorting them out, as is suggested by the popular version of relativism that Double Click brandishes as a threat whenever someone refuses to judge everything by the standard of his straight talk. It implies only that the sorting out will have to take place, from now on, on a level playing field, contingent on precise tests, and we shall no longer able to endow ourselves with the astonishing facility of asserting that these particular beings exist for sure while those others are, at best, mere "ways of speaking." We see why the expression "to each his own (truth)" not only has the relativist tonality people often grant it; it also implies the daunting requirement of knowing how to speak of each mode in its own language and according to its own principle of veridiction.

If this is such a strong requirement for anthropology, it is because there is, among the Moderns, a very strange feature that consists in defending certain values by saying at the outset that, quite ⊙ IN ORDER TO SPEAK OF VALUES WITHOUT BRACKETING REALITY. obviously, they have no real existence! Or that, if they do exist, they do so phantasmatically, in thought, in the mind, in language, in symbols. It takes great impertinence to claim to respect those to whom one is speaking while complacently asserting that, naturally, one has to "bracket" all questions about the reality of what they are saying.

I am well aware that anthropologists dealing with remote beings have accustomed us to the necessities of such a "bracketing." It was probably the only way for them to absorb worlds whose composition differed so much from their own. But it was *among the others*, precisely: among those whose differences had all been sheltered once and for all under the umbrella word "culture," making it possible to protect all their strange aspects through a somewhat hypocritical respect for "REPRESENTATIONS," while being careful not to specify too precisely the relations that these representations maintained with "reality." But the anthropologists of the Moderns are going to have to learn to address "us." Consequently they can no longer so easily ignore the real existence of the values to which we hold. They can no longer practice the mental restriction that has served "intercultural dialogue" for so long—especially now that we are well aware that we are not just in a "culture" but also in a "nature." To speak well of something to someone is first of all to respect the *precise ontological tenor* of the VALUE that matters to him and for which he lives. This is surely the least one can ask of an investigator.

LANGUAGE IS WELL ARTICULATED, LIKE THE WORLD WITH WHICH IT IS CHARGED, ⊙ The consequence of the foregoing is that, to continue the inquiry, we need to proceed to a different distribution of tasks between reality and speech. However inaccessible it may appear, the goal is not out of reach, since we have already been able to recognize in matter the cause of a speech impediment whose origin we have pinned down: it respected neither the requirements of knowledge nor those of persistence, nor, finally, the interlocutory situation in which it was engaged, since it put an end to discussion without an attributable test. What prevents the inquiry from proceeding in the same way, step-by-step, for all the other modes?

Let us recall first of all what is unique about the division of labor between on the one hand an unarticulated world—but which is as obstinate as it is obtuse—and on the other an articulated language—but which is as arbitrary as it is changeable. The division is all the stranger in that, to obtain transfers of indisputable necessities, we had already crossed the two (very awkwardly) with the invention of "facts that speak for themselves." The famous barrier between questions of ontology and questions of language may not be all that insurmountable, since it was so easily crossed with the *res ratiocinans*—although in a contraband operation, as it were. The commerce between the two would already become more regular (in all senses of the word) if we used the term ARTICULATION to designate both the world and words. If we speak in an articulated manner, it is because the world, too, is made up of articulations in which we are beginning to identify the junctures proper to each mode of existence.

Our anthropologist probably jotted this down in her notebook: "Definition of the Moderns: they believe in language as an autonomous domain that is carrying on in the face of a mute world." And later on she must have added a codicil that makes the first curiosity noted even more curious: "A world probably *made* mute for *argumentative* reasons!" And this observation, too: "To silence their adversaries, they have preferred to silence the world and to deprive themselves of speech even as they let facts speak on their own!" And this: "They are frightened of the 'silence of those infinite spaces,' even though they themselves were the first to learn how to make them speak at last!"

Perhaps she has even been astonished to see the amused seriousness with which Magritte's painting *This Is Not a Pipe* is treated. As if the painted pipe were at once too close to a real pipe—it looks like the real thing—and too remote—it is not the real thing, whereas no one is astonished at the fact that the real, solid pipe for its part is also *preceded* by its antecedents and *followed* by its consequents, which are also at once very *close* to it—they look like it, since it persists—and very far away; these are the n - 1 and n + 1 stages in the trajectory of its existence; and the pipe would not have persisted without going through them. If we replace the image with a real pipe decked out with the sign "this really is a pipe," Magritte's gentle irony would lose all its savor. And yet the metaphysical truth would continue to have a meaning, for this is what should be inscribed on its label: "This is not a pipe *either*—just one of the segments along the path of a pipe's existence." The articulation of the pipe with itself, then the articulation of this first articulation with the word "pipe," then the articulation of these first two articulations with the picture of the pipe would also deserve to be painted—even if the experience would not provide an occasion for the knowing, blasé chuckle of the critical connoisseur.

Wherever there is a HIATUS, there is an articulation. Wherever one can define antecedents and consequents there is DIRECTION. Wherever one has to add *absent beings* that are necessary to the comprehension of a situation, there are SIGNS. If we want to define a sign by what stands "in the place of something else whose place it takes," then we can say it about the pipe and about all beings. All of them *pass by way of* others

⊙ PROVIDED THAT WE TREAT THE NOTION OF SIGN WITH SKEPTICISM.

in order to exist. In other words, the word "sign" has no contrary to which it can be opposed—and especially not the word "thing." If the abyss between the world and things appears immense, from one articulation to the next it is no longer an abyss at all. The word "dog" may not bark, but it takes only a few hours of training before the summons "Fido" brings to your feet the warm ball of fur that you have designated by that name and that has gradually taken on reality despite the supposed chasm between words and things.

MODES OF EXISTENCE
ARE INDEED AT STAKE,
AND THERE ARE MORE
THAN TWO OF THESE, ⊙

It is precisely in order to give up the sign/thing distinction completely that I have chosen to speak of "mode of existence," a term introduced into philosophy in a masterful way by Étienne Souriau. We are going to be able to speak of commerce, crossings, misunderstandings, amalgams, hybrids, compromises between modes of existence (made comparable in this sense according to the mode of understanding that we have already recognized under the label "preposition" [PRE]), but we shall no longer have to use the trope of a distinction between world and language.

What counts in this argument, moreover, is not so much the choice of terms we use on either side of the distinction as the fact of managing at last to *count beyond two*. Are we going to be able, in the course of the inquiry, to push ontology to take into account more than two genres, two modes of reality? Dualism has its charms, but it takes the anthropologist only a few months of fieldwork to notice that dichotomies do not have, among the Moderns in any case, the extraordinary explanatory virtue that the anthropology of remote cultures so readily attributes to them. The raw and the cooked, nature and culture, words and things, the sacred and the profane, the real and the constructed, the abstract and the concrete, the savage and the civilized, and even the dualism of the modern and the premodern, do not seem to get our investigator very far. It might almost make us doubt the ultimate light that such plays of contrast are supposed to shed when they are applied to the "others."

But are we quite sure that we have really counted beyond two? To get there for certain—we are already at seven!—we would still need to be sure that each of these modes has been credited with its share in reality. Without this equal access to the real, the jaws of dichotomy will snap shut and we shall once again find ourselves facing the distinction between a single world and multiple modes of interpretation. Fortunately, in the correspondence between knowledge and the beings of the world (as we have restaged it in the preceding chapters) we have a fine example of the way in which one can gradually charge the comings and goings of forms with reality.

Go into a laboratory and take it first of all at its beginnings. (It will have become clear that the LABORATORY serves as the Archimedean point in this inquiry, the spot where we place our lever.) The research group may project some bold hypotheses into the world, it may order expensive instruments and assemble a competent team, but it is still quite unable, at this stage, to transport anything resistant, consistent, real, through the intermediary of what it says. Visit the same site a few years later: it will have established (if it is well run) with some of the beings of the world—beings that have been mobilized, modified, disciplined, formed, morphed—such regular commerce, such efficient transactions, such well-established comings and goings that reference will be circulating there with ease, and all the words that are said about the beings will be validated by those beings through their behavior in front of the reliable witnesses convoked to judge them. Such access is not guaranteed, but it is possible; there is no shortage of examples.

What matters in the laboratory is that it can serve as a model for seeing how one might charge *other* types of realities thanks to networks *other* than that of reference. This is when we can really speak about modes of "existence" and address at last, in their own languages, those who hold to these values without bracketing the reality of what they are talking about. There would be BEINGS, yes, real beings, that leave in their wake the passes, unique to each case, of modes. To each mode there would correspond a singular local ontology, just as original in its productions as the invention of objective knowledge. The hypothesis is obviously astounding at first glance: it would be necessary to take into account beings of law, beings of politics, and even beings of religion. As Souriau says good-humoredly: "If there is more than one kind of existence, it means the world is pretty vast!" The Moderns protest: "That makes for too many beings, far too many beings! Give us a razor . . ." This reaction is not without piquancy, coming from those who have always boasted that they have conquered unknown worlds and who, at the very moment when they were taking the inventory of the universe, after their fashion, tried to make everything it contained fit into *two* kinds of existence! Do they really believe it's feasible, the anthropologist wonders, to arrange all the treasures they have discovered in just two categories, Object and Subject? An astonishing, formidably rich paradox that has to be called a state of metaphysical famine.

⊙ A FACT THAT OBLIGES US TO TAKE THE HISTORY OF INTERMODAL INTERFERENCES INTO ACCOUNT.

But before we go on, we probably have to make honorable amends to the divine Plato for treating the idea of the Idea with too little respect, and for recalling with too much indignation how he used it to dismiss the Sophists. He was so successful that straight talk has rendered crooked talk incomprehensible to the very people who discovered it and refined its turns. If we allowed ourselves this impiety, it was because, to measure the influence one value has on the others, one has to know how to weigh the role of the *moment* in each case. In the beginning, in Athens, after years of turmoil, this promise of intelligibility must have seemed so dazzling that the price to be paid, the neglect of common sense, seemed very slight. It is only today, many centuries later, that its cost appears too high to us, and its promises untenable. This is what it means to have an embarrassment of riches: after inventing both demonstrations (proofs) and democracy, the Greeks have never explained to us how we could preserve both without mixing them up in the impossible amalgam of a politics that dreams of going straight by way of unchallengeable demonstrations (a "political science"!). We see why it is so problematic to say that the Moderns inherited Reason: this legacy hides a treasure, but one crippled by debts that we are going to have to honor. If it is true, as Whitehead says, that all philosophy consists of a series of footnotes to Plato, it must not be so hard, after all, to introduce into it this slight *rectification*, this *erratum*, this regret: different ways of speaking divide the false from the true with razors sharpened differently from the razor of straight talk—which may be Occam's, but without a handle and without a blade.

As we see, it takes time to learn to talk. The anthropologist of the Moderns slowly resumes speaking, catches her breath, resists the temptation of straight talk, multiplies the forms of veridiction, manages to stop distinguishing the question of words from the question of the world, undoes invasive matter, and hears in a completely different way what is usually called articulated language. But just as things are getting clearer, they turn out to become terribly complicated. For we now have to take another step, by exploring the doubt introduced by the Moderns not about the link between the world and words, this time, but about the link between construction and truth. Once this new obstacle has been overcome, we shall finally be able to get to work for real: experience will have become expressible, and we shall at last be able to use it as a guide for understanding what has happened to the Moderns and what they can decide to inherit.

Chapter 6

CORRECTING A SLIGHT DEFECT IN CONSTRUCTION

The difficulty of inquiring into the Moderns ⊙ comes from the impossibility of understanding in a positive way how facts are constructed, ⊙ which leads to a curious connivance between the critical mind and the search for foundations.

Thus we have to come back to the notion of construction and distinguish three features: ⊙ 1. the action is doubled; ⊙ 2. the direction of the action is uncertain; ⊙ 3. the action is qualified as good or bad.

Now, constructivism does not succeed in retaining the features of a good construction.

We thus have to shift to the concept of instauration, ⊙ but for instauration to occur, there must be beings with their own resources, ⊙ which implies a technical distinction between being-as-being and being-as-other ⊙ and thus several forms of alterity or alterations.

We then find ourselves facing a methodological quandary, ⊙ which obliges us to look elsewhere to account for the failures of constructivism: ⊙ iconoclasm and the struggle against fetishes.

It is as though the extraction of religious value had misunderstood idols ⊙ because of the contradictory injunction of a God not made by human hands, ⊙ which led to a new cult, antifetishism, ⊙ as well as the invention of belief in the belief of others, ⊙ which turned the word "rational" into a fighting word.

We have to try to put an end to belief in belief ⊙ by detecting the double root of the double language of the Moderns ⊙ arising from the improbable link between knowledge and belief.

Welcome to the beings of instauration.

Nothing but experience, but nothing less than experience.

THE DIFFICULTY OF
INQUIRING INTO THE
MODERNS ☉

THE INQUIRY PROPOSED TO THE READER RESTS ON THE POSSIBILITY OF ESTABLISHING A DISTINCTION BETWEEN THE MODERNS' EXPERIence of their values and the account they give of these values. An account that they have transposed to the rest of the world—until that world decides that it is not at all a "remnant"! The Moderns' new situation of relative weakness offers an excellent opportunity to set aside the official account, to redefine their values differently—through the now well understood notion of mode of existence—and begin to offer alternative accounts (which will be the object of a still faltering diplomacy later on). By starting with the deployment of networks [NET], specifying them through the detection of prepositions [PRE], then undoing the confusion of the two modes amalgamated in the notion of matter [REP · NET], and finally making explicit why straight talk conflated two values that should be respected equally [REF · POL], we have given the investigators basic equipment to allow them to rediscover the thread of experience independently of the official versions. If it is true that "we have never been modern," it is going to become possible to say in a more precise way what has happened to "us" and what "we" really care about.

We still have to be able to explain this continuing distance between practice and theory. Why is it so hard to follow experience? The gap is too large to be attributed simply to the customary distance between the intricacies of daily life and the limits of vocabulary. Among the Moderns, this gap has become a major contradiction, one that accounts simultaneously, moreover, for their energy, their enthusiasm, and their complete opacity. We cannot do their anthropology if we settle for speaking of illusion or false consciousness in this regard. Why have they put themselves in the untenable position of defending values without giving themselves the means to defend them? Why have they cast doubt on the mediations necessary for the institution of these values?

As always, in this first Part, we need to start with the laboratory—for laboratories are to metaphysics what fruit flies are to genetics. By proposing the laboratory as the model of an original bond between words and things, as the prime example of what is to be understood by ARTICULATION, we have made our task somewhat easier, for we have proceeded as though everyone could agree that, in this case at least, the *artificiality* of the construction and the *reality* of the result went hand in hand; as if the ethnologist finally had the key that would allow her to exit from that persistent opposition in principle between language and being; as if the description of *experimentation* would indeed put us on the path of *experience*.

⊙ COMES FROM THE IMPOSSIBILITY OF UNDERSTANDING IN A POSITIVE WAY HOW FACTS ARE CONSTRUCTED, ⊙

This would be to forget what effect the field of science studies has had on the scientific public. Whereas this field claimed to be describing scientific practice at long last, the practitioners themselves judged on the contrary that the practice was being drained of its real substance! If our investigator has the slightest talent for her work, she has to take very seriously the cries of indignation uttered by those who felt they were being attacked. If it is impossible to link manufacture and reality without shocking the practitioners, it is because there is in the very notion of CONSTRUCTION (and in the scholarly theme of CONSTRUCTIONISM) something that has gone very wrong. How can one do justice to the sciences if the deployment of their chains of reference looks like a scandal to those charged with setting them up? More generally, how could we be capable of doing justice to the different trajectories proper to each mode if doing so amounted to discrediting each of their segments, each of their mediations? We lack an adequate tool here; we shall have to forge one, no matter the cost, if we want to succeed later on in instituting experience.

After all, nothing has prevented anyone from cherishing chains of reference. Everyone can agree that facts are facts. Apart from some hasty generalizations and perhaps an immoderate taste for "indisputable facts," nothing would have been lost for lovers of the sciences if we had offered them these chains fully developed at the outset, warmly clothed, richly veined, in short, sheathed in their networks, just as the history and sociology of the sciences present them today. And yet the fact remains, it can't be avoided, it happens every time: as soon as you draw up the list of the ingredients necessary for the production of objectivity, your interlocutors feel—they can't help themselves, it's too much for them—that objectivity is *diminished* rather than increased. Talking won't help; you can carry on as much as you like, there is no positive version that lets you say about some bit of knowledge in a single breath that it is proven *and* that it depends on a fragile progression of proofs along a costly chain of reference that anything at all may interrupt. Eventually, someone will always ask you to take a stand: "Yes, but is it objective, or does it *depend* on the interpretation of an instrument?" Impossible to reply: "Both! It is *because* the results of an instrument have been well interpreted that the fact turns out to be *proven*." You might just as well try to put a marble on top of a hill: it will roll into one of the valleys ("you're a realist!") or the other ("you're a relativist!"), but one thing is certain: this Sisyphean stone will not stay balanced on the crest.

It is as if the essential tool for understanding the role of MEDIATIONS had been broken or at least cracked; it can only be used incorrectly; it smashes what you want to protect; far from transforming everything into gold, like King Midas's hand, it jinxes everything it touches. Even if one avoids its most banal uses—what is called "social" construction, or even more trivially, the claim that "the illusory causes of our actions have real consequences"—one cannot use it as a basis for acceding to truths in a sustainable way. Whatever we do, we'll always be injecting doubt at the same time. "If it is constructed, then it's likely to be fake." Even without the addition of the adjective "social," even in small doses, the appeal to the notion of construction always remains a tool for critique.

It is true that the establishment of chains of reference poses a delicate problem of follow-through, since there is no resemblance between one FORM and the next, and because, as we saw in Chapter 4, one must always pay for the continuity of the constants by the discontinuity of the successive materials. But this is a minor paradox, in the end; only habit and a taste for observation are lacking; ethnographers have disentangled much more mixed-up rituals in faraway lands. In any case, the difficulty of following the chains cannot account for the extent of the misunderstanding about knowledge, the ineradicable attachment people have to a theory of Science that is so contrary to the interests of the advocates of objectivity [REF]. Of course, we have just seen what temptations a political use of Reason can offer: Double Click thought he had found certainties in Science that no artifice could contaminate, no deliberation could slow down, no costly instrumentation could weaken: "The facts are there, whether you like it or not." With this sort of fist-pounding on a table, it's true, one could put an end to any discussion in the agora [REF · DC]. But from here to depriving ourselves of all the means for fabricating the same facts? That's pretty unlikely. No, there must be another, much more important motive than the interests of knowledge, stronger even than political passions. Otherwise there would be no way to explain why mediations have been so discredited.

The ethnologist can hardly turn to critical thought to make the notion of construction more positive. If it has become impossible to say in a single breath "This is *well* constructed, it is therefore *true*, and even *really* true," it is because the negative view of mediations (even though they alone are capable of establishing the continuity of networks) is shared both by those who want to defend values and by those who want to undermine them. Or rather, the thought that is always trying to reveal *behind* the institutions of the True, the Beautiful, the Good, the All, the presence of a multiplicity of dubious manipulations, defective translations, worn-out metaphors, projections, in short transformations that cancel out their value—this thought has become "CRITIQUE." The position that in effect justifies, *a contrario*, in the eyes of their adversaries, the invocation of a substance that would maintain itself, for its part, without any transformation at all. The two parties are thus in agreement on this point: "As long as we do not possess pure and perfect information, let everyone abstain from speaking of real truth." The advocates of the absolute have found the enemies they deserve. As a result of this combat, in order to belittle the demands of Reason (exaggerated demands, to be sure), the Moderns have destroyed,

⊙ WHICH LEADS TO A CURIOUS CONNIVANCE BETWEEN THE CRITICAL MIND AND THE SEARCH FOR FOUNDATIONS.

one after another, the mediators necessary to the diversified advances of truth, even though these mediators are reason itself, the only means for subsisting in being. And it is finally Derrida, the Zeno of "differance," who was right always to preface the notion of construction with the preposition "de": constructivism is always in fact *de*-construction. We shall never be able to reconstruct the pathways of truth, but only deconstruct them. The ideal is still the same, except that the idea is not to learn how to reach it, but how to live stoically in the *disappointment* of not reaching it. We can see why there is no question, if we want to restore meaning to constructivism, of giving ourselves over to the temptations of the critical spirit.

The temptation becomes all the more dangerous in that FUNDAMEN-TALISM and all its dangers now has to be added to the mix. The quest for foundations that no interpretation, no transformation, no manipulation, no translation would soil, that no multiplicity would corrupt, that no movement along a pathway would slow down, could have remained at bottom a fairly innocuous passion—since at all events, in practice, the Moderns have always done just the opposite and multiplied mediations, whether in science, politics, religion, law, or elsewhere. But everything changed a few decades ago, when their inability to present themselves politely to the rest of the world led the others to believe what the Moderns were saying about themselves! "We" know perfectly well that we have never been modern. Necessarily: we would not exist otherwise. But we seem to have kept this secret to ourselves. As long as we were the strongest, this hardly mattered to us. On the contrary: we were exporting the drug and keeping the antidote for ourselves. The more or less failed modernization of the other cultures was a source of amusement for us, actually: spot a camel in front of a petrochemical factory, and here we have another "land torn by contrasts," we would say with a little smile, "between modernity and tradition." As if we ourselves were not torn, we too, between what we say about ourselves and what we do!

But today everything has changed; "the others" have absorbed enormous doses of modernization, they have become powerful, they imitate us admirably, except that they have never had the occasion to know that we have never been . . . Suddenly, our amused smile has frozen into a grimace of *terror*. It's a little late to yell: "But no, not at all, it's not that! You've got it all wrong, without a path of mediation you can't access any foundations, especially the True, but also the Good, the Just, the Useful, the Well Made, God too, perhaps . . . " We have become so fragile it's frightening. Terrorized? Yes, one would be terrorized for much less. We were able to pretend we were modern, but only so long as we were not surrounded by modernizers who had become *fanatics*. What is a fanatic? Someone who can no longer pronounce this benediction, this veridiction: "*Because* it is *well* constructed, it may *therefore* be *quite true*." But who taught the rest of the world that speaking that way is a sacrilege? Who taught the others to secularize an expression attributed to Islam: "The doors to interpretation must be closed," attributing it *first* to the sciences, technologies, economics? And they claim to be secularized! How poorly they know themselves!

It is one thing to recognize the *differences* between modes of existence by jealously retaining the *multiplicity* of types of veridiction; it's quite another to go back to the source, heaping discredit on *all* construction. To protect the diversity of truths is civilization itself; to smash the pavement laid down on the paths that lead to truth is an imposture. If there is a mistake that, for our own salvation, we must not commit, it is that of confusing respect for the various alterations of the modes of existence with the resources of critical thought.

Thus to escape from critique, in the end, we have to go back over this whole affair of constructivism and understand by what accident the requisite tools now lie broken in our shed. How do we reequip ourselves with new ones? First, by reclaiming everything that can be recuperated from meticulous attention to mediations—and which we forgot to include in the instruction manual for modernity when we exported it. To say that something—a scientific fact, a house, a play, an idol, a group—is "constructed," is to say at least *three different things* that we must manage to get across simultaneously—and that neither the formalists nor their critics can hear any longer.

THUS WE HAVE TO COME BACK TO THE NOTION OF CONSTRUCTION AND DISTINGUISH THREE FEATURES: ☉

First of all, it is important to stress that we find ourselves in a strange type of *doubling* or splitting during which the precise source of action is lost. This is what the

☉ 1. THE ACTION IS DOUBLED; ☉

French expression *faire faire*—to make (something) happen, to make (someone) do (something)—preserves so preciously. If you make your children do their vacation homework assignments, you do not do them yourselves, and the children won't do them without you; if you read in your Latin grammar that "*Caesar pontem fecit*," you know that the divine Julius himself did not transport the beams that were to span the Rhine, but you also know for certain that his legionnaires would not have transported them without his orders. Every use of the word "construction" thus opens up an *enigma* as to the author of the construction: when someone acts, *others* get moving, *pass into action*. We must not miss this particular pass.

⊙ 2. THE DIRECTION OF THE ACTION IS UNCERTAIN; ⊙   Second, to say of something that it is constructed is to make the *direction* of the vector of the action uncertain. Balzac is indeed the author of his novels, but he often writes, and one is tempted to believe him, that he has been "carried away by his characters," who have forced him to put them down on paper. Here we again have the doubling of *faire faire*, but now the arrow can go in either direction: from the constructor to the constructed or vice versa, from the product to the producer, from the creation to the creator. Like a compass needle stymied by a mass of iron, the vector oscillates constantly, for nothing obliges us to believe Balzac; he may be the victim of an illusion, or he may be telling a big lie by repeating the well-worn cliché of the Poet inspired by his Muse.

We find the clearest instance of this oscillation pushed to an extreme with marionettes and their operators, since there can be no doubt about the manipulator's control over what he manipulates: yes, but it so happens that his *hand* has such autonomy that one is never quite sure about what the puppet "makes" his puppeteer do, and the puppeteer isn't so sure either. The courts are cluttered with criminals and lawyers, the confessionals with sinners whose "right hand does not know what the left hand is doing." There is the same uncertainty in the laboratory: it takes time for colleagues to decide at last whether the artificial lab experiment gives the facts enough autonomy for them to exist "on their own" "thanks to" the experimenter's excellent work. A new oscillation: to receive the Nobel Prize, it is indeed the scientist herself who has acted; but for her to deserve the prize, facts had to have been what made her act, and not just the personal initiative of an individual scientist whose private opinions don't interest anyone. How can we not oscillate between these two positions?

We can break out of this oscillation by identifying the third and most decisive ingredient of the composite notion of construction. To say of a thing

⊙ 3. THE ACTION IS QUALIFIED AS GOOD OR BAD.

that it is constructed is to introduce a *value judgment*, not only on the origin of the action—double trouble, as we have just seen—but on the *quality* of the construction: it is not enough for Balzac to be carried away by his characters, he still has to be *well* carried away; it is not enough for the experimenter to construct facts through artifices; the facts still have to make him a *good* experimenter, well situated, at the right moment, and so on. Constructed, yes, of course, but is it *well* constructed? Every architect, every artist, even every philosopher has known the agony of that scruple; every scientist wakes up at night tormented by this question: "But what if it were merely an artifact?" (In this respect, at least, who doesn't feel like a scientist?)

Here is an astonishing thing, which proves how hard it is, when one lives among the Moderns, not to be mistaken about oneself: *none* of these three aspects shows up in the use of the word "construction," as it is commonly deployed in critical moves.

NOW, CONSTRUCTIVISM DOES NOT SUCCEED IN RETAINING THE FEATURES OF A GOOD CONSTRUCTION.

When someone asks the question "Is it true *or* is it actually a construction?" the implication is usually "Does that exist *independently* of *any* representation?" or, on the contrary, "is it a completely arbitrary product of the imagination of an *omnipotent* creator who has pulled it out of his *own* resources?" The doubling of the action? Lost? The oscillation as to the direction of the vector? Gone. The judgment of quality? Out of the question, since all constructions are equivalent. In the final analysis, the term "constructivism" does not even include something that the humblest craftsman, the most modest architect, would have at least recognized in his own achievements: that there is a huge difference between making something well and making it badly! With constructivism used this way, we can understand why the fundamentalists have become crazed with desire for a reality that nothing and no one has constructed.

What is astonishing is that the Moderns all live surrounded by constructions, within the most artificial worlds ever developed. Saturated with images, they are savvy consumers of tons of manufactured products, avid spectators of cultural productions invented from A to Z; they live in huge cities all of whose details have been put in place one by one, and often recently; they are dazzled with admiration for works of imagination. And yet their idea of creation, construction, production, is so strangely bifurcated that they end up claiming they have to *choose* between the real and the artificial. Anyone who thinks at all like an anthropologist can only remain dumbstruck before this lack of self-knowledge: how have they managed to last until now while being so badly mistaken about their own virtues? If we take the "fundamentalist threat" into account, we have to wonder about their chances of survival.

WE THUS HAVE TO SHIFT TO THE CONCEPT OF INSTAURATION, ⊙

How can we decant into a different word the three essential aspects I have just listed, which the word "construction" no longer seems to be able to contain? When one wants to modify the connotation of a term, it's best to change the term. Here I turn to Souriau once more: let us borrow the term INSTAURATION in the sense he gave it.

An artist, Souriau says, is never the creator, but always the instaurator of a work that comes to him but that, without him, would never proceed toward existence. If there is something that a sculptor never asks himself, it is this critical question: "Am I the author of the statue, or is the statue its own author?" We recognize here the doubling of the action on the one hand, the oscillation of the vector on the other. But what interests Souriau above all is the third aspect, the one that has to do with the quality, the excellence of the work produced: if the sculptor wakes up in the middle of the night, it is because he still has to let himself do what needs doing, so as to finish the work or fail. Let us recall that the painter of *La Belle Noiseuse* in Balzac's short story "The Unknown Masterpiece" had ruined everything in his painting by getting up in the dark and adding one last touch that, alas, the painting *didn't require*. You have to go back again and again, but each time you risk losing it all. The responsibility of *the masterpiece to come*—the expression is also Souriau's—hangs all the heavier on the shoulders of an artist who has no model, because in such cases you don't simply pass from power to action. Everything depends on what you are going to do next, and you alone have the competence to do it, and *you don't know how*. This, according Souriau, is the riddle of the Sphinx: "Guess, or you'll be devoured!" You're not in control, and yet there's no one else to take charge. It's enough to make anyone wake up at night in a cold sweat. Anyone who hasn't felt this terror hasn't measured the abyss of ignorance at whose edge creation totters.

The notion of instauration in this sense has the advantage that it brings together the three features identified above: the double movement of *faire faire*; the uncertainty about the direction of the vectors of the action; and the risky search, without a pre-existing model, for the excellence that will result (provisionally) from the action.

But for this notion to have a chance to "take," and to be invested gradually with the features that the notion of construction ought to have retained, there is one condition: the act of instauration has to provide *the opportunity to encounter beings capable of worrying you.* BEINGS whose ontological status is still open but that are nevertheless capable of *making* you *do* something, of unsettling you, insisting, obliging you to speak well of them on the occasion of branchings where Sphinxes await—and even whole arrays of Sphinxes. *Articulable* beings to which instauration can add something essential to their autonomous existence. Beings that have their own resources. It is only at this price that the trajectories whose outlines we are beginning to recognize might have a meaning beyond the simply linguistic.

⊙ BUT FOR INSTAURATION TO OCCUR, THERE MUST BE BEINGS WITH THEIR OWN RESOURCES, ⊙

On this account, the statue that awaits "potentially" in the chunk of marble and that the sculptor comes along to liberate cannot satisfy us. Everything would already be in place in advance, and we could only alternate between two bifurcated descriptions: either the sculptor simply follows the figure outlined in detail in advance or else he imposes on the shapeless raw material the destination that he has "freely chosen." No instauration would then be necessary. No anxiety. No Sphinx would threaten to devour the one who fails to solve the riddle. This ontological status is hardly worthy of a statue, at least not a statue of quality; at most, it would do for molding a set of plaster dwarfs for a garden. No, there have to be beings that escape both these types of resources: "creative imagination" on the one hand, "raw material" on the other. Beings whose continuity, prolongation, extension would come at the cost of a certain number of uncertainties, discontinuities, anxieties, so that we never lose sight of the fact that their instauration could fail if the artist didn't manage to grasp them according to their own interpretive key, according to the specific riddle that they pose to those on whom they weigh; beings that keep on standing there, uneasy, at the crossing.

⊙ WHICH IMPLIES A TECHNICAL DISTINCTION BETWEEN BEING-AS-BEING AND BEING-AS-OTHER ⊙

As there is no commonly accepted term to designate the trajectories of instauration—Souriau proposes "anaphoric progression"!—I shall introduce a bit of jargon and propose to distinguish BEING-AS-BEING from BEING-AS-OTHER. The first seeks its support in a SUBSTANCE that will ensure its continuity by shifting with a leap into the foundation that will undergird this assurance. To characterize such a leap, we can use the notion of TRANSCENDENCE again, since, in uncertainty, we leave experience behind and turn our eyes toward something that is more solid, more assured, more continuous than experience is. Being rests on being, but beings reside elsewhere. Now the beings that demand instauration do not ensure their continuity in this way. Moreover, they offer no assurance regarding either their origin or their status or their operator. They have to "pay for" their continuity, as we have already seen many times, with discontinuities. They depend not on a substance on which they can rely but on a SUBSISTENCE that they have to seek out at their own risk. To find it, they too have to leap, but their leap has nothing to do with a quest for foundations. They do not head up or down to seat their experience in something more solid; they only move out *in front of* experience, *prolonging* its risks while remaining in the same experimental tonality. This is still transcendence, of course, since there is a leap, but it is a *small* TRANSCENDENCE. In short, a very strange form of IMMANENCE, since it does have to pass through a leap, a hiatus, to obtain its continuity—we could almost say a "*trans-descendence,*" to signal effectively that far from leaving the situation, this form of transcendence deepens its meaning; it is the only way to prolong the trajectory. (We shall often return to this distinction between bad and good transcendence.)

In fact, this jargon has no other goal but to shed light on the central hypothesis of our inquiry: from being-as-being we can deduce only one type of being about which we might *speak* in several ways, whereas we are going to try to define how many *other forms of alterities* a being is capable of traversing in order to continue to exist. While the classic notion of CATEGORY designates different ways of speaking of the same being, we are going to try to find out how many distinct ways a being has to *pass through* others. Multiplicity is not located in the same place in the two cases. Whereas there were, for Aristotle for example, several manners of speaking about a being, for us all those manners belong to *a single mode*, that of knowledge of the referential type [REF]. The being itself remains immobile, *as* being. Everything changes if we have the right really to question the alteration of beings in several keys, authorizing ourselves to speak of being-as-*other*. If it is right to say, as Tarde does, that "difference proceeds by differing," there must be several modes of being that ensure their own subsistence by selecting a *distinct* form of alterity, modes that we can thus encounter only by creating different opportunities for instauration for each one, in order to learn to speak to them in their own language.

⊙ AND THUS SEVERAL FORMS OF ALTERITY OR ALTERATIONS.

What is interesting here is the anthropological consequence of an argument that would otherwise remain overly abstract: why have the Moderns restricted themselves to such a small number of *ontological templates* whereas in other areas they have caused so many innovations, transformations, revolutions to proliferate? Where does this sort of ontological anemia come from? Beings in the process of instauration: these are precisely what we have trouble finding among the Moderns, and this is why it is so hard for the Moderns to encounter other COLLECTIVES except in the form of "CULTURES."

In the previous chapters, we have seen why: either the Moderns find themselves face to face with obtuse raw materiality, or they have to turn toward representations that reside only in their heads. And what is more, they have to choose between the two: in theory, of course, since in practice they never choose; but this is exactly the split that interests us here: why is what is necessary in practice impossible in theory? In the wake of what events has the very civilization whose continual practice has been to transform the world so radically agreed to acknowledge in theory only two domains of reality?

WE THEN FIND OURSELVES FACING A METHODOLOGICAL QUANDARY, ⊙ To answer this much too vast question, we find ourselves in a dangerous situation, since the challenge is to bring to light, using our previously established method, the CROSSING between two modes of existence—one, to which I have already alluded, though too rapidly, religion, noted [REL]; and another, still buried, which we shall meet only in the following chapter, and which has something to do with the power of idols. Now, to do justice to these two modes, we shall need to have *already answered* the question that preoccupies us here, since in both cases it is a matter of entering into contact with beings that remain totally resistant to articulation, expression, instauration, as long as we lack another path that would allow access to alterity. We are caught up in a contradiction, a catch-22, from which I propose to escape abruptly by imposing a solution—a totally hypothetical one, of course, until its fruitfulness can be tested in the following chapters.

We have to adopt a radical position with regard to the old chestnut of Western history according to which we would be, if not Judeo-Christians, then at least Judeo-Greeks. It is actually the connection between these two questions—how to understand chains of reference, how to understand creation—that has thrust the Moderns into such a passion for obscurity and compromised everything they have asserted about the illumination their Enlightenment was supposed to shed on the world.

⊙ WHICH OBLIGES US TO LOOK ELSEWHERE TO ACCOUNT FOR THE FAILURES OF CONSTRUCTIVISM: ⊙ If European history didn't tell us this clearly enough, anthropology would already have put us on the track: only *religion* can explain that we are mistaken at this point on the precise role of the mediations necessary for the establishment of knowledge. To put it bluntly, it is the "religion of knowledge" that we must now take on in order to free our inquiry at last. The theory of Science has been only the "collateral victim" of a discredit that had a different origin and was aimed at a different, even more formidable target, which we must now approach with fear and trembling. If the Moderns had confused only the sources of the Science, their anthropology would be doable, if not easy; but they have also confused the sources of religion, or rather they have mixed them in frightfully with the sources of Science. Behind every question of epistemology lies another question: what to do with the idols, or FETISHES? This is the most striking feature of the anthropology of the Moderns: they believe that they are *anti-idolators* and *antifetishists*.

So we are going to act as though the Moderns had been victims of an *accident of constructivism!* Their ancestors behaved more or less normally, instituting a multiplicity of potential beings as best they could, when they set out to bring to light a new CONTRAST, a contrast that they hadn't thought they could kindle without breaking with the rhythm of instauration. This accident, this event of events, has a multiplicity of names—all mythical, by definition—and because it was so daring, so improbable, so contingent, it has not ceased to play out over and over in the course of European history, but it can almost certainly be held responsible for the discredit into which mediations have fallen. The event in question can be called the "Mosaic division" (Jan Assmann's name for it), "antifetishism," "iconoclasm," "critique," it hardly matters. To cover this whole business that makes us all the happy elect of an ever-so-sanctified history, I have proposed the term ICONOCLASH—deliberately blending Greek with English. If there is indeed something that defines, ethnographically, the fact of being Western, European, Modern, if there is at least one history that belongs to us, it is that we are descendants of those who overturned the idols—whether this meant destroying the Golden Calf, toppling the statues of the Roman emperors, chasing the moneylenders out of the Temple, burning the Byzantine icons, looting the Papist cathedrals, beheading the king, storming the Winter Palace, breaking the "ultimate taboos," sharpening the knives of critique, or finally, more sadly, taking the dust fallen from the ruins of postmodern deconstruction and further pulverizing it, one last time. Dear reader, go back up your genealogical tree: if you don't have an iconoclast among your ancestors, you must be neither Jew nor Catholic nor Protestant, neither revolutionary nor critical, neither an overthrower of taboos nor a deconstructor; if that is the case, it is not to you, in your blissful innocence, that this discourse is addressed...

⊙ ICONOCLASM AND THE STRUGGLE AGAINST FETISHES.

At this stage in our progression, what counts is to realize in what aporia those who bring the charge of falsity against idols are going to be plunged. Or rather, since among the Moderns everything always happens "doubled," we are going to find ourselves confronting two multiplied aporias: one bearing on the quality of the beings that we claim to reveal by destroying idols; the other on the meaning that we should have given the idols, had we not taken it upon ourselves to destroy them. (Still, we shall be able to give a positive twist to these two arguments later on in this book, once we have recognized their particular trajectories, in Chapter 7 for the "idols" and in Chapter 12 for the "gods.")

Let us hypothesize that, in order to extract the contrast from a new mode of existence—one very poorly characterized but nevertheless instituted in history under the name "monotheism"—we shall find ourselves obliged to misunderstand idols and idolators. Here is something strange and properly—or holily—diabolical: there is always something false in idols, but this falsity too is doubled. It comes not only from the fact that idols manufacture deceitful divinities (deceitful in the eyes of the religious), but from the fact that they are at risk of being struck down carelessly, indiscriminately. It is as though they incited us to misunderstand them, to get them wrong, mistake them. The Moderns engage in the phony history of believing themselves to be antifetishists, and this blinds them to themselves and especially to the others, in the long run. We can hardly be surprised, for we have become familiar with this kind of conflict between values. We saw in the previous chapter how the emergence of objective knowledge had inadvertently attacked as an imposture the ever-so-subtle and essential veridiction of politics [REF · POL]. Well, we are going to presume that the same thing happened, in a mythical past, to another value, it too essential, namely, the religion of the living God, which—also mistakenly—laid low the unfortunate idols that were directed, as we shall soon see, toward an entirely different type of veridiction, one that was nonetheless equally essential to our survival. The whole problem arises from this pileup of category mistakes that threatens to make us lose the thread—and even to despair of the inquiry.

The whole affair arises from the collision and entangling of two contradictory injunctions, each of which is marked in turn by a fundamental contradiction in the form of an aporia. Let us begin with the first aporia, which is easy to understand even though it constitutes, as we shall see, a category mistake of incalculable consequence.

⊙ BECAUSE OF THE CONTRADICTORY INJUNCTION OF A GOD NOT MADE BY HUMAN HANDS, ⊙

Those who counter idolators by affirming that "we, unlike the others, don't make images of our God" have obviously positioned themselves, to speak somewhat euphemistically, in a *false situation*: for at once, of course, inevitably, obviously, fortunately, necessarily, they are going to have to find *other* images, *other* mediations, *other* recipients, *other* temples, *other* prayers, *other* conduits to achieve the instauration of this God. Let us recall that, without a path of alterations, without instauration, there is no possible subsistence. But here we are, it's done, the path has been taken, it's already too late: we *shall never again be able to admit it*, because we have embarked on the impossible enterprise of presenting ourselves before a nonmanufactured God. From here on, we shall have to carve out an *absolute* difference between what the left hand is doing ("We do not manufacture images") and what the right hand is doing ("Alas, we cannot not manufacture images"). Iconoclasm has become our cult. Impossible to turn back. This is what Claudel has Mesa say: "A knife is a narrow blade, but when it cuts through fruit, the pieces will never be put back together."

And yet, of course, at once, we have to do the opposite of what we have said. As soon as the Golden Calf has been overturned, someone has to build the Tabernacle with its sculpted cherubins; Polyeucte has just destroyed Zeus's temple, and someone is already erecting an altar on the same spot with the relics of *Saint* Polyeucte; Luther has paintings of the Crucifixion taken down, and Cranach is already painting the "mental image" of the Crucifixion as it emerges from one of Luther's sermons in the minds of believers; and the list goes on. Malevitch exhibits his big black paintings, but he is said to have painted the edge of an icon. In the *Iconoclash* catalog, we have begun to inventory this sumptuous and overwhelming legacy, showing that those who have inherited the "Mosaic division" have to maintain a gulf that no good sense can ever fill between the prohibition of these images and their necessity. "If only we could get along without images, we could at last reach God, the True, the Good, the Beautiful!" they say out of one side of their mouths, while out of the other they sigh: "If only we had images, we could at last reach God, the True, the Good, the Beautiful!" Whether it is a matter of destroying idols, renewing works of art, or making chains of reference disappear, the same "image wars" continue for religions, the arts, morality, and the sciences. The sense of instauration—in the sense in which one says "the sense of sight"—has been completely severed, since it will never again be possible to prolong a trajectory in the same tonality.

If we have been speaking from the start about the double language of the Moderns, it is clear by now that there were good reasons to do so. It is by no means simply a matter of a guilty conscience, of a veil thrown modestly over practices, but of a fundamental crack in what the theory of action can absorb, assimilate, take in, from any practice whatsoever. How could one live at ease under the grip of such a contradictory injunction? It is enough to drive people crazy, for sure.

⊙ WHICH LED TO A NEW CULT, ANTIFETISHISM, ⊙   Here we discover one of the factors that kept constructivism from succeeding: what cracked the tool that made the instauration of beings possible was that hammer-blow repeated throughout history and always accompanied by the cry "Idols must be destroyed!" But what was destroyed was not only the idol, it was also the hammer—without forgetting the head of the hammerer who was hit on the rebound ... Antifetishism is the religion of Europeans, the one that explains their piety, the one thanks to which, whatever they may claim, they are impossible to secularize: they can decide on the weapon, the target, and the divinity to which they are going to sacrifice their victims, but they cannot escape the *obligation* to worship this cult, that of iconoclasm. This is the only subject on which there is perfect agreement among the most religious, the most scientific, and the most secular of persons.

The anthropologist of the Moderns thus has to get used to living in a cloud of dust, since those whom she is studying always seem to live amid ruins: the ruins they have just toppled, the ruins of what they had put up in place of the ones they toppled, ruins that others, for the same reason, are preparing to destroy. Mantegna's *Saint Sebastian* in the Louvre, pierced by arrows, his corpse already stony, upright on a pedestal at the foot of which lie the idols of the gods that he has just sacrificed, the whole framed by the arch of one of those Roman ruins so admired during the Renaissance— this is the sort of emblem we confront when we approach the tribe of taboo-breakers. Or the astonishing film that ran in a continuous loop in the *Iconoclash* exhibit, showing the consecration of the church of Christ the Savior in Moscow, performed by popes and patriarchs in clouds of god and incense (this was from the early days of cinema), then the destruc- tion of the church of Christ the Savior by the Bolsheviks in clouds of dust, followed by the construction of a Soviet swimming pool, followed at once by its destruction, in new clouds of dust, which permitted the construc- tion of a facsimile of the church of Christ the Savior, once again conse- crated, a century later, by Orthodox bishops, once again gleaming again with gold and precious stones . . . Have you ever come across a critical mind that was secularized? Go ahead and unveil the divinity to which he has just presented the broken limbs of his victim—swimming pool or church. (Anyone who accepts the Moderns' claim to have had, at least, the immense merit of having done away with the taste for human sacrifice must not follow the news very closely, and must know very little about twentieth-century history.)

What interests us for the moment is not the linguistic predicament in which we are going to find ourselves when we try to speak of this God not made by human hands (we shall return to the issue at

⊙ AS WELL AS THE INVENTION OF BELIEF IN THE BELIEF OF OTHERS, ⊙

length in Chapter 11) but the second aporia, the one through which we are going to attribute to the shattered idols a function that they cannot have had. It is as though, to bring to light the figure of this nonconstructed God (with the right hand, but necessarily supported by the left), we had believed ourselves obliged to accuse the idols of a supplementary crime that they are completely incapable of committing. We are on the razor's edge here, but we have to be able to designate as a category mistake what impels people to strike out at idols unjustly: it is *the belief in the* BELIEF *of others* that defines rather precisely what can be expected of the Moderns when they believe they are disabused. These blind folks are intriguing in that they never blind themselves as much as when they think they have their eyes wide open. This is why the urgency of doing their anthropology is always underestimated ...

A glance at the history of the First Contacts between the Europeans and the "Others" (contacts that we shall look at quite differently in Part Three) should suffice to highlight the misunderstanding: none of the "pagans" who saw their idols destroyed, their fetishes broken, their altars overturned, their acolytes exterminated, ever really understood the fury of the Christian iconoclasts—especially given that the latter wore medals of the Virgin or the saints, set up altars, celebrated the truth of their Book, and organized the most subtle, the most lasting, the most imperial of institutions. But then none of the same faithful servants of the same Catholic Church ever understood the fury of those who wanted to put an end to their worship, empty their churches, get their priests to forswear their faith, dissolve the orders—especially given that those who sent them to the scaffold worshipped Reason, developed the State of Terror, and wore the Phrygian cap with pride. The same misunderstanding has been reproduced in all the iconoclastic crises, within religions and then against them, in secularized critical thought and even in the worn-out forms of "taboo breakers." And what if no one, ever, had had the sort of *adherence* to taboos that the *breakers* of taboos presuppose? What if the maleficent strangeness of the taboo were not in the idolator's mind but instead in the *anti*-idolator's?

⊙ WHICH TURNED THE WORD "RATIONAL" INTO A FIGHTING WORD.

The question is worth asking, since the antifetishist is probably the last to hear the complaint that has been arising for centuries from the smoking ruins where the idols lie at the bases of their pedestals: "You're on the wrong track, completely mistaken: we do not *hold* at all *in that way* to the thing you want to wrench away from us; you and you alone are living in the illusion to which you would like us to put an end." Blind, they are also deaf, *like the idols* whom they proclaim, quite wrongly, to be deaf as well as mute ... This category mistake of iconoclasm, an astonishing error that says a great deal about the Moderns' capacity for self-knowledge, is also the mistake that has been ignored the most constantly, the most stubbornly, for the longest time. A Modern is someone who knows that the others are plunged into belief, even when the others affirm that they are not. More precisely, a Modern is someone who, facing this denial on the part of believers, placidly affirms that the latter cannot bear to "face the truth," at the very moment when it is he, the courageous critic, who is *denying* the protests of all the believers who do not believe, who have never believed—in any case not *in the way* in which the Moderns believe that the others believe. In short, a Modern is someone who, in his relations with the other, uses the notion of belief and believes furthermore that he has an obligation to undeceive minds— while deceiving himself on what can be deceptive! How many crimes have we not committed in seeking to shed our naïveté?

We are going to have to get used to thinking that the "pagans" are clearly innocent of at least one of the accusations brought against them, the charge that idolators worship an object made of wood, clay, or stone without "knowing" that it has also been manufactured by them or by other humans. The accusation does not hold up for a second, since the avowal of the object's fabrication is on the contrary unanimous: "But of course we made them, fortunately, and they are even *well* made!" The only ones who imagine that there is an absolute obstacle here, a shibbo-leth capable of drawing a line between rational humanity and irrational humanity, are precisely those who have appealed to the tribunal, those who, as good "constructivists," have lost the very sense of what "instaura-tion" can mean. These are the ones of whom it must be said that they see the speck in their neighbor's eye but not the log in their own (Luke 6:42). The difficulty is not here; on this point we see clearly, speck and log have already been removed; instauration is not construction (and thus not deconstruction, either). The kind of beings whose instauration the idols achieve still has to be grasped—but this will be for the next chapter.

After this detour, we understand why our inves-tigator had to agree to hear the scientists' complaints, which she thought she had described faithfully: this is the only way she can detoxify herself after   WE HAVE TO TRY TO PUT AN END TO BELIEF IN BELIEF ⊙

imbibing the poisonous resources of antifetishism. Someone will object that she is using exactly the same critical resource in claiming to put an end to that long error of antifetishism, by disabusing minds of their millennial illusion. Not entirely. "Putting an end to belief" is indeed the goal of rational inquiry, with the understanding that it is a matter of putting an end to *belief* and not to *beliefs*. Belief, in the present inquiry, has a precise meaning: it is, in the literal sense, to take one thing for another, to interpolate two or more modes of veridiction. Whereas "beliefs" have *the truth* as their opposite, *belief* has as its contrary the explicit determination of *prepositions*. There is belief when the iconoclasts believe that the idolators believe, whereas the latter are only following the path of a different mode of veridiction—which we still have to define. There is belief when critics who believe they are secularized populate the world with naïve believers because they have not grasped the interpretive key in which what the latter do must be understood. There is belief again when Double Click interpolates the paths of knowledge with the paths of reproduction. In other words, belief never targets a precise object, and thus—a crucial consequence—one cannot expect objective knowledge ever to reveal the real target of belief. (We shall see, moreover, in Chapter 11, that one cannot even use it to respect the religious contrast.)

If we claim to be heirs of the rational project, we too can thus affirm that it is indeed a matter of *putting an end to belief*, if we mean by this proud slogan that we seek to free ourselves of the *very concept* of belief in order to interpret the instauration of those beings to which the Moderns are right to want to cling. This is the only way we can inherit from the Enlightenment—without the immense shadow it has cast up to now over the rest of history by making the whole expanse of the planet semi-willing victims of the necessary *illusion*.

⊙ BY DETECTING THE DOUBLE ROOT OF THE DOUBLE LANGUAGE OF THE MODERNS ⊙ Here we are concerned only with the completely unanticipated telescoping between what we discovered in the previous chapter and this adventure of antifetishism—which would warrant detailed study on its own account. If it is true, as I have argued, that the extraction of chains of reference had the unexpected consequence of destroying political value; if it is true, as I have just suggested, that the emergence of religious value could happen only when people spoke ill of (I don't dare say "committed blasphemy against"!) idols; if it is thus true that, in both cases, a different value, an innocent victim, had to pay the price; still, nothing, absolutely nothing—nothing but the contingencies of Western history—prepared the way for the amalgamation, for the chimera (in the biological, teratological sense of this term) of these *two misunderstandings*.

We are now in a position to propose a definition of the double source of the double language of the Moderns. The prerequisite was the impromptu encounter (it is up to the historians to sort out the exact circumstances of this meeting) between two distinct events that by ill fortune were both going to deprive constructivism of *all its means*—that is, of its mediations. Curiously, one arose from belief, the other from knowledge. We believe that we know. We know that the others believe.

A simple mistake of constructivism would not have sufficed: it would have resulted in pious people who would merely have been mistaken about the others when they attributed abominable beliefs to them—an attribution so badly aimed that it would have left the so-called idolators completely indifferent. From another standpoint, a simple category mistake, such as the one we encountered earlier concerning knowledge, would not have sufficed either: while chains of reference would have been deprived of any even slightly extended use, remote beings would still have been engaged in the destiny of objective truth—but this access would never have become the sovereign right of anyone wearing a lab coat to disqualify all other access routes. It is *the conjunction of the two that makes us modern*; this conjunction alone explains the astonishing "We" that allows "us," through opposition, to define "Them," the Others, absolutely rather than relatively: "We are those who do not construct our gods"; and "We are those who know how to speak literally and not just figuratively." The first claim strikes the idols of wood and stone to the benefit of a nonmaterial God; the second strikes the ever-so-material forms of confirmed knowledge to the benefit of an ideal form that no materiality will come to corrupt any longer. If we splice the two injunctions together, then, yes, the "Others" become truly different from "Us": "They" adore the gods that they manufacture although they don't dare admit this to themselves; "they" speak figuratively in confusing their phantasms with the order of the world.

It becomes clear why it is hard to follow the anthropologists who define Westerners as a bloc by their "NATURALISM" and their "rationalism" while acknowledging a few contradictions in practice—or at least paying lip service to the possibility. No, the Whites are much quirkier than this, much more interesting; their tangled objects deserve one or two ethnological museums of their own. Recalling what was said earlier about

BIFURCATION, following Whitehead, the reader will forgive me for using and perhaps abusing the idea that the Whites are of the cleft genre. Yes, unquestionably, the Moderns are serious bifurcators. When they speak indignantly about "chasing out the demons," we would do well to look at their feet!

⊙ ARISING FROM THE IMPROBABLE LINK BETWEEN KNOWLEDGE AND BELIEF. Here we probably have the junction point, itself cleft, the articulation that makes it possible to give meaning to this anthropological feature that is no longer, let us hope, a vain accusation of illusion about the self, or of false consciousness. Let us recall that the first Bifurcation we encountered came from a difficulty concerning knowledge: as knowledge produces assured—and in some cases even apodictic—results, the Moderns tried to explain this miracle by acting as though the things known had in themselves a resemblance of form to the FORMALISMS that authorized access to them. The hypothesis was so contrary to experience, so much in contradiction with the self-evidence of common sense (Mont Aiguille does not go around in the world the way the map allows access to Mont Aiguille) that it could be maintained only by being fused with another hypothesis, it too a risky one, but more heavily invested with passion. We now see what this was: worship rendered to a God not made with human hands who required the sacrifice of all the idols, and, in passing, of all realism. But this hypothesis in turn would never have held up for long, in any case would not have become our seemingly insurmountable horizon, if the Moderns had not tied it to the passion for knowledge.

Without the link between these two enthusiasms, the stronger supported by the weaker and vice versa, the Bifurcation could never have been maintained: the *chains* of reference, like the *processions* necessary for the emergence of the living God, would have appeared clearly [REF · REL]. But the two impossibilities, combined, produced a single, invincible history, and—this is the fascinating part—both are true at once, that is, the impossibility and the invincibility: the impossibility is always there, since people can never literally "talk straight" and since they are after all obliged to achieve the instauration of their own Gods; the invincibility comes from the fact that people nevertheless have managed— at least until quite recently—to act as if these two impostures had *no consequences* for practice. It is impossible, obviously, to give up on practice; suicide would have to ensue. The stupefying solution has been to dig a bottomless gulf between theory and practice, to install a source of irrationality and blindness at the very heart of the project of unveiling. Everything is to be unveiled, everything except that abyss.

It is clear that I was not so far off base earlier when I distinguished in the modern Constitution what I called procedures of "hybridization" (below) and of "purification" (above). The terms were too simplistic, but the diagnosis was accurate: at the core of the Moderns there is a source of foundational irrationality, since they must never be able to draw the consequence between on the one hand the search for substance and the search for a God not made by human hands, a search that they have made the origin of all virtue, and on the other the practice that obliges them not to take that project into account. The source of their formidable energy has indeed been localized: how many powers they are going to be able to launch, since they will never have to follow the consequences of their acts and harmonize theory with practice! After all, the cliché may well be true: we are indeed Judeo-Christians. Contrary to the pious legend, Paul's preaching on the Acropolis may not have been in vain: far from leaving Athens in disgust, as the Acts of the Apostles asserts, he may have set up shop there. The search for substance and the God not made by human hands: this is the debt-scarred legacy whose hidden treasure we must learn to extract.

The collateral victim of this double accident is, of course, EXPERI-ENCE: despite the volumes written on empiricism, if there is one thing one could not require of the Moderns, it is that they be faithful to what is given in experience, since they cut off *twice over* any continuity with the two articulations on which what they cherished the most depended. From this point on it was impossible to follow in any explicit way the threads of the mini-transcendences necessary to the subsistence of beings in the process of instauration. The threads would have become visible. The whole business would have fallen apart. Without belief, knowledge would have remained one-eyed; without knowledge, belief would have remained simply vision-impaired. With both, one runs the risk of total blindness...

It is true that a startled blind person can hurl himself fearlessly ahead, unaware of danger (this is the hubris of the Moderns). But if he begins to hesitate, he ends up discouraged (this is postmodernism). If he is truly frightened, the most insignificant terrorist can terrorize him (this is fundamentalism). Three centuries of total freedom up to the irruption of the world in the form of the Earth, of Gaia: a return of unanticipated consequences; the end of the modernist parenthesis.

WELCOME TO THE BEINGS OF INSTAURATION.

At the end of this lengthy Part One (but perhaps its length will be forgiven, owing to the difficulties of the task), the reader will now have grasped the goal of the inquiry: it is impossible to escape the weakening of constructivism, impossible to retain the notion of instauration, if we are not resolved to repopulate the world that the Moderns actually inhabit but that they believe they have been obliged to depopulate in advance (it is true that they have become expert exterminators...).

Are we now in a position to replace the irreparable crack between what is constructed and what is true by deploying trajectories that distinguish among the various modes of veridiction? If I am to believe thirty years of debate and quite a few heated disputes, the answer can only be "No!" In which case the fundamentalists will always win, and their critics, too, in alternation. And yet once the opposition between theory and practice has been diagnosed, its origin situated in the unforeseen conjunction between two contradictory requirements, the one concerning manufactured nonmanufactured facts, the other concerning the constructed nonconstructed God, perhaps we can come back to the beings in need of instauration. In any case, we have definitively left behind the opposition that required us to choose between representations and things.

Up to now, the Moderns thought they had to refer to "fictional beings," "gods," "idols," "passions," "imaginings," as *real things* only *out of charity*—critical rather than Christian charity. It was understood that with such things it was only a matter of "REPRESENTATIONS" "taken for beings" by people whose "convictions" had to be respected, to be sure, but whose "phantasms" had to be "feared" at the same time, and one had to protect oneself above all against "an always possible return of the irrational and of archaism." The real nature of these beings devoid of existence "quite obviously" came from elsewhere, since it could not lie in "material" things. Which does not mean, as we have understood, that the Moderns were "materialists," since we now know how *res extensa, cogitans et ratiocinans* is to be understood among them. The "materialists" never proposed anything invigorating since matter, in the modern scenography, arose from an a priori confusion between modes of knowledge—whose networks had been lost from view—and the reproduction of things—which the materialists had completely neglected to follow [REP · REF]! What passed for materialism was a phantom added to these phantasmatic representations—we can understand that it hardly matters, then, whether the materialism in question is subtle or crude. To become materialists for real, we are going to have to instill in materialism a bit of ontological *realism*, counting on *many* beings, well-nourished, fattened up, plump-cheeked. To push this inquiry forward, we are going to have to go through ontological *fattening* therapies. Our anthropologist doesn't want to have anything more to do with "representations," those succubi engorged with wind that disappear like Dracula at the break of day, leaving you bloodless. Materialism is still a thought of the future.

But aren't we already partway there? From the start, haven't we managed to allow several modes of existence to run, flow, pass, each one appearing indeed to possess its own conditions of truth and falsity and its own mode of subsistence? If we consider the *libido sciendi*, we now know how to recognize the branching that allows us to stop confusing the chains of reference [REF] it has to establish in order to ensure knowledge with the leaps that things have to make to maintain themselves in existence [REP]; here we surely have two distinct modes, each articulated and each real in its kind. I have alluded several times to law [LAW], to its passage, its processes, its procedures; nothing prevents us from speaking from now on about "beings of law," those beings that wake a judge up at night and force him to ask himself "Did I make the right decision?" We have also seen that the deployment of the surprising heterogeneous associations we have called networks [NET] allowed us to unfold an entirely different landscape from that of the *prepositions* [PRE]. Do we not have here two modes of existence, both also distinct, both also complete, each in its own way? Similarly, we have figured out that we ought to be able to detect something true or false in the political, something that Double Click [DC] missed for sure [POL]. We have just introduced the religious [REL], still so inadequately grasped if we limit ourselves to one of its divagations, namely, the destruction of idols, while the latter must have something to do, themselves, with another mode developed further on as [MET]. Thanks to Souriau, we have been introduced to beings of FICTION that it would surely be unjust to classify under imagination alone, since they emerge, they survive, they impose themselves "from the outside," even though they do not resemble the other modes of subsistence (we shall learn to spot these, noted [FIC], later on). If I am not mistaken, *this already makes ten.*

So it is not unfeasible, after all, to throw ourselves into the enterprise of *capturing* the modes of existence thanks to which the Moderns could get a new grip on themselves and discern for each mode the category mistakes that they risk committing when they confuse one mode with another. Perhaps we shall then be able to assemble them in quite a different way. Isn't this a more productive enterprise than trying to fix the cracks of constructivism?

NOTHING BUT EXPERIENCE, BUT NOTHING LESS THAN EXPERIENCE. To sum up this Part, we could take up William James's argument again: we want *nothing but* experience, to be sure, but *nothing less* than experience. The first EMPIRICISM, the one that imposed a bifurcation between PRIMARY AND SECONDARY QUALITIES, had the strange particularity of removing all relations from experience! What remained? A dust-cloud of "sensory data" that the "human mind" had to organize by "adding" to it the relations of which all concrete situations had been deprived in advance. We can understand that the Moderns, with such a definition of the "concrete," had some difficulty "learning from experience"—not to mention the vast historical experimentation in which they engaged the rest of the globe.

What might be called the *second empiricism* (James calls it *radical*) can become faithful to experience again, because it sets out to follow the veins, the conduits, the expectations, of relations and of PREPOSITIONS—these major *providers of direction.* And these relations are indeed in the world, provided that this world is finally sketched out for them—and for them all. Which presupposes that there are beings that bear these relations, but beings about which we no longer have to ask whether they exist or not in the manner of the philosophy of being-as-being. But this still does not mean that we have to "bracket" the reality of these beings, which would in any case "only" be representations produced "by the mental apparatus of human subjects." The being-as-other has enough *declensions* so that we need not limit ourselves to the single alternative that so obsessed the Prince of Denmark. "To be or not to be" is no longer the question! Experience, at last; immanence, especially.

·

PART 2

HOW TO BENEFIT FROM THE PLURALISM
OF MODES OF EXISTENCE

Chapter 7

REINSTITUTING THE BEINGS OF METAMORPHOSIS

···

We are going to benefit from ontological pluralism ⊙ while trying to approach certain invisible beings.

There is no such thing as a "visible world," any more than there are invisible worlds ⊙ if we make an effort to grasp the networks [NET] that produce interiorities.

Since the autonomy of subjects comes to them from the "outside" ⊙ it is better to do without both interiority and exteriority.

Back to the experience of emotion, ⊙ which allows us to spot the uncertainty as to its target ⊙ and the power of psychic shifters and other "psychotropes."

The instauration of these beings has been achieved in therapeutic arrangements ⊙ and especially in laboratories of ethnopsychiatry.

The beings of metamorphosis [MET] ⊙ have a demanding form of veridiction ⊙ and particular ontological requirements ⊙ that can be followed rationally, ⊙ provided that the judgment of Double Click [DC] is not applied to them.

Their originality comes from a certain debiting of alteration, ⊙ which explains why invisibility is among their specifications.

The [REP · MET] crossing is of capital importance, ⊙ but it has been addressed mainly by the other collectives; ⊙ thus it offers comparative anthropology a new basis for negotiations.

WE ARE GOING TO BENEFIT
FROM ONTOLOGICAL
PLURALISM ⊙

I N PART TWO WE ARE GOING TO TRY TO BENEFIT FROM THE PLURALITY OF THE ONTOLOGIES THAT WE HAVE JUST RELEASED FROM THE CRUSHING division between Object and Subject—a division whose origin we shall ultimately have to explain in some way beyond merely calling it a "mistake." So that we can interrogate these beings appropriately and extract templates for existence that will suit them, rather than the ones every other mode seeks to impose on them, I propose to use the term "SPECIFICATIONS," a term used in project management. For each type of being, we shall ask what specifications their ontology must respect and what their "essential requirements" are. This useful term from the world of standardization designates specifications about which there must be agreement in international negotiations to define norms *at a minimum*, even if all the parties involved are otherwise ready to allow great latitude in the way each one commits to meeting these requirements.

We have learned to recognize a mode every time we realize, in a test (most often a test constituted by a category mistake), that a certain type of continuity, a trajectory, is outlined through the intermediary of a discontinuity, a hiatus, a new and original one in each instance. We also know how to recognize a mode when it has its own, explicit, self-referential way of qualifying the difference between its felicity and infelicity conditions. And in addition, for each of these declensions, we can discern the right way and the wrong way to grasp the mode. A mode of existence

is thus always both a version of BEING-AS-OTHER (a debiting of discontinuity and continuity, difference and repetition, otherness and sameness) and also its own regime of veridiction.

Since in this second Part we can expect to profit from the ontological pluralism authorized by the inquiry, let us try to hone our skills, as it
were, on two contrasting examples. It is often claimed that the Moderns
have put an end to the irrationality of superstitions and discovered
the effectiveness of technologies. This is in any case the way they have
presented themselves to those whom they have encountered during
the process of endlessly extending the pioneering front, the war front of
modernization. To discover exactly what to make of this frontier, we are
going to use the same QUESTIONNAIRE for the beings that seem to be the
most nonmaterial of all, and then, in the following chapter, for those that
seem the most material. Through this sort of warm-up exercise, we shall
attempt to rediscover the thread of two experiences, one overly negative
and the other overly positive, whose oscillation very largely determines
the anthropology of the Moderns.

Let's begin with the charge of superstition.   ⊙ WHILE TRYING TO APPROACH
Even a superficial reading of ethnographic litera-   CERTAIN INVISIBLE BEINGS.
ture would suffice to convince any inquirer of the
abyss that exists between on the one hand the enormous work done by
the COLLECTIVES known as "traditional" to capture, situate, institute, and
ritualize "invisible beings" and on the other the continual defense of the
societies known as "modern" *against* these beings, so as to *prevent* them
from securing their positions. Institution on one side, destitution on
the other. This is even what allows a Modern to declare himself as such:
he, at least, doesn't believe in "all that nonsense." He has "battled" those
monsters; he has exposed the snares of magic and, when he conquered
the world, it was in the style of *Tintin in the Congo*: destroying fetishes,
delivering peoples from their ancestral terrors, putting an end to the
power of sorcerers and charlatans. We can tell that it's hard for the investigator to compare such different collectives, where one set *comprehends*
and encompasses invisible beings that are totally absent and totally
incomprehensible among the others. On one side, these beings exist fully;
on the other, not at all. Here is a good opportunity to see whether the
notion of *mode* of existence allows us to see the picture more clearly.

If our anthropologist can't start from such a sharply delineated position, it is because she already knows to what extent one must be skeptical, dealing with Whites, of what they call the "visible world." In Part One, we saw that the "healthy materialism" of the Moderns' "good old good sense" captures the experience of knowledge as poorly it does that of the reproduction of existents. Perhaps the Moderns lack an adequate way to grasp those beings to which the other cultures seem to pay so much attention. How can we trust those who have been so mistaken about the visible when they head off to war against the "occult powers"? Better, then, to be skeptical about what they call "illusions" and "phantasms," in contrast to what they call "materiality," and against which they feel obliged to go on fighting. And it is truly a matter of wars, here—even of massacres. The bonfires are still smoking with the witches burned alive at the time of the scientific revolution; the ashes are not yet cold after the auto-da-fés in which lay and religious missionaries alike piled up fetishes (and sometimes the fetish-makers) every time they came in to "deliver the tribes from their archaic superstitions." In the face of such violence, it would be most imprudent of our investigator not to be skeptical of everything her informants tell her about the nonexistence of such beings. If the beings are deceptive, it is perhaps not for want of existence but because there is a risk of being mistaken about the precise value of existence that they should be granted—and a risk of tragic self-deception.

THERE IS NO SUCH THING AS A "VISIBLE WORLD," ANY MORE THAN THERE ARE INVISIBLE WORLDS ⊙

She is all the more skeptical given that her informants insist on situating the unique origin of all the "irrationality" displayed by other cultures in what they call "the psyche of human subjects." All these beings whose elaboration forms an essential and hair-raisingly complex part of ethnographic research could never be found among us, to tell the truth, they say, except through the intermediary of PSYCHOLOGY. We must not look outside, but inside, into the mind, even into the brain. Now, if the investigator has been so doubtful about the direction she should take when she was hearing talk about objects and objectivity, it would be prudent on her part not to rush too quickly toward what the Moderns designate as "subjects" and "subjectivity." The two directions are linked, since they have no definition except that of mutual opposition. If the one—exteriority—leads only to dead ends, the same will surely be true of interiority. Indeed, we are going to discover that psychology plays the same role among the Moderns as epistemology, but in inverted fashion: whereas the latter exaggerated the outside world, the former overplays the inside world. This could put our whole investigation on the wrong track.

Faithful to our methods, we shall thus have to work at determining whether among the Moderns there are networks for the production of "interiorities" and "psyches" endowed with some materiality, traceability, solidity similar to those of the networks for the production of "objectivities" that we have already identified. To be sure, we have to acknowledge that there is no *positive* institution that allows us to welcome invisible beings, as other peoples can, but it is quite possible that the Moderns deceive themselves when they declare that they are entirely freed from (or deprived of, depending!) such arrangements and such handholds. The very violence with which the informants strip invisible beings of all external existence and insist on locating them only in the twists and turns of the self, the unconscious, or the neurons reveals such a deep discomfort, such an intense anxiety, that it really demands a closer look.

⊙ IF WE MAKE AN EFFORT TO GRASP THE NETWORKS [NET] THAT PRODUCE INTERIORITIES.

Someone is sure to object that this externalization—or rather, to give it its true name, this deliverance!—is more or less plausible for objective knowledge, which is always strongly equipped, collective, materialized, and "networked," but what about what quite incontestably makes up interiority, that is, the passions of the soul? The anthropologist has trouble explaining her project. "You really don't expect to throw those out as well, toss them *outside* while you're cleaning up the sanctuary of subjectivity? Knowledge, especially in our day, after three hundred years of science, in fact has something public, instituted, attributable, about it; but psychology? the depths of the self? the secret folds of the human mind? You're not really going to argue . . . The Moderns can be mistaken about themselves, but not to that point. You can localize, historicize, anthropologize as much as you like, cultures, customs, technologies, even the sciences, if that appeals to you, but there is still an unchallengeable foundation, a flowing spring, a primordial origin, a cavity from which the self surges forth, everywhere and from the beginning, the indisputable *ego*, the common property of universal humanity."

Here, in any event, is a branching point that our investigator does not want to miss: to be anxious whereas there is actually *nothing* outside; or, on the contrary, not to realize that, if one is frightened, it is because there really is *something* that provoked the fright, something pressing! Projection or encounter? anguish or fear? She is going to have to choose.

Now, very quickly, an ordinary approach focused on networks allows her to gather up in her net a fairly large number of indices regarding the apparatus necessary for the production of interiorities.

For, after all, she is studying people who at all hours of the day and night address themselves to beings that they do take, but in practice only, to be forces that transcend them, oppress them, dominate them, alienate them; people who have developed the largest pharmaceutical industry in history and who consume psychotropic drugs, not to mention illegal hallucinogenic substances, to a hallucinating extent; people who cannot subsist without psycho-reality shows, without a continuous flood of public confessions, without a vast number of romance magazines; peoples who have instituted, on an unprecedented scale, the professions, societies, technologies, and works of psychoanalysis and psychotherapy; people who delight in being scared at horror movies; people whose children's rooms are filled with "Transformers," or "transformable characters" (to back-translate from French, where the label is particularly apt), people whose computer games consist essentially in killing monsters or being killed by them; people who cannot go through an earthquake, experience an automobile accident, or deal with a bomb explosion without calling in "psychological service providers"; and so on.

Even if they appear less continuous, less equipped, less standardized, and thus less easily traceable than the arrangements that produce rectified knowledge, nothing would seem to prevent us from following the networks that could be designated PSYCHOGENIC, since in their wake they leave interiorities on the basis of the outside (as opposed to reference, which could be said to engender exteriorities on the basis of the inside of its networks).

SINCE THE AUTONOMY OF SUBJECTS COMES TO THEM FROM THE "OUTSIDE" ⊙ And yet, if the ethnologist questions one of her informants, the latter will assert unhesitatingly that he does not have to "engender" his psyche, since he possesses a self that is native, autochthonous, primordial, authentic, aboriginal, individual. And if she points out the immense apparatus that seems required for the manufacture—or the instauration—of his interiority, he will look at her *without understanding*, not seeing the relation between his "self" and those psychogenics. "All that has nothing to do with *me*; there's no connection. Me? No, not at all; I'm 'clean.'" Her informants even grow furious that anyone might accuse them of "the practices of savages." This is something really intriguing: what puts them "beside themselves" is that someone designates the source of what is agitating them as something outside themselves!

More surprisingly still, they will make condescending fun of the "other tribes" or the "bumpkins" who are "still" obliged to believe in sorcery, to protect themselves with fetishes or amulets, to send for a spell-breaker, or to go through a shaman to interpret their dreams. Yes, the subscriber to romance magazines, stuffed with downers, stretched out on the couch, who for good measure may well have added to a long string of home health aides a few soothsayers, gurus, imported fetish-makers, osteopaths, seers, zen masters, and various charlatans (on the pretext that "it can't hurt")—this modernist believes that the others believe in beings external to themselves, whereas he "knows perfectly well" that these are only internal representations projected onto a world that is in itself devoid of meaning . . . And, to prove it, he will make fun of the abracadabras of dark-skinned charlatans, even as he accepts the voluble mumbo jumbo of psychology, which has "become scientific," as a "radical epistemological break" in the history of "Western Reason." He will even assert that he ought to work very hard on himself so as "finally" to become authentic, by plunging deep inside himself to reach the "truth of the subject" and to "bring out the 'ego' where the 'id' had been in charge."

Even in Freud we find this astonishing statement: "Demons do not exist any more than gods do, being only the products of the psychic activity of man." Our ethnologist, who is somewhat familiar with the literature, wonders whether everything is not topsy-turvy in such a statement: the words "demon," "gods," "man," "activity," "psychic"— even the word "product" (and let us note the "any more than," so typical of de-constructivism), not to mention the verb "exist," the least respected verb in the entire language. Here is a truly capital category mistake: if there were ever a case of one thing being *mistaken for* another, it is surely that of the psychogenic networks being mistaken for "a product of the human mind." How can one not be stunned by such a lack of self-understanding?

The analyst finds herself facing the same type of paradox as when she had to grasp the material and practical means of objectivity of the [REF] type. Just as the connection between the results and the means of knowledge remains invisible to the Moderns, so the infrastructure that authorizes them to possess a psyche seems to escape them completely. It is as though they hadn't succeeded in defining comfortably either the "external world" or the "internal world." The symmetry is so fine that she can't resist recording it in her notebook: "My informants are doubly shocked by what I say about them, when I propose to accompany the beings of psyches *outside* and to accompany the beings of knowledge *inside* their networks. They say that I am mistaken about both the outside and the inside. But what if they are the ones who are doubly mistaken?"

⊙ IT IS BETTER TO DO WITHOUT BOTH INTERIORITY AND EXTERIORITY.

To explore interiority, psychology does not seem to be a more reliable guide than epistemology is for advancing in exteriority. Because of their theory of objects, the Moderns do not seem to have any place to put the effect of the psychogenic networks. Everything happens "in their heads," because of the radical, essential, distinction between Object and Subject. If the outside was filled up too quickly by a clumsy gesture expanding the chains of reference and confusing them with that to which they had access, the *inside*, too, is perhaps only the result of another clumsy gesture, a simple problem—one hardly dares to suggest this—of *storage*, or in any case of logistics. "Sorry, our *res extensa* is already full; go find yourselves lodging somewhere else!" Unable to discharge or even to situate a set of phenomena that are quite real, quite objective, quite fleshed-out, but that do not resemble what was anticipated in the too-quickly-saturated vessel of exteriority, the Moderns could be said to have gotten rid of them, there's no other way to put it, by calling them "internal to the subject."

This does not mean that subjects lack cavities, but that any such must always be *dug out* by an effort of mining; it must be cleared out, shored up, equipped, instituted, maintained. To keep it from filling in, one has to keep drying it out at great cost with machinery of increasingly enormous size. The Modern psyche resembles a subterranean city, a material infrastructure, an artificially pressurized sphere. No one would think of saying that a secret military base or the catacombs or sewers of Paris are "intimate" spaces on the pretext that the light of day is never seen in them. There is thus no reason, either, to confuse the practical nuance between above and below, public and private, free access and restricted access, with this radical break between the intimate and the extimate, interiority and exteriority, the subjective and the objective, the material and the nonmaterial, the personal and the instituted. As every urbanist well knows, both the visible city and its invisible infrastructure must always be taken into account.

There is thus no reason at all not to follow the networks that would allow us to empty out and dig down, to equip, illuminate, maintain, and move subjects around. The Moderns have big egos, it's true, but if we listen carefully, we hear the regular humming of the exhaust pipes that maintain the void in their ever-so-precious interiority. These pumps, too, are of respectable size and very costly. We shall thus have to learn to follow the manufacture of interiority, to follow along those paths, here as always, along particular networks, and learn to discover the exact tenor of what they transport. The construction site is immense, but it is open: we are going to have to pay the price of interiorities in hard cash as we have learned to pay for—or rather to pave over—the paths of reference.

Here is where things get complicated and where we have to go further to capture, not the networks in all their heterogeneity, this time, but the original BACK TO THE EXPERIENCE OF EMOTION, ⊙ experience that defines their specific mode of extension. Without that experience, indeed, we run the risk of committing with regard to the Moderns the same mistake we often make when we see someone talking wildly to himself and making sweeping gestures—before we notice that he is talking to *someone* through the intermediary of a portable phone. "Ah, so I was wrong about the setup: he isn't crazy, he is using a device to talk to someone else!" These are the questions we now have to raise: *who* is being addressed by those who assert that they are only talking to themselves, and *what apparatus* serves as their go-between?

So we find ourselves confronting the same type of question as with equipped and rectified knowledge: how to bring out an experience that the accounts that have become official succeed only in squelching? Here, however, our inquirer is even more astonished at the distance between theory and practice. She readily acknowledges that scientific networks pass unnoticed: they interest few people, they are encumbered with instruments, they excite only the *libido sciendi*; she even understands that the networks of law have drawn little attention: after all, apart from practitioners, most of us go from cradle to grave without stepping into the office of a lawyer or a judge. She sees clearly that the networks of reproduction are too speculative to interest anyone but the most seasoned metaphysicians and a select few poets. But the networks that gnaw, burrow, and shore up psyches? Not a single one of us escapes, not for a second of our days or nights, and they go on, they proliferate, they bug us and nag us starting even before we're born, and they keep on going even after we die. And yet the Moderns persist in describing the "self" as if it were an island surrounded by sharks and inhabited by natives dressed like Eve.

What happens if we begin to direct our attention toward the entities described by psychology as endowed with an authentic self? It doesn't take the analyst long to notice that the same informants who claim to be free of all superstition are nevertheless unstoppable when it comes to describing an experience that seems to them to have come from the "outside." This experience is the test of feeling oneself "targeted by" an *emotion*, that is, any kind of setting into motion—a hurtful word, a shocking attitude, an untoward gesture, sometimes even qualified by the term "evil eye"—that gives the impression that one is being taken over, overcome by an uncontrollable force—"I don't know what *got into me.*" Let us call this experience a *crisis*. And since we are always "in the grip" of something in a crisis, let us call it a *gripping* crisis.

⊙ WHICH ALLOWS US TO SPOT THE UNCERTAINTY AS TO ITS TARGET ⊙

But, at the same time, at the heart of the crisis, a suspicion creeps in, a suspicion that always causes our anthropologist to prick up her ears, and also the patient, the person in the grip of an emotion—the suspicion that there is something else, something *other* in this trial, that he or she has made a mistake as to the attribution, the target, the goal:

"I'm not the one targeted," "I didn't hold a grudge against you," "that's not getting to me." In other words, the vague feeling, almost as powerful as the emotion itself, that there has been a mistake of address. One feels quite odd, or rather quite *other*, as if there were indeed something other in play. But what? Crisis in crisis. Anxiety about anxiety. Cry behind the cry.

There ensues—sometimes, not always (there are always people whom nothing shakes, nothing moves)—through the trouble, the tears, the screams, piercing cries that traverse the soul, convulsions, shocks, upsets, something like an *exit* from the crisis, a transformation that is the inverse of the previous change: you see, you get it, you realize your mistake, in a quite particular form: "But no, of course, I *wasn't the one* who was the target." You have the vague impression, sometimes comic, always stupefying, that *something quite different* has happened—and that you *knew this very well*, at the very heart of the crisis: "There, it's over, I'm OK, whew, that was rugged, it was *awful*, really terrifying, I almost *disappeared*."

And now, from the other side of this transformation that was as sudden as that of the crisis, floods of new energy carry you away. But it is not at all the same *carrying away*. It has changed signs, as it were, from negative to positive. From black to white. From moral, morbid, criminal, it has now become tonic, active, inventive. You are transported as if by the same wave that, a second earlier, was going to "devour" you, drown you. You have the impression of being *transformed*, yes, of having undergone a sort of *transmutation*. A gesture, a word, a memory, a ritual, something unnameable, something elusive and decisive has *made you pass* from one side of this trial to the other.

Sometimes, on the contrary, you don't come out of it at all, you feel that you are *going under*, and there, through more convulsions, more cries, more turmoil, more tears, you yourself become a raging force, a thirsty monster, a tornado: you're no longer in control. The suspicion, the hesitation of a moment ago gradually disappear—and with them, any hope of "pulling out of it," of "getting through it." Now, a certainty: "I am really the one targeted? Well, I'm going to get my revenge!" From this point on, terror runs rampant through the world and nothing can staunch it. The "I" becomes another for real. "I'm not myself any longer." "Stop me or I'm going to do something terrible." The diagnosis is made, language has the words for it: I am *alienated*, yes, *possessed*. I am an *alien*, one of the living dead, a zombie. Except that of course there is not, there is almost no longer, there is no longer at all any "I" subject of the verb "to be," since the "I" and the gripping "other" are now one and the same. An impersonal form. It rains. It goes. It kills. A vampire has come out, taken over, overcome. Terror. Misery. Destruction. Tailspin. Fall into rack and ruin. Rewind.

⊙ AND THE POWER OF PSYCHIC SHIFTERS AND OTHER "PSYCHOTROPES."

Isn't this how the informants declare that they subsist in good years and bad, from crisis to crisis? Isn't it their common lot, the lot of mortals? Subsistence dearly bought. A little like the continually agitated balance, in meteorology, between the anticyclone and what is called (the metaphor works pretty well) a *depression zone*. Pressure and counter-pressure, the subtle atmosphere of their moods. To exist, for a self (the word is still uncertain; we shall define it later on), is first to resist successive waves of fright, any one of which could devour us; but each could also be shifted off course by an attraction, a snare, a device, a gimmick, any sort of *artifice* thanks to which we suddenly discover that the beings that were deceiving us are on the contrary helping us to exist, finally, that we can surf on them, through them, with them, thanks to them, by dint of skill, as on a wave that carries you away but that isn't targeting you personally. As if there were many beings that have to be called PSYCHOTROPES in the sense that they modify who you are almost completely; they get you all mixed up by making your soul spin.

These ruses, these skills, are depicted for us by mythology; rituals manifest them for us; the confidential revelations of romance magazines delight in them. "He told me, then I told him, and then he let me have it; don't get me wrong; that doesn't get to me; I didn't mean what I was saying; he's got it in for me; he had it coming; this always gets to me; I wasn't after *you*." Is there a single moment when we escape from this tension, this twisting, these ruses, the obligation to protect ourselves, to reverse the forces, to put an end to the tensions, to resist the low-pressure areas, the depressions? A single moment when we don't benefit from the formidable energy of what seems to *transit* in us? This obstacle course through the flux of fears and terrors resembles the itinerary of a salmon going back upstream—if it lets itself go, even for a second, it is swept downstream, wiped out, cleaned up, undone, depressed, carried away, consumed, morbid, dead, rotten—but if it holds on, if it insists, if it beats its fins vigorously enough, then it goes toward what allows it to be, to *come*, and to reproduce. A state that is designated by a happy conjunction of the verbs "to be" and "to have": "We've been 'had'"—that's it: "We've been *possessed; carried away; taken over; inhabited.*"

But by whom are you possessed? Who carries you away? Who traverses you? Who has begun to dwell in you? As in the case of the false madman who seems to be speaking to himself as long as we don't see his phone, the analyst can't help but ask the question: to whom are you speaking? And here the informants, unstoppable up to now, fall silent, hesitate: they don't want to be taken for fools who talk to themselves; they do feel that they are in contact with owners, conveyors, beings in transit, particularly demanding inhabitants, but they seem unable to find the right expressions to account for them.

It's possible, then, that our anthropologist may look at them with a somewhat disappointed air: when she thinks about what all the other collectives have developed to enter into commerce with such "owners," such "inhabitants," this sudden speech impediment demonstrates a sort of lack of culture, almost rudeness. It is as though there were no relation between the search for a complete, autonomous, authentic, true self and the swarming of entities necessary to the self's constant fabrication, its continual mutations. As if there were something truly pathological in the systematic avoidance of the forces in which other peoples are said to "believe too easily"?

Is it possible to discover, at the heart of the modern collectives, an approach that will allow the analyst to go further in the recognition of beings that we feel don't come from within—everyone acknowledges this—even if we can't yet qualify their exteriority, that is, the type of objectivity that would suit them? It's tempting to choose *therapeutic* arrangements to grasp their complex logic and their specific rationality; by taking the term in a broad sense that would cover everything that attempts to offer care and treatment, from the awkward expressions of an overwhelmed lover, through the psychoanalyst's couch or the pharmacologist's laboratory, to rituals of exorcism. Isn't it in these arrangements, at least among the Moderns, that we verify day

THE INSTAURATION OF THESE BEINGS HAS BEEN ACHIEVED IN THERAPEUTIC ARRANGEMENTS ⊙

by day what liveable relations we can maintain with these beings whose positioning strikes us as so strange?

As if the procedures of treatment offered the beings that devour, carry away, and possess what laboratories offer the beings of the afore-mentioned Nature: a privileged site in which we might grasp their instauration. In relation to the rituals, cosmologies, and philosophies developed by the other cultures, it is obviously a matter of a weakened form of existence: therapeutic arrangements are the only ones that have escaped, in practice if not in theory, from the generalized denial of invis-ible beings.

⊙ AND ESPECIALLY IN LABORATORIES OF ETHNOPSYCHIATRY.

Especially if we approach them after the fashion of ETHNOPSYCHIATRY. Indeed, our analyst would have despaired of making her investigation empirical, if she hadn't had the bright idea of concentrating her attention on one of the rare sites capable of establishing a comparison between the so-called modern technologies of healing and those said to be "archaic or primitive." This is where these transformational beings can best be identified. Just as situated cognition (and science studies) had made it possible to make it attributable, that is, rational, to follow this component, one among others, of the train of thought that stems from chains of reference—to the point that an ordinary brain becomes objec-tive and scientific by branching out along these chains—so ethnopsychi-atry allows us to make perceptible another component, even more diffi-cult to trace, that of the "owners" who are charged with "possessing" us. With ethnopsychiatry—certainly in Tobie Nathan's fruitful version—the self is no longer a madman who talks to himself in search of authen-ticity: one speaks to that self, it answers, it has an apparatus at its disposal, it may even understand what people are telling it in a language that is at first foreign.

But how are we to understand the language of those beings whose instauration is at stake? By acknowledging, first of all, that we have to pass through artificial arrangements, rituals, if we are to become familiar with them. And then by recognizing that, if they can metamor-phosize "us," it is because they "mistake us for others," for *aliens*, without "aiming" at us in any way at us as PERSONS; we shall see why later on. Moreover, this explains why we can often trick them. Since they do not take us as selves, we may be able to "get them to mistake something else for us." An overly simplistic explanation that lacks depth? It seems that no one has ever found anything better that corresponds to the requirements of these particular creatures. Black magic and white magic are separated only by a slight difference: shall we be devoured or, on the contrary, shall we catch what threatened to devour us by making it shift to another target that it has mistaken for us and that we are going to be able to straddle in order to transform ourselves? Ulysses managed in the end to tie up even Proteus.

Each of us preserves a more or less treasured anthology of our contacts with invisible beings like these: words of white magic that have saved us—"in fact, it was nothing; I felt targeted, but it wasn't me, it was something else"; wounds that won't scar over caused by the black magic that has put us, as we say, "in a state," and that we gnaw on for the rest of our lives; words that kill whose semiconscious messengers we have been—"that's not what I meant to say, I'm really sorry"; yes, but, the thing is, what I said hit its target anyway, and killed quite effectively. These forces can be "made to pass elsewhere" (something that psychoanalysis has expressed very well with the idea of *transfer*). They pass, we transform them, transfer them and, in passing, if we go about it right, they give us the energy to go on, to go farther in quite a different way, as a river activates the wheel of a mill—provided that the mill resists the pressure. A vast scholarly or popular encyclopedia, on public display or kept private, complacent or reserved, depending.

All the scholarly apparatus, sometimes heavily equipped, that ethnopsychiatry has taught us to respect among ourselves as well as among the others (the "charlatans," precisely) are actually arrangements of "transactions" that make it possible to *compromise* with beings that are infinitely more powerful than we are but that are blind to our persons, as it were, so that we can treat them like Chronos and make them mistake a stone for one of their children in order to mislead them, make them shift their formidable power elsewhere, or, on the contrary, we can install them at last, "seating" them, instituting them, turning them into tutelary divinities.

Therapeutic setups offer the investigator an opportunity to sketch out the specifications of these   THE BEINGS OF METAMORPHOSIS [MET] ⊙

entities and their very particular mode of presence or absence. If they are so *gripping*, it is because they can transform us at any time; they can inexplicably turn us into monsters, yes, make us *alienated*, *possessed*, or, depending on circumstances, make us *ingenious*—which is not always more reassuring. Whether we are dreaming or awake, from simple anger to the monstrous crimes we sense we are capable of committing, from "black looks" to stormy passions, there is no one among us who is not in constant touch with these transformational beings. In our relations with them, it is as if we were some fragile envelope constantly bombarded by an incessant rain of beings that bear psyches, each of which is capable of influencing us, moving us, messing us up, upsetting us, carrying us away, devouring us, or, on the contrary, making us do something we didn't know we were capable of doing, something that inhabits and possesses us from then on. Like Cassavetes's heroine, we all live "under the influence," positive or negative, according to the turn taken by these "psychotropes." We still have to find out what they are, what they want, what language they speak, and how to "seat" them.

It is clear that this experience—of shock, alienation, detection, transformation, installation—is at once very common and very difficult to qualify within the narrow framework of psychology. The continuity of a self is not ensured by its authentic and, as it were, native core, but by its capacity to let itself be carried along, carried away, by forces capable at every moment of shattering it or, on the contrary, of installing themselves in it. Experience tells us that these forces are external, while the official account asserts that they are only internal—no, that they are nothing. Nothing happens. It's all in our heads. One thing is certain: we have here a form of continuity that is obtained by leaps, by passes, by hiatuses through a dizzying discontinuity. And thus we have, as a first approximation, an original mode of existence that we shall note from here on as [MET], for METAMORPHOSIS.

⊙ HAVE A DEMANDING FORM OF VERIDICTION ⊙ The reader will perhaps acknowledge that these beings have a certain exteriority—experience says so, popular wisdom as well—but that the TRAJECTORIES they sketch out cannot have the type of veridiction required by the writing of specifications. Everyone knows perfectly well that this is all just a bunch of nonsense, at best old bonesetters' formulas, at worst mental manipulations. "We're certainly not going to rely," everyone will say, "on just anything, on magical 'passes,' on sleight-of-hand." The word "magic" may be debatable, but the word "pass" suits us very well! For this is what is at issue: how to make something *pass* that otherwise is going to aim, strike, bury, crush, possess, devour?

Now here we find a demanding form of veridiction, because one really can't say just anything at all about these beings: one ill-chosen word, one misunderstood gesture, one carelessly performed ritual, and it's all over: instead of freeing, they imprison; instead of taking care, they kill. It's not that everything plays out on a hair's breadth, a razor's edge; it's not that the razor isn't sharp. On the contrary, as we expected, as the rational—the search for reasons—requires, there is a subtle casuistics that will allow us to distinguish the expression capable of *speaking well* (turning away well, deflecting well, installing well) from the one capable of *speaking badly*, or cursing (poisoning, or—to put it bluntly—enchanting, in the strong sense).

If there is one distinction that must not be lost, must not be crushed by accusations of irrationality, magic, charlatanism, it is certainly this crucial distinction between *enchanting* and *disenchanting*. It is not because these felicity conditions are most often mute; it is not because they have been shunted off to gloomy suburban dispensaries around the edges of pharmacology labs or to horoscopes in tabloids and romance magazines, that we must ignore their capacity to *discriminate* between good and bad, true and false. These felicity and infelicity conditions define, on the contrary, a terribly demanding form of veridiction. So demanding that people rush to the office, the temple, the convent—to the laboratory of the rare souls who are capable of speaking about this *effectively*, that is, who are capable, through what they say and do to you, of *taking care* of you. These practices may be encumbered with all the false mystery and clutter of occultism, but we must surely not confuse this theatrical darkness with the real mystery of the difference between a ritual that makes things better and one that makes them worse. This difference, on the contrary, must be preserved, cherished, and certainly protected against accusations of irrationality, which miss precisely the *ratio*, the judgment it manifests.

This was the third feature, let us recall, of INSTAURATION: the difference between *well* and *badly* made. The trashy opposition between charlatanism and psychology, ultimately rational, must not be allowed to obscure the only contrast that counts: the small but crucial *nuance* between the good and the bad charlatan, the good and the bad psychoanalyst, the good and the bad neuropsychiatrist, the good and the bad dosage of a medication. We must not be forced to choose between the rational and the irrational when souls in trouble are simply, desperately, looking for a *good therapist*.

⊙ AND PARTICULAR
ONTOLOGICAL
REQUIREMENTS ⊙

But to grasp this form of articulation we still have to be able to define the beings that it is a matter of addressing well or badly. What are these *aliens* capable of simultaneously attacking the self, casting doubt on whether the target is the right one or not, letting themselves be deflected by a gesture of treatment or, on the contrary, successfully contaminating another self that takes itself henceforth for what was targeted and becomes *alien* in turn by transforming itself from top to bottom thanks to *those beings* that obsess and inhabit it from then on? It is impossible, as we have seen, to assert calmly that all these *aliens* come "from me," reside "in me," and are nothing but mere "projections" of my mind, my desires or my fears, onto objects that are themselves devoid of all existence and thus of all danger. Such a diagnosis would amount, strangely enough, to *helping* them reside in me and devour me; it would amount to conspiring with them against myself! As if ordinary psychology ("Look, it's just you, it's nothing, you're projecting, you're delirious, don't worry, it's all in your head") had had the astonishing result of further facilitating the residence of succubi and devourers... No, we have to give them consistency, externality, their own truth. But what consistency?

⊙ THAT CAN BE FOLLOWED
RATIONALLY, ⊙

In modernism, one cannot avoid approaching them by asking this one question: "But finally, those beings, are they real or not?" In answering too quickly, we would be falling back again into psychology. And yet it is impossible to *suspend* that question by saying that it doesn't arise for this type of subject since it is appropriate to "bracket" the existence of invisible entities in order to speak freely about them. As if it were a matter of being freed from them! An agreeable method that would allow us to avoid dealings with those we claim to grasp. No more than we could avoid the great quarrel of RELATIVISM when it was a matter of equipped and rectified knowledge, we cannot avoid it when it comes to addressing the invisible beings in their own languages [MET · REF]. In this inquiry, all *beings* insist equally in the expectation of receiving from us their exact ontological pasture.

To have a chance to move forward, we have to hold strictly, despite appearances, to the meaning we gave the adjective RATIONAL at the outset. The continuous following of a network to which one adds its interpretive key, its PREPOSITION, is rational. In what *tonality* are we to hear what follows when we say that there are "beings of possession" that try to devour you, but that do not target you since your little person is not at all in question—it is without importance for them, and moreover without content? And what do we really mean by asserting that therapy—a material, collective, and highly instituted material technical arrangement—can intervene in such a way that these voracious beings, instead of *taking you for another, take another for you* while allowing you thus to *become quite other* if you know how to figure out what they want and can find the means to install them properly? That one can, in other words, not only *deceive* them, trick them, deflect them, since they are already after the wrong target, not aiming at you, but because they can—because they are not aiming at you—become the energy source that is going to transform you for real. At last, "I is an other"!

Let us first recall, and there is no doubt about this, that if you apply to such an arrangement only the template proposed by Double Click, people will inevitably exclaim: "So you believe there are devils and demons, just as there are stones and tables!" In

⊙ PROVIDED THAT THE JUDGMENT OF DOUBLE CLICK [DC] IS NOT APPLIED TO THEM.

other words, the anthropologist searching for the ontological weight of these beings will be a madwoman in an insane asylum [MET · DC]. But let us not forget that, in the eyes of this same Double Click, the most austere scientists ought to be wearing straitjackets themselves, since the "stones" of geology deposited in the admirable museum of the School of Mining don't resemble the stones of good sense any more than the beings we want to designate [REF · DC]—as for tables, we shall find out what to think of them in the next chapter; they, too, have some surprises in store.

Double Click's mistake about psychic beings has something tragicomic about it: when he does everything in his power to imagine "transports without transformation" (believing he is imitating the IMMUTABLE MOBILES necessary to reference), he ends up with the exact opposite of what therapy is striving to do: ensuring that there is *no* transport *without* a radical transformation. Double Click thus tries to avoid all opportunities for metamorphosis. It is as though you were advising a destroyer pursued by a submarine to go straight ahead instead of making twists and turns to avoid the torpedoes . . . Boom! It will sink for sure. Double Click is not satisfied with deceiving, as usual, by turning away the sources of veridiction; this time, he kills.

The anthropologist has one advantage over this Evil Genius (and perhaps also a capacity not to curse and thus not to kill) quite simply because she can now take advantage of *several* ontological templates, rather than just two. When she declares that it is necessary to recognize the existence of psychogenic beings, while seeking to determine their exact weight of being, she no longer needs to confuse them with "material things." The reason has become clear: "MATTER" can no longer serve as the standard for any being, since it conceals rather badly the category mistake imposed by Double Click [REP · DC]. On this account, "material things" no longer fit, either, in this Procrustean bed, since what this label designates no longer amalgamates two modes, reproduction and reference [REP · REF]. Consequently, when the anthropologist tries to externalize these beings so as to respect the weight of their existence (yes, they are indeed outside me and this is what may perhaps treat me by forcing me to make room for them), she is not referring at all to "NATURE." Which avoids the temptation of classifying this new race of beings in a "supernature": without Nature, there is no supernatural, either. (As we shall see later on, this *subtraction* will be welcome in our quest for immanence.)

If we agree to answer the question of specifications this way, it must be possible to grant them a certain externality, without having to credit them with the same continuity in being, the same ISOTOPY as tables or chairs. These beings are no longer representations, imaginings,

phantasms projected from the inside toward the outside; they unquestionably come from elsewhere, they impose themselves.

But how? They have their weight of being, their ontological dignity.
But which?

If therapeutic arrangements allow an initial
identification of such beings, it must be possible
to go further in the specifications by adding a new
item to our questionnaire and asking each time
what ALTERATION is involved. In fact, now that we are beginning to free
ourselves from the scenography of Subject and Object, the question
becomes essential: if there are several ways to exist, and not just two, we
can no longer define the one simply as the opposite of the other. Every
time, the analysis must hazard a diagnosis on the *manner of being* proper
to this mode, on the positive way it has of inventing a new way of *altering
itself.* In our questionnaire, from now on, we must add to the hiatus, to
the felicity conditions, to the trajectory thus traced by these beings, the
alteration that each mode will, as it were, *debit* from being-as-other. We
are still echoing Tarde here: "Difference proceeds by differing."

THEIR ORIGINALITY COMES
FROM A CERTAIN DEBITING
OF ALTERATION, ⊙

We can already define them by starting from the therapeutic question alone. Their name may be Legion, but Proteus for sure, and probably
Morpheus of the thousand dreams. This proteiform character is familiar
to all of us, since we touch on it in dreams during nearly half of our existence and in our waking moments the rest of the time. What would we
do without them? We would be always and forever the same. They trace
thoughout the multiverse—to speak like James—paths of alteration
that are at once terrifying (since they transform us), *hesitant* (since we
can deceive them), and *inventive* (since we can allow ourselves to be transformed by them). As soon as we begin to recognize them, consequently,
we simultaneously measure the gulf into which they pull us, the means
of pulling out, and the formidable energy that would be available to us if
we only knew "how to go about it." It is only if we are afraid of them that
they start deceiving us cruelly. This is why the word "metamorphosis"
designates at one and the same time what happens to these beings, what
happens to the humans who turn out to be attached to them, and what
happens during the therapies that allow us to spot them and sometimes
to install them.

⊙ WHICH EXPLAINS WHY INVISIBILITY IS AMONG THEIR SPECIFICATIONS.

It is moreover this capacity for metamorphosis that explains why we speak about these beings as *invisible* entities. Their invisibility does not depend at all, as we pretended to believe at the beginning of this chapter, on a supposed contrast with the "visible world." It has to do with the fact that these entities correspond to an entirely different template for existence. There is nothing strange about this statement, nothing that could encourage rejection or, on the contrary, some morbid appetite for mystery. We are simply declaring that these beings undergo metamorphoses such that one can surely not attribute to them the mode that would make them *persistent* entities, on the same basis as tables and stones. We may choose to call them *occult*, but as one says of planets or asteroids that they are undergoing "occultation," for these beings, too, appear and *disappear*. They are peculiar aliens in that one can say, from one second to the next, "I thought it was *nothing*, but then I turned around and there it was, terrifying"; or, conversely, "I thought it was *something*, but then I turned around, and there was *nothing* there after all." Curiously, the countless "special effects" of horror movies may well be the best expression of this particular metaphysics. As Rilke suggests, perhaps all the monsters in our lives are only lovely young girls asking for help.

The invisibility of these beings is not irrational, supernatural, or mysterious; it comes from their precise form of articulation: *we take them for others because they take themselves and they take us for others*, thereby giving us the means to *become other*, to deviate from our trajectories, to innovate, to create. If we never see them the same way twice, it is because they are transformed—and so are we, as a result of the arrangements that allow us to capture them. To learn to follow them is thus not to succumb to the irrational but rather to explore one of the paths of objectivity—a multi-modal term if there ever were one, to which it is appropriate to add an abbreviated notation to make it quite clear that OBJECTIVITY is in question [MET].

THE [REP · MET] CROSSING IS OF CAPITAL IMPORTANCE, ⊙

Having reached this point, it would be useful to venture a hypothesis that would give these beings greater weight than that of therapy alone. If our ethnologist started from therapeutic treatment, it is because that was the only institution that could give her a handhold. Now, the other collectives have given these beings versions that are not only therapeutic but also cosmological. Would it be possible to propose an identification that does justice to such ontologies and that would allow us to begin negotiating on a basis other than that of Bifurcation between the "inner world" and the "outer world"—which cannot capture, as we now understand, either the Moderns or the others?

This is not impossible if, in order to qualify them, we approach the crossing they form with the beings of reproduction [REP · MET]. Let us recall that the latter manage to extract from being a particular value of ALTERATION: they re-produce precisely themselves, beings identical to themselves, or at least *almost* identical to themselves. The advantage of speaking of being-as-other is that we can count on a *reserve* of alterities on which many other forms of differences can draw. Being, we might say, explores the variety of its differences as if every civilization offered a particular version of its contrasts. Let us suppose, now, another form of alteration that no longer explores the resources of reproduction but rather those of *transformation*. This time it is no longer a question of paying for continuity in being with a *minimum* of transformations—persistence—but with a *maximum* of transformations, one might say—with metamorphosis. Everything can, everything must, become something else.

If this is the case, we understand that we are seriously reducing the scope of such existents by capturing them only through therapeutic arrangements. They are infinitely more inventive. Like the ⊙ BUT IT HAS BEEN ADDRESSED MAINLY BY THE OTHER COLLECTIVES; ⊙ beings of reproduction, they *precede* the human, infinitely. If the ones ensure persistence, the others multiply metamorphoses, the entire set crossruling being-as-other, giving it its own rhythm, made of the combination of the two, a sort of matrix or kneading process from which the "human" can later take nourishment, perhaps, can in any case branch out, accelerate, be energized, but that it will never be able to replace, engender, or produce. Two modes of existence that are not "primitive" but on the contrary prodigiously advanced, even if to us latecomers they appear primordial, or at least prehuman. We would be grasping them very poorly if we qualified them as "PRELINGUISTIC," when there is no proposition, no articulation, no *predication* except in that they are already responsible for the proliferation of reproductions and metamorphoses that the other modes are going to be able to *modulate* by different choices of alterations and renewal. Rather, they form the basso continuo without which no music would be audible.

⊙ THUS IT OFFERS COMPARATIVE ANTHROPOLOGY A NEW BASIS FOR NEGOTIATIONS.

Such a hypothesis would allow us to understand the total misunderstanding created by the **MODERNIZATION FRONT**. By requiring that the beings of reproduction be taken for the "external world" and the beings of metamorphosis be taken for an "inner world," the Moderns could not understand themselves any better than they could understand those whom they claimed they were "freeing from their superstitions." This neglect of the invisible entities is all the more strange in that they offered the Moderns a unique opportunity to enter into contact, on an equal footing for once, with all the other peoples. In fact, the thing the least well distributed in the world is not reason—in any case not Cartesian reason, it is surely not Science, but rather the subtle elaborations invented by all the collectives to explore the crossing between the beings of reproduction and those of metamorphosis. Now, this crossing, which the other cultures have practiced quite systematically, has no form of official existence among the Moderns. It is not surprising that the colonizers kept going from one surprise to another, and that the colonized, startled, wondered: "But unless they deal with such beings, how do they manage to keep on subsisting?"

By a reversal that should not surprise us by now, the Whites presented themselves to the others as people who were "finally" in possession (this is the word for it!) of a rational psychology, a native subjectivity, an authentic, nonfabricated self, which was going to be able to extend the benefits of subjectivity to the whole planet—with the unconscious as a bonus and medications as an option. "The old Adam still harbors phantasms that Reason is going to dissipate by showing you that *that is nothing*; nothing has happened to you; there are no divinities; it's all in your heads." *Totius in mente.* Here we rediscover the premature universalism of a particularly local form of ethnocentrism that we have already encountered under the auspices of "matter."

The result is that the Moderns, although swimming more than any other people in the diffuse institutions of psychology, remain *without treatment* in the face of the "forces of Evil." They indeed possess the UNCONSCIOUS, but unfortunately there isn't much they can *do* with it, since it is not composed of beings whose energy could be deflected by tailored artifices. A monster, yes, but one that no longer gives access to any cosmology. As if there were something *diabolical* in the Moderns' insistence on the internal origin of their emotions: this *division* between the most constant of their experiences and what they allow themselves to think about it. Whence the anxiety of our anthropologist: aren't the Moderns dangerously *alienated*? Wouldn't that explain a large part of their history? As if there were a madness of the Subject after that of the Object. And yet she cannot keep from believing in them: "If there were one area where you need to learn something from 'the other cultures,' as you say, it is surely this one. Once freed from Bifurcation, nothing will keep you from reconnecting with existents with which you have in fact never ceased to interact."

Chapter 8

MAKING
THE BEINGS OF
TECHNOLOGY VISIBLE

....

The singular silence imposed on technologies ⊙ and on their particular form of transcendence ⊙ requires, in addition to an analysis in terms of networks [TEC · NET], ⊙ the detection of an original mode of existence ⊙ different from reproduction [REP · TEC].

We need to return to the experience of the technological detour, ⊙ which is hidden by Double Click and the form/function relation.

By drawing out the lessons of the [REP · REF] crossing on this point ⊙ we shall no longer confuse technology with the objects it leaves in its wake.

Technology offers a particular form of invisibility: ⊙ the technological labyrinth.

Its mode of existence depends on the [MET · TEC] ruse ⊙ as much as on the persistence of the beings of reproduction [REP · TEC].

The veridiction proper to [TEC] ⊙ depends on an original folding ⊙ detectable thanks to the key notion of shifting.

The unfolding of this mode gives us more room to maneuver.

LET ME REASSURE THE READER: BY TRYING TO DEFINE THE RELA-TIONS THAT THE MODERNS MAINTAIN WITH THE BEINGS OF MET-AMORPHOSIS—RELATIONS LIMITED TO BEINGS THAT PRODUCE psyches but that the other collectives have ballasted quite different-ly—we have not manifested a dubious penchant for phantoms, witch-es, or demons. We have simply tried to understand why the cosmology of the Moderns remains so hard to retrace, so contradictory to the les-sons of experience. To continue our warm-up exercise and really benefit from the ontological pluralism that will allow the inquiry to count more than two modes, we are now going to try to capture beings that enjoy an entirely different status in modernism. "The modernization front al-lowed the Moderns to represent themselves," we speculated, "as the peo-ple who put an end to superstitions and finally discovered the effective-ness of technologies." We now know what to think of the first part of this claim; what are we to make of the second?

THE SINGULAR SILENCE IMPOSED ON TECHNOLOGIES ⊙

When informants insist on the nonexis-tence of certain beings, they make them proliferate, but when they emphasize—and so proudly!—the massive presence of other existents, we can scarcely make them out. This is the case with the beings of technology (noted [TEC]). The transition from beings of magic and charms to beings of tech-nology is by no means unheard-of; Gilbert Simondon had already broken the path in his book *On the Mode of Existence of Technical Objects*, a text as famous as it is little read. In passing from one mode to the other, we are going to add more depth to what has already been said more than once about the not-so-very material MATTER, that "idealism of materialism" of the Moderns. In turning toward the craftsmen, the ingenious engineers who actually build engines and machines, we shall be able to clarify the strange notion of construction to which "CONSTRUCTIVISM" does not seem to be particularly faithful.

In the eyes of our ethnologist, one of the most astonishing aspects of the Moderns is not the way they treat divinities, knowledge, or gods, but the fact that they grant so little room to what has defined them most sharply in the eyes of all the others since the era of the great discoveries began: the art and manner of deploying technology. Those who pride themselves on being "solid materialists" do not seem to have given a second thought to the solidity of MATERIALS (a word that we shall use to intensify the contrast with matter). Dismissing religion with scorn is understandable: that figure has not managed to hold its ontological ranking in the face of competition from the sciences. Nothing is more natural than being skeptical about the beings of metamorphosis and their tamperings: they always contaminate those who manipulate them in rather dangerous ways. But tools? Robots? Machines? The very landscape that they have ceaselessly turned over and plowed for hundreds of thousands of years, the inventions that have disrupted our lives more than all the other passions over the last three centuries, the systems of production on such a massive scale that they now weigh heavily on the whole planet?

And yet for a thousand books on the benefits of objective knowledge—and the mortal risks that challenging it would entail—there are not ten on technologies—and not three that signal the mortal danger one risks by not loving them. Even political philosophy, less prolix than epistemology, can still flatter itself that it has engendered more books than the philosophy of technologies; we could count the latter on our fingers. The proof of this decline is that in the word epistemology we still hear knowledge about knowledge, whereas in the word technology, despite the efforts of André Leroi-Gourhan and his disciples, we fail to remember that some sort of reflection on technology lies imprisoned. We don't hesitate to say about the most modest washing machine full of chips that it is an instance of "technology"—even "modern technology"—but we don't expect to learn any lessons from it. We ask a "technician" only to come repair our machine; we don't ask him for an in-depth reflection on it. What would we do with his philosophy? Everyone knows that technology is nothing but a heap of convenient and complicated methods. There is nothing to think.

Even if the ethnologist is no longer surprised at how hard it is to find reliable informants on questions so central to those she is studying, because of the consubstantial distinction between theory and practice, she continues to be astonished that there is no legitimate institution to shelter technologies, any more than there is one to teach us how to come to terms with psychogenic beings. How have the Moderns managed to miss the strangeness, the ubiquity, and yes, the spirituality of technology? How could they have missed its sumptuous opacity? Our ethnologist begins to note: "We don't really know what they *make*. Father, forgive them, they don't even know what they *do*."

⊙ AND ON THEIR PARTICULAR FORM OF TRANSCENDENCE ⊙

If we can hesitate over the mode of existence of reproduction [REP] (because its persistence hides its gaps), if we hesitate again over that of chains of reference [REF] (because once we reach remote entities, we are in danger of ultimately forgetting the instruments that have allowed us this access), we must hesitate also over the HIATUS introduced into each course of action by the detour and delegation proper to the technological trajectory. One little error of inattention, and we risk missing its own characteristic form of mini-TRANSCENDENCE.

All modes can be said to be transcendent, since there is always a leap, a fault line, a lag, a risk, a difference between one stage and the next, one mediation and the next, n and n + 1, all along a path of alterations. Continuity is always lacking. There is nothing more transcendent, for example, than geodesic reference points with respect to the readings jotted down by a surveyor-geometrician in his notebook—forms in the first and second meanings given in Chapter 4 [REF]; nothing more transcendent than the question of a single line of text proposed to the jury in a trial in relation to the thousands of pages of a heavy dossier rolled on a trolley all the way to the court reporter [LAW]; nothing more transcendent than the relation between the lukewarm character of a perfunctory prayer and the gripping effect of grasping its meaning for the first time [REL]; nothing more transcendent than the relation between the papier-mâché stage setting and the exuberance of the characters that seem to emerge from it [FIC] (whom we shall learn to encounter later on); nothing more transcendent than the distance separating what you were from what you have become after being seized by a psychogenic being [MET]. Transcendences abound, since between two segments of a course of action there is always a discontinuity of which they constitute, as it were, the price, the path, and the salvation.

These are what we must learn to name, every time. Do we need to recall that, in the eyes of the anthropologist, who has been forced to become something of a metaphysician in order to succeed in her quest, there are no longer two WORLDS, the first one immanent and full, *above* and *beyond* which another has to be added—the supernatural—and *beneath* which, for good measure and in order to house "representations," still another—interiority—has to be carved out? There is no longer anything before her but *subnatural* beings—Nature included!—that are all slightly transcendent in relation to the previous stage of their particular paths. They form networks, and these networks most often pay no attention to one another, except when they intersect and have to come to terms with one another by avoiding category mistakes insofar as possible. What appears most lacking is IMMANENCE, or rather, immanence is not native but secondary, and it too depends on a very particular mode of existence, as we shall see in Chapter 10. The world, or rather the multiverse, is thus full of—or rather, no, the multiverse is constantly *emptied out* by—circulating transcendences that dig down into it, all along a subtle *dotted line* formed by the leaps and the thresholds that have to be crossed one after another in order to exist a little longer. In short, an obstacle course.

The reader must now see why the investigator could not possibly do justice to technologies with the two patterns of "Objects" and "Subjects" as her only resources. In comparing modes of existence she is complicating her task, to be sure, since she has to multiply the templates of beings to be taken into account, but, in another sense, she is *simplifying* the task, since she finally has on her workbench a large number of different instruments for determining the weight of each mode by *comparison*.

Simondon's genius lay in seeing that one could not specify the mode of existence of technological beings without *titrating* them thanks to the beings of magic, religion, science, and philosophy. Every handyman knows that his skill increases if he has at his disposal not just a few rudimentary tools but a panoply of screwdrivers and wrenches, saws and pliers. This is the only rational use that can be attributed, as I have already said, to Occam's proverbial razor. We use it clumsily if we start making random cuts to limit the number of beings arbitrarily. On the contrary, it should be used like a workbench where tools of various sizes can be used to cut out, following the articulations of the creature itself, all the modes of existence, without condemning any one of them to the cutting-room floor...

⊙ REQUIRES, IN ADDITION TO AN ANALYSIS IN TERMS OF NETWORKS [TEC · NET], ⊙

To advance on this new construction site, our ethnologist can rely on what are now often called *socio-technological networks*. This is of course a mild euphemism for describing the surprising heterogeneity of the material arrangements according to the well-identified mode of networks [NET]. As if a nuclear power plant, a drone, an eel trap, or a metal saw could be content to maintain itself in existence with the help of elements from two domains, the "social" and the "technological"— and these two alone. The ethnologist has already learned this at her own expense: even though what historians call "technological systems" do exist on the local level, they are no more made *of* technology than law is made *of* law or religion *of* religion. What complicates the analysis is that there is no domain at all that can be mistaken for that of "technology" (there is no domain of "the social," either, but that is another matter).

To follow even the smallest course of action, she is going to have to record the various segments that belong, for example, to dozens of scholarly disciplines, to the economic arbitrations performed by groups of experts, to international standards, tests of the resistance of materials (often contested, moreover), social laws, as well as gear drives, chemical reactions, or electric currents, all this very quickly mixed up with questions of patents, breakdowns, pollution, or organization. It is no accident that the very notion of network (this time in the sense of the ACTOR-NETWORK THEORY) cut its teeth, as it were, on the foregoing DOMAIN of technology. This is because we have to add various things for any technology to start working [TEC · NET]!

The tools for our inquiry are well known, and their fruitfulness no longer needs to be demonstrated. All we have to do is reconstitute the path taken by the smallest innovation, follow the slow process of learning a previously unknown skill, come across an object whose meaning completely eludes the archaeologists or the ethnologists, in order to record the countless discontinuities that are necessary for the continuity of any action whatsoever. But it is with *controversies* that the heterogeneity of technological systems appears most clearly. An accident, a breakdown, an incident of pollution, and suddenly the "system," by dint of polemics, trials, media campaigns, becomes as unsystematic as possible, multiplying the unforeseen branchings that delight sociologists of technology.

It was surprising to learn that an investigative commission finally decreed that the catastrophic explosion of the spaceship *Challenger* had come about owing to the resistance to cold of a small rubber o-ring, but also to the distribution of decision-making responsibilities in NASA's complicated flowchart. It is surprising to note, thanks to violent polemics, that, in order to limit the proliferation of green algae on the beaches of Brittany, the mayors of municipalities dependent on tourism have to pay as much attention to elections in agricultural unions as to the reactions of nitrates or to the enticing propositions of garbage collection equipment salesmen, not to mention bringing in the minister of the environment along with the laboratories of the French Research Institute for the Exploitation of the Sea. The more one studies technological arrangements, the more one considers their ins and outs, the less chance one has of unifying them in a coherent whole.

If there is one area in which the results of science studies and technology studies can be considered robust, it is indeed in the vertiginous deployment of the heterogeneous elements necessary for the maintenance of technological arrangements. These studies can always be criticized when they bear on the sciences because of the key question of relativism—we have spent a fair amount of time on this—but the deployment they have made possible for technological systems no longer poses any problems except that of access to the terrain. When we talk about a "technological infrastructure," we are always designating a more or less patched-together mix of arrangements from more or less everywhere that others seek to render *irreversible* by protecting it from analysis, making it a carefully sealed and concealed BLACK BOX. It may be hard to penetrate these places that have been made secret, but it is never because we would come across chains of indisputable necessities. There is no area of technology whatsoever that would send us back all sanctimoniously to the sole ineluctable fate of "materiality." Here, at least, the advantages of constructivism are clear: everything that has been *set up* can be broken down.

⊙ THE DETECTION OF
AN ORIGINAL MODE
OF EXISTENCE ⊙

And yet, even if we are becoming familiar with that literature, we are not necessarily coming closer to a mode of existence other than that of networks. There is indeed discontinuity, there is indeed heterogeneity, there are indeed surprises all along each course of action, every time we discover the *other* components that "spring into action" to complete a given course of action, but in the end we are simply following the logic of this already well-demarcated mode of existence. Even if he does no more than look around in his own vicinity, the reader will notice without difficulty that he cannot make a gesture without *passing through* one or the other of those ingredients, whose mediation, intervention, and translation are indispensable to its achievement. However lazy he may be, even if he is just shifting position in his hammock, it is *through* this hammock that he must pass to keep himself up in the air, keep himself away from the stinging nettles or the ticks on the ground . . . It is indeed on the solidity of this weaving and these ropes that *he rests*. It is to them that he delegates the task of *holding him*. Is there in this *detour* and this *delegation* something more than the surprising linkage between beings on which every existent depends for self-maintenance—and that we already know how to note as [NET]?

This supplement is not so easy to capture. Not only because there is no identifiable domain or institution of technology, but because if we begin to follow the list of beings necessary to the maintenance of any being at all then *everything*, on this basis, becomes technology. Not just the hammock but also the two solid tree trunks to which it is attached! They too depend for their existence on a multiplicity of beings with which they have "learned to connect" and which they have "turned away" from their initial goals—yes, translated, enrolled and twisted— as surely as the strands of fine wool have been taken from the fleece of a sheep. All beings, to maintain themselves, have turned others away from their own paths.

Might we be dealing then with the beings of reproduction [REP]? This is definitely the case when we talk about the "evolution" of trees, the "invention" of photosynthesis, the "discovery" of leaves during the long "history of life." It is as though we were imagining that the beings of reproduction had had "to solve problems" and that they had "chosen" one particular branching rather than another. The hammock and the tree to which it is solidly attached would then share the same inventiveness, the same capacity to enroll other beings: sheep, for the first, nodules and bacteria for the second. In short, all existents would stem from the same technicity.

⊕ DIFFERENT FROM REPRODUCTION [REP · TEC].

And yet we have a clear sense that by assimilating the hammock with the tree this way, by assimilating the beings of technology with the beings of reproduction [REP · TEC], we would be making a diagnostic error as surely as if we confused them with networks [TEC · NET]. For with the beings of technology, we are dealing with something new in the order of alteration. If we can imagine a technological history of trees—for example, the "invention" of photosynthesis—it is because we imagine that the tree could have *started over several times*; that it could have benefited from *several chances* to persist in being by combining in different ways those beings it had at its disposal. Now, it is precisely the opportunity to start over that is absolutely unavailable to the beings of reproduction.

This is even what defines them: they throw themselves into the hiatus of existence without any possibility of turning back. As we saw above, this is their harsh, devastating felicity condition: to be or no longer to be. We can of course read the evolution of living beings—and even of inert ones—according to the mode of technological beings [TEC], but here we are dealing with a feature that we shall come across many times during our inquiry: each mode grasps all the others according to its own type of existence—and misunderstands each of them in a particular way each time. To undo the [REP · TEC] crossing, we would still need to be able to define what is truly original about the alteration proper to technological beings.

WE NEED TO RETURN TO
THE EXPERIENCE OF THE
TECHNOLOGICAL DETOUR, ☉
Let us try, as we have each time, to approach experience while setting aside the hope of making it coincide with a domain. To learn to speak appropriately about it, let us recognize that technological TRAJECTORIES are not easy to grasp and that they do not go straight—no more than do the beings responsible for the establishment of chains of reference [TEC · REF]. Everything in the practice of artisans, engineers, technicians, and even weekend putterers brings to light the multiplicity of *transformations*, the heterogeneity of combinations, the proliferation of clever artifices, the delicate setups of fragile skills. If this experience remains difficult to register, it is because to remain faithful to it we would have to accept its scarcity, its dazzling invisibility, its deep constitutional opacity. For it is always oscillating between two lists of contradictory elements: rare and ordinary, unforeseeable and predictable, fleeting and constantly begun anew, opaque and transparent, proliferating and controlled.

This experience seems to bring us to grips with what can be defined at first as a dazzling zigzag. Thanks to unpredictable detours, beings very far away in the order of reproduction become the missing pieces in a puzzle that requires an unexpected degree of ingenuity. Through a long series of detours, each cleverer and less predictable than the next, we find atomic physics turning up, for example, in a hospital wing for cancer treatment. By another detour, wood and steel are mutually implicated in the grip of a well-balanced hammer. By still another, the successive layers of a program, a compiler, a chip, a radar, manage to complicate each other and align themselves to the point of replacing the solid couplings that up to now had connected the cars of an automated subway system, ending up with wholly calculated "nonmaterial couplings." Moreover, it is not always worth the trouble to look far and wide for brilliant innovations so as to grasp the detours they have taken, their total originality. We can find the same flashes in the humble gesture of the tinkerer who finds a wedge to keep a door from closing too quickly, or in the minuscule discovery of a designer who shifts the placement of the handle on a handbag or the cap on a medicine bottle. "There's a trick to it," just waiting to be found; that's the whole thing in a nutshell.

Like Zorro, the technological being traces a fiery Z in a lightning stroke! Let's try to follow this zigzag. Nothing more common, more ordinary: you were heading for your office, getting into your car, and suddenly, without quite grasping what's going on, you find yourself in a garage, trying somehow to understand what a mechanic in work clothes is muttering as he crouches under the chassis, seeming to point with his hand dirtied by the oil leaking out to a part whose name and function escape you completely, except that (you are beginning to get it) you are starting to "expect miracles" from the availability of the spare part and from the skill of the mechanic, knowing that "you're going to have to go through this" if you want to find the path to your office again—and that, in addition, when it comes time to pay the bill, you're "going to feel it." A cascade of indubitable detours.

There, you have felt the breath of technology pass over you, but—here is the whole difficulty—*only for a brief moment.* As soon as you have paid the bill and left the garage, the purring under the hood will make you *forget everything* right away—even if you continue to grouse about the bill for a while. It is this strange presence and absence that makes the beings of technology in fact so difficult to grasp. Like the beings of metamorphosis, would these also be beings of occultation, then? Would they also depend on a particular "pass," even a magical sleight of hand? No doubt about it, to write up the specifications of these beings the analyst is going to have to take into account, simultaneously, the *detour*, this zigzagging course, the *delegation* that makes the action reliant on other materials, and the *oblivion* that the beings leave behind once the new composition has been established. In this, they differ from the beings of metamorphosis, which cannot be forgotten for a second: if you forget them, they will "get" you around the next corner. Technology, for its part, seeks to be forgotten. Definitely, it is about technology rather than nature that we can say "it likes to hide."

What is interesting, in the case of technology, is that this zigzag that ought to be so easy to grasp, given that the experience is so common, in fact totally disappears, for two related reasons: the ⊕ WHICH IS HIDDEN BY DOUBLE CLICK AND THE FORM/FUNCTION RELATION. habitual ravages of Double Click on the one hand and on the other the confusion that is always made between technology, or the technical, and the things left in its wake. Contrary to the title of Simondon's book, it isn't the mode of existence of the technological *object* that we must address but the mode of existence of technology, of *technological beings* themselves. (Let us recall that, in this inquiry, we are shifting from the question "What is the *being* or the *identity of* X or Y?" into a different question: "How are we to address *beings* or alterities, the *alterations* X or Y?")

Let's begin with Double Click [TEC · DC]. He has come, as always, to propose his services and calibrate a manner of being for which he is even less apt than for judging the path followed by facts, demons, angels, or legal means. But, as usual, instead of rejecting such a manifestly inadequate template, he has chosen to bring technology, too, into this Procrustean bed. Whereas the whole experience rebels against such a mutilation, he has acted as though technology, too, *transports* mere information, mere forms, *without deformation*. It is true that the engineers haven't protested; they go to great lengths to resemble the image of stubborn and somewhat dopey characters that has often been attributed to them! Double Click strikes everywhere: knowledge, yes, psyches, yes, but also, but especially, matter. If we want to measure the gulf that the Moderns are capable of digging between practice and the account of their practice, we must look not only into epistemology, psychology, or theology but also into TECHNOLOGY (used here in the sense of reflection *on* technology).

How could we impose a transport without transformation on a technological act when everything points to the opposite? It suffices to add *utility, effectiveness,* or, to use a more technical term, *instrumentality.* Effectiveness is to technology what objectivity is to reference: the way to have your cake and eat it too, the result without the means, that is, without the path of appropriate mediations (we shall see later on that the same thing is true for Profitability, the third Grace of this archaic mythology). All the technological whirlwinds and troublemakers can be forgotten all at once if you say that you are only transporting through the technological *object* the *function* that it must content itself with fulfilling *faithfully.* If you succeed in seeing in all technology a preexisting form that it applies to a hitherto inert and formless matter, then you are going to be able, by sleight of hand, to make the material world disappear even while giving the impression that you are populating it with objects whose materiality would have the same phantasmatic character as that of Nature! Here is where *Homo faber* comes on stage, shaping his needs through tools by "effective action on matter." Four little words as completely innocent as they are inadequate to grasp such a zigzag: there is no matter, one does not act "on" it, the action is not "effective" (it will be, perhaps, but later on), and, finally, as we shall see, it is not at all certain that this is an "action," at least not the action of "someone."

Give me needs and concepts, the form will arise from them and the matter will follow. An automobile? It "corresponds" exactly to the "need for transportation," and each of its forms "follows" from the needs of drivers. A computer? It "fulfills effectively" the function for which it was conceived. A hammer? It too derives from a reflection on the "best way" to balance an arm, a lever, wood, and steel. Molière made fun of the "sleep-inducing virtue" of the poppy invoked by doctors; the humorist is not far behind when, to a child's series of questions about one technological arrangement or another, he responds disingenuously: "That's what it was made for!" If there is an unworthy way to treat technologies, it lies in believing that they are means toward ends.

What do they become in this case? Thought applied to matter, itself conceived as FORM (in the third sense of the term), so that, once again, form and thought repeat each other, and this repetition arouses the same enthusiasm among the rationalists as *adequatio rei et intellectus*. We have lost the materials, we have lost the technological detour, we have lost the clever artifice. When people say of technologies that they are neither good nor bad, they forget to add: nor neutral.

Fortunately, the investigator knows how to undo this amalgamation, since she finds herself situated at the same branching point as the one that had already almost made her miss the worth proper to the sciences. The scorn with which people view technologies comes from the fact that they are treated according to the same model that we saw used to misunderstand the work of reference. Just as there was, in epistemology, a theory of objectivity as "correspondence" between map and territory, there is in technology a theory of effectiveness as *correspondence* between form and function. Technology is believed to be an action stemming from a human being—most often male, moreover—that would

BY DRAWING OUT THE LESSONS OF THE [REP · REF] CROSSING ON THIS POINT ⊙

then bear "on" matter itself conceived through confusion between geometry and persistence [REP · REF]. Technology then becomes an application of a conception of science that is itself erroneous!

If there really is one thing that materialism has never known how to celebrate, it is the multiplicity of materials, that indefinite alteration of the hidden forces that enhance the shrewdness of those who explore them. Nothing is less proper to technologies than the relation between the end and the means, since ends and means are invented simultaneously. It is a grievous misunderstanding to claim to see technologies as mere "applications of Science" and mere "domination of Nature"—we now know how to counter the weight of the mistakes borne by those two proper nouns.

As we can see, it is not only psyches that suffer from being misunderstood; technicians fare no better. The Moderns view them as scientists, but of lower rank—and they are mistaken about both groups. But it is not technology that is empty, it is the gaze of the philosophy of being-as-being, which has deliberately emptied itself of all contact with its own experience. In the finest dam, this philosophy doesn't manage to see anything original with regard to Being. "Mere beings," Heidegger would say, thus repeating and reinforcing the universal movement that obscures the scientific enterprise. Science is merely an avatar of Technology, after the latter has already been misunderstood as *Gestell*: a masterful misunderstanding about mastery; a fine case of forgetting being as technological; a quite cruel lack of ontological generosity!

The idea that one could *deduce* all the twists and turns of technological genius by always-well-formed a priori principles has always made engineers laugh—although not out loud. Isabelle Stengers had the idea of undertaking a radical thought experiment to reduce all technological inventions to the "basic principles" recognized by scientists and presented to students as their "incontestable foundations": reduced to the Carnot cycle, locomotives would immediately stop running; limited to the physics of lift, airplanes would crash; brought back to the central dogma of biology, the entire biotech industry would stop culturing cells. What have to be called the *invisibles* of technology—deviations, labyrinths, workarounds, serendipitous discoveries—would vanish, reducing the efforts of the sciences to nothing [TEC · REF]. No more invisibles; no more domination. For Vulcan the Lame doesn't care a whit about Athena's claim that she can impose her laws on him.

For ingenuity, everything in materials is food for thought. How have we lost this contrast to the benefit of a dream of control and domination? How have we been able to neglect the MATERIOLOGY that was honored by an entire, admittedly somewhat obscure, tendency in French philosophy, from Diderot through Bergson and of course Simondon to François Dagonet? The loss is as serious for a civilization, as we shall see, as is the loss of the religious or the political modes. Just as tragic an inversion, since technologies follow such twisted paths that they leave in their wake all sorts of other invisibles: danger, waste, pollution, a whole new labyrinth of unanticipated consequences opened up under our feet and whose very existence continues to be denied by those who think they can go directly ahead, without mediations, without running the risk of a lengthy detour, "straight to the goal." The "magic bullet," the "technical fix." A strange blindness on the part of the Moderns toward the most precious source of all beauty, all comfort, all efficiency. What a lack of politeness toward their own genius! It is awfully late to speak out of the blue about the *precautions* that should be taken to love technologies with all the delicacy required.

But the difficulty we have grasping the beings of technology arises, too, from the fact that the term "technological object" leads analysis astray, because we can clear up the misunderstandings of Double Click only by focusing on the hiatus itself and not on what it leaves behind *after* it has sketched out its mark in the form of a lightning stroke. We shall never find the mode of technological existence *in* the object itself, since it is always necessary to look *beside* it: first, between the object itself and the enigmatic movement of which it is only the wake; then, within the object itself, between each of the components of which it is only the temporary assemblage. And the same thing applies to the skillful gestures that the artisan eventually makes habitual, after long practice: when we began to establish them, they required the presence of a technological detour—which was painful and strenuous; but once these gestures become assured, routine, regulated, adjusted, we no longer feel them, any more than we feel the presence of the mechanic in

⊙ WE SHALL NO LONGER CONFUSE TECHNOLOGY WITH THE OBJECTS IT LEAVES IN ITS WAKE.

the purring of the engine under the hood. Despite what is often said of cold, smooth technology, in it there is never anything but *breaks in continuity*; things never quite connect. And even if we forget technology and let the thing created live its life, as soon as the thing in question needs to be maintained, restored, revised, renewed, other ingenious approaches will be required; we shall have to invoke the spirit of technology once again to maintain it in being. There is nothing more "heteromatic" than a robot, an AUTOMATON.

TECHNOLOGY OFFERS
A PARTICULAR FORM
OF INVISIBILITY: ⊙

The technological object is opaque, and—to put it bluntly—incomprehensible, the ethnologist concludes, in that it can only be understood provided that we add to it the invisibles that make it exist in the first place, and that then maintain, sustain, and sometimes neglect and abandon it. To learn to enter into relations with the beings of [TEC], we thus have to avoid, as always, the temptations of Double Click and go backward from the things to the movement that has transformed them and of which they are never anything but a provisional segment along a trajectory whose signature is singular.

This is why it is important to talk about OPERATIONAL SEQUENCES, as technologists following André Leroi-Gourhan have done, to try to determine the trajectory proper to technologies that leave objects in their wake, of course, but that cannot be reduced to those objects. The test of this encounter with such sequences is easy to administer: it suffices to stand idly in front of a "gadget," a "gimmick" whose meaning completely eludes you—perhaps a gift you have received, or an apparatus whose purpose is unclear, or a rock from the Châtelperronian period with cut marks made by someone who disappeared forty thousand years ago: everything is there, and yet *nothing is visible*. As if the object were only the print of a trajectory whose direction escapes you and that you have to learn to reconstitute, a fragment at a time.

Which leads our investigator again into more than one quarrel. "Mind in machines? No, really, here you're exaggerating. Invisibles again? It's a mania, an obsessional tendency to add irrationality even at the heart of the most material, the most rational effectiveness!" "And yet without the invisibles, no object would hold together, and in particular no automaton would achieve this marvel of automation." "Ah, you mean that there are technicians, engineers, inspectors, surveyors, intervention teams, repairmen, regulators, *around* and *in addition to* material objects? In short, humans, and even a 'social context'?" "No, I didn't say anything of the sort, for the good reason that technologies *precede* humans by hundreds of thousands of years. I am simply saying that if you are capable, you Moderns, of leaving out the paths of reference when you speak of objective knowledge, you are perfectly capable of leaving out what is responsible for the instauration of technological objects on the pretext (which is also true) that they hold up on their own once they are launched. Except that they can never remain alone and without care—which is also true! It is only the flow of operational sequences that allows us to sketch them." Technology is better hidden than the famous *aletheia*.

If the investigator is determined to speak of invisibles, it is not owing to a taste for the irrational, it is in order to follow the thread of this labyrinth rationally—the real labyrinth, the one that the architect Daedalos built for Minos. If nothing in technology goes in a straight line, it is because the logical course—that of the *episteme*—is always interrupted, deflected, modified, and because in following it one goes from displacement to deviation: in Greek, a *daedalion* is an ingenious detour away from the direct route. This is what we mean, quite banally, when we assert that there is a "technological problem," an obstacle, a snag, a bug; this is what we are referring to when we say of someone that "he's the only one with the technical ability" to solve a given problem: "he has what it takes," "he has the knack." We need to see "**Technique**" and "**Technology**" not in their noun forms but as adjectives ("that's a technical issue"), adverbs ("that's technically/technologically feasible"), even sometimes, though less often, in verb form ("to technologize"). In other words, "technology" does not designate an object but rather a difference, an entirely new exploration of being-as-other, a new declension of alterity. Simondon, too, made fun of substantialism, which, here again, here as always, failed to grasp the technological being. To borrow from Tarde one of the fine words that he opposed to the exclusive search for *identity*: what is the *avidity* proper to the mode of technological existence?

⊙ THE TECHNOLOGICAL LABYRINTH.

ITS MODE OF EXISTENCE
DEPENDS ON THE
[MET · TEC] RUSE ⊙

If technology, like the metamorphoses we have just been studying, like all the other modes, explores alterity, it must do this in its own way. But which way? Without question, it's a matter of a leap, a fault, a break, even, a rupture in the course of things, something we cover over rather too hastily with the term *invention*, whether modest or brilliant, it hardly matters. We can grasp this first of all as a derivation of two of the modes we have already recognized. As if technology relied on the power of metamorphoses [MET] to extract from the beings of reproduction [REP] unknown new capacities.

To convince ourselves of this, all we have to do is look around. If you begin to think about the materials that went into the objects that surround you, you have to think in terms of many metamorphoses. The stones of which your house is built lay in a distant quarry; the wood in your teak furniture was doing its thing somewhere in Indonesia; the sand from which your crystal vase is made was sleeping deep in some river valley; the hammock where you are snoozing as you read this book was still wool on the back of a sheep; and so on. Yes, there is magic in technology—all the myths tell us this, and Simondon grasped it better than anyone. Look around you again: you will have a lot of trouble establishing any *continuity* between the quarry, the tropical forest, the sandpit, the sheep, and the forms they managed to suggest to their manufacturers as they became some of the components of your home. There has thus been transmutation, transformation. And it is no accident that we speak, with reference to technology, about ruse, skill, indirection, cunning. There are many HARMONICS between the subtlety necessary to interact with the beings of metamorphoses and the subtlety that has to be put to work to find the "trick." This is why myths so often bring together the lessons of these two types of ruse and deviation. At all events, both types *hedge*. There's an admirable popular expression for this in French: *C'est qu'il y a toujours moyen de moyenner*. There's always a way to muddle through, to manage. If Ulysses is "crafty," if Vulcan limps, it is because, in the vicinity of a technological being, nothing goes straight, everything is done on the bias—and sometimes, even, everything goes askew.

But at the same time, my table, the walls of my house, my crystal vase *persist* after their transformation. Unlike the beings of metamorphosis, once they have been radically transformed the beings of technology *imitate* those of reproduction through their persistence, their obstinacy, their *insistence*. It is as though technology had dragged some of the secrets out of reproduction [REP · TEC] and of metamorphoses [MET · TEC] by *crossing* the two species of modes of being. Technology appears in a first approximation as a mixed mode: proteiform speed on one side, persistence on the other. It's hardly surprising that Prometheus's fire has been seen as something that liquefies all things and at the same time gives them new durability, solidity, consistency. No archaeologist worthy of the name fails to be moved by the pottery she digs up, which, even in shards, will last as long as our Earth.

⊕ AS MUCH AS ON THE PERSISTENCE OF THE BEINGS OF REPRODUCTION [REP · TEC].

As we can see, the adjective "technological" does not designate in the first place an object, a result, but a *movement* that is going to take from inert entities and from living ones—including the body of the artisan, which becomes more skillful by the day—what is needed to hold together in a lasting way, to freeze, as it were, one of the moments of metamorphosis. Neither the wall nor the table nor the vase—nor the car nor the train nor the computer nor the dam nor the culture of domesticated bacteria—is "technological" once it is left to its own devices. What is lasting and persistent in these things depends on the presence of composites that have been drawn out by metamorphoses [MET] from the persistence of beings of reproduction [REP], each of which lends certain of its virtues, of course, but most often without leaving us the possibility of profiting from its initiative and its autonomy in a lasting way. The ingredients of these blends always remain foreign to one another. They "lend themselves," as the expression puts it so well, to being translated, deflected, disposed, arranged, but they nevertheless remain "themselves," ready to let go at the slightest pretext. If we are not careful, the wall falls down, the wood is eaten away by worms and crumbles into dust, the crystal cracks, the car breaks down, the train derails, the bacteria culture dies, the hammock's ropes fray; as for the computer, it malfunctions through a sort of malevolent depression. What is certain is that the technological detour leaves behind a differential, a gradient of resistance, a whole leaf-pile or layering of diverse materials that holds up "on its own" and at the same time can be dispersed. The expression *traduttore, traditore* applies much better to technologies than to texts.

What mode goes further in alteration than this one? The risk of reproduction is admirable, of course, but the beings of reproduction never cross existents in as dizzying a fashion as the components of the humblest technology [REP · TEC]. This is what reduces the paleontologist in the Olduvai Gorge to tears when he is lucky enough to come across a stone carefully worked on both sides. In a Museum of Natural History we can be very impressed by the profusion of living beings (a fine example of [REP · MET] crossing), yes, but the series of bicycles in the Museum of Arts and Crafts, or the electric locomotive gliding noiselessly into the train station along its shiny rails, ought to move us just as much. Through technology, the being-as-other learns that it can still be even more infinitely *altered* than it thought possible up to that point.

THE VERIDICTION PROPER TO [TEC] ⊙   But is there a veridiction proper to the beings of [TEC], the ethnologist wonders, as she continues to complete the SPECIFICATIONS for this mode? At first glance, to speak of the truth and falsity of technology seems to make even less sense than speaking of the truth and falsity of the beings that produce psyches. How could they, too, have FELICITY AND INFELICITY CONDITIONS?

And yet our investigator will not hesitate long, provided that she begins to count the number of times her informants affirm that they *judge* a tool, a function, a utensil for its good or poor quality, and provided that she notes the subtle way they seem to creep along the gradient that goes from the least effective, the least useful, to the *most* effective, the *most* useful, the *best* adapted. The least skillful project manager working for the Paris subway system explores all the possible solutions, one by one, to make sure that the nonmaterial couplings that connect the cars of the automated metro system succeed in *holding to* the constraints of the design office. The clumsiest cook rummages in her drawer until she finds the right knife, the one best *adjusted* to the job. The most sybaritic of sleepers punches his pillow until he finds the right *harmony* between his head and the cushion. How much time does it take a dog trainer to learn how to *correspond* with the animal and end up learning from it?

Don't look at the most skillful artisan, but rather at the apprentice who is seeking skill through a slow trajectory and finds himself corrected at every step by his master. Don't try to grasp the movement of a technology that's "working," but rather the gropings of innovation, precisely where something is not yet working and obliges the artisan to *start over* several times, going from one obstacle to the next. "Judgment," "adjustment," "rectification," "fresh start," no question about it, here we find ourselves confronting difference—often mute, perhaps, but always extraordinarily subtle—between the true and the false, the well made and the poorly made.

It is this displacement, this translation, completely original every time, that artisans, architects, engineers practice day after day, and that Double Click no more manages to grasp than he does chains of reference [REF]; and for the same reason he mistakes the final result—yes, it is adjusted, yes, it works, yes, it does what it's "made to do," yes, it "holds together"—for the movement that led to that result [TEC · DC]. This sideways, crablike motion, this perpendicular movement of rummaging around, exploring, undulating, kneading, which so obstinately misses the relation between form and function and the relation between ends and means, is precisely the motion that will perhaps (but not necessarily) produce forms or means corresponding to functions or ends. To say that technologies are effective, transparent, or mastered is to take the conclusion for the pathway that led to them. It is to miss their spirit, their genesis, their beauty, their truth.

What can we call this spirit that we would miss entirely if we were to make the mistake of limiting technology to the objects left in its wake without reproducing its ever-so-particular movement? How can we qualify its mode of being more precisely? In other words, what is the equivalent for the zigzags, the brilliant flashes, the detours and discontinuities, of *chains of reference* for objective knowledge [REF], *processions* for religion [REL], *means* for the passage of law [LAW], *persistence* for the beings of reproduction [REP]? We shall call it *technical* FOLDING. We could have used the word PROJECT, as opposed to "object," but we would have needed another mode, that of organization, a crossing that we shall not learn to master until much later [TEC · ORG].

⊙ DEPENDS ON AN
ORIGINAL FOLDING ⊙

The term "folding" will allow us to avoid the blunder of speaking of technology irreverently as a piling up of objects or as an admirable example of mastery, transparence, rationality, that would prove "man's dominion over matter." Technology always entails folds upon folds, implications, complications, explanations. Its canonical representation, thoroughly studied by the sociology of technologies, sketches it in the form of a series, often a very long series, of nested TRANSLATIONS, a labyrinth. There is technical folding every time we can bring to light this second-level transcendence that comes to interrupt, bend, deflect, cut out the other modes of existence, and thus by a clever ploy introduces a *differential* of materials.

Once we have been freed of our obsession with matter, nothing prevents us from identifying the diversity of these differentials. We can talk about technical folding with respect to the delicate establishment of muscular habits that make us, through apprenticeship, competent beings endowed with a high degree of skill, just as well as to talk about the molten iron that spews out of the Mittal blast furnaces, or to designate the distinction between a software program and its compiler, or to celebrate the legal "technology" that makes it possible to link a somewhat more durable text with a dossier that will be less durable, or to support an argument over a somewhat heavier and more cumbersome metaphor by using what is rightly called "literary technology." What counts, each time, is not the type of material but the difference in the relative resistance of what is bound together. Curiously, there is *nothing material* in technology. Where there is differential resistance and heterogeneity among the components, technology is found as well.

⊙ DETECTABLE THANKS TO THE KEY NOTION OF SHIFTING. It is by insisting on the notion of SHIFTING that we shall succeed in qualifying these gradients of resistance more accurately. There is a great temptation, in fact, to think that if there are technologies, it is first of all because there are technicians! If we gave in to this view, we would be firmly placing the origin of technological beings in thought, or at least in the gestures of *Homo faber*. The spirit that we are invoking would simply be the inventive spirit of humans, the creator that has to precede all creation, or so we are told. This is indeed what we were supposing, in the fantasy sketched out above, when we were claiming to grasp the beings of reproduction in the mode of technological invention [REP · TEC]. We were then imagining a manufacturer—"Mother Nature"—whose wisdom and inventiveness would have made it possible to solve the problems posed by another character, "Natural Selection." If we pretended to grasp living beings as technological inventions, it was only by making them stand out against the background of a giant factory animated by the spirit of an ingenious creator.

Now, by using the metaphor of shifting that comes from the mechanics of gear drives, SEMIOTICS may have put us (not necessarily on purpose) on the trail of an entirely different way of grasping technological beings. Let us recall that for semiotics, shifting—we shall come back to this in the next chapter—makes it possible to grasp a quadruple transformation starting from a zero point. To start with, displacements in time, in space, and in the type of actor. This is what we mean when we talk about *reprise*: others, elsewhere, before, after, go into action.

These three instances are easily recognizable in any technological detour whatsoever: when you are resting in the hammock, it is indeed the hammock that takes over—and it does not resemble you, others have woven it for you; when you entrust yourself to an aspirin tablet, it is the tablet, another actor from elsewhere, manufactured by others, to whom you have entrusted or delegated the work of treating your headache—and the tablet doesn't resemble you in the least, either; when a shepherd, tired of watching over his sheep, entrusts to a fence and to his dogs the task of protecting the flock against wolves (or perhaps stray dogs), those who are now standing guard are the fence posts, the barbed wire, and the dogs, each with its own history, its own fidelity, and its own fragility. With the folding of technological beings, a *dislocation* of the action emerges into the world and makes it possible to differentiate between *two levels*, the starting level and the one toward which you have precisely shifted gears by installing in it other actors who possess different resistances, different durations, different degrees of solidity. It was moreover this dislocation that interested us in the definition provided above for CONSTRUCTIVISM. Whatever the technological detour may be, this is in fact what makes it possible, not to do something, but to *have something done*.

But we mustn't forget the fourth agency involved in any operation of shifting. When an artisan, an explosives manufacturer, or an engineer goes into action, others do too, of course, but this also means that *the one who manufactures is himself also shifted down*. A new dislocation, which this time does not go ahead, toward level n + 1, but rather *falls short* of the starting point. This level n - 1 is presupposed, *implied* by the action, and it is this level that begins to give weight and shape to the virtual author of the action. If we always have to maintain the ambiguity of constructionism without ever believing in the assured existence of a builder, it is because the author *learns from what he is doing* that he is perhaps its author. In the case of technological beings, this general property is of capital importance, since technologies have preceded and generated humans: subjects, or rather, as we shall soon name them, QUASI SUBJECTS, have sprung up little by little from what they were doing. This is why we had to be so suspicious of the concept of "action on matter," which threatened to place the point of departure in the depths of a human subject instead of waiting for this human subject to emerge *from his works*—though the possessive adjective is quite unwarranted, because the human subject does not master "his" works any more than he possesses them.

Instead of situating the origin of an action in a self that would then focus its attention on materials in order to carry out and master an operation of manufacture in view of a goal thought out in advance, it is better to reverse the viewpoint and bring to the surface the encounter with one of those beings that teach you what you are when you are making it one of the future components of subjects (having some competence, knowing how to go about it, possessing a skill). Competence, here again, here as everywhere, *follows* performance rather than preceding it. In place of *Homo faber*, we would do better to speak of *Homo fabricatus*, daughters and sons of their products and their works. The author, at the outset, is only the *effect* of the launching from behind, of the equipment ahead. If gunshots entail, as they say, a "recoil effect," then humanity is above all the *recoil* of the technological detour.

THE UNFOLDING OF THIS MODE GIVES US MORE ROOM TO MANEUVER. By freeing the beings of technology from their association with matter; by localizing their effectiveness a little bit ahead of the fabricating subjects; by completely abandoning the notion of mastery and transparence; by letting these beings explore the entire gamut of materials; by no longer obliging them to remain confined in the narrow prison of means and ends, the inquiry could not only become compatible with the long history of technologies and the slow anthropogenesis that they have allowed, it could also open up less uneven interactions with the other collectives, since all humans are the children of what they have worked on.

We would no longer base the comparison on the somewhat wobbly expression "material culture," which technologists use as if the term "culture" remained problematic—given how much the so-called symbolic, aesthetic, and social dimensions of technology can vary— whereas the term "matter" would hardly pose any problems since everyone "clearly sees what is involved." Now, as we well know by now, nothing is less widespread throughout the world than the notion of matter or even that of production. The Australian aborigines whose toolbox contained only a few poor artifacts—made of stone, horn, or skin—nevertheless knew how to establish with technological beings relations of a complexity that continues to stun archaeologists: the differentials of resistance that they arranged were located rather in the tissue of myths and the subtle texture of kinship bonds and landscapes. The fact that their materiality was slight in the colonizers' eyes tells us nothing about the inventiveness, the resistance, and the durability of these arrangements. To keep open opportunities for negotiation over the successors of the contemporary production arrangements, it is crucial to restore to the beings of technology a capacity for combination that liberates them entirely from the heavy weight of instrumentality. Freedom of maneuver that is indispensable in order to invent the arrangements to be set up when we have to dismantle the impossible MODERNIZATION FRONT. If the verb "ECOLOGIZE" is to become an alternative to "modernize," we shall need to establish quite different transactions with technological beings.

We learn (we used to learn) in the catechism that the Letter of Scripture remains inert without the Spirit that blows where it will. This is even truer of the bleached bones of the technical object that are waiting for the spirit of technique to raise them up, recover them with flesh, put them back together, transfigure them—resuscitate them, if the word is not too strong. The ethnologist has tears in her eyes in front of this Valley of the Dead that she sees in a sudden revelation. What? Are we going to have to restore to the Moderns not only the beings that engender psyches but also resuscitate for them the technologies of which they are at once so proud and so ignorant? "But why," she sighs at the prospect of such a mission, "why have they not known how to celebrate them with appropriate institutions?"

SITUATING THE BEINGS OF FICTION

Multiplying the modes of existence implies draining language of its importance, ⊙ which is the other side of the Bifurcation between words and the world.

To avoid confusing sense with signs ⊙ we have to come back to the experience of the beings of fiction [FIC].

Beings overvalued by the institution of works of art ⊙ and yet deprived of their ontological weight.

Now, the experience of the beings of [FIC] invites us to acknowledge their proper consistency ⊙ an original trajectory ⊙ as well as a particular set of specifications.

These beings arise from a new alteration: the vacillation between raw material and figures, ⊙ which gives them an especially demanding mode of veridiction.

We are the offspring of our works.

Dispatching a work implies a shifting ⊙ different from that of the beings of technology [TEC · FIC].

The beings of fiction [FIC] reign well beyond the work of art; ⊙ they populate a particular crossing, [FIC · REF], ⊙ where they undergo a small difference in the discipline of figures ⊙ that causes the correspondence to be misunderstood.

We can then revisit the difference between sense and sign ⊙ and find another way of accessing the articulated world.

THE READER MAY BE BEGINNING TO DISCERN THE BENEFIT THAT CAN BE DERIVED FROM THE PLURALISM OF MODES OF EXISTENCE, IF for each mode we cut out a template capable of respecting the singular experience that each one offers. This is what we have done with the beings of metamorphosis, to which we denied objective existence, and then with those of technology, whose empire we stretched well beyond their reach. And yet, even if we are managing to multiply the modes, we are still far from having the flexibility we need for this inquiry. We do not yet benefit from the habits of thought that would allow us to take the measure of the true *existence* of these beings whose instauration we are claiming to follow. Inevitably, we risk falling back on the idea that there is, on one side, that which exists, and, on the other, "representations" of that which exists. In this view, existence would always be a unity; representations alone would be multiple.

To make real progress in the inquiry, then, we have to approach the other side of the BIFURCATION. While we have seen how to undo the amalgamation that has rendered the world mute and filled it with RES EXTENSA, we still do not know how to treat—in any sense of the word— the symmetrical obsession with a "speaking subject." Is it possible to discern in LANGUAGE another amalgamation, different, to be sure, from the one that engendered the notion of MATTER but that also mixed several modes together? If we could manage to disamalgamate this one as well, we might be in a position to offer the Moderns a wholly different version of what they care about.

Curiously, the difficulty is almost greater on this side of the Bifurcation than on the other. At bottom, it was not so difficult to relate objective knowledge to the networks that make sense of it: it is so costly, so heavily equipped, so collective, that we could readily see how much labor is needed to accede to remote entities [REP · REF]. Its very successes, in the course of recent history, have only multiplied the handholds that allow us to reembody it, to reinsert it, to refold it within itself. It is a wholly different story with the other two resources that aspire to the position of judge external to all networks: "SOCIETY" and "LANGUAGE." It is fairly easy to get around "NATURE's" claim to hold the position, so striking is the gap between the distinct requirements of the beings of reproduction [REP] and those of reference [REF], requirements that the idea of matter had merged. But the two offstage voices of "Language" and "Society" seem impossible to circumvent, for it appears self-evident that all activities that are "not solely material" must then "bathe in" language and "be situated" in a "social context." All our efforts to extirpate ourselves from "Nature" will thus be vain if they only end up entrusting the judgment of last resort to its two substitutes, since it is from the collaboration of these three that the notion of *representation* arises—mental, social, or collective representations confronting a world that is itself inarticulate (except, let us recall, in the single case of "FACTS THAT SPEAK FOR THEMSELVES" . . .).

As it is clear that we are dealing with two aspects of a single problem, nothing prevents us from using the same method for the other side of the Bifurcation. The reign of matter—an amalgam ⊙ WHICH IS THE OTHER SIDE OF THE BIFURCATION BETWEEN WORDS AND THE WORLD. of reproduction [REP], reference [REF], and also politics [POL]—over the realm of the PRIMARY QUALITIES has obliged the Moderns to group together somewhere all the SECONDARY QUALITIES on which SUBJECTIVITIES, feelings, meaning, in short, "lived experience," rely. But this realm, once it has been reduced to a bare minimum, turns out to be deprived of any ontological reality, since "real reality" is located on the other side, in the primary qualities—which are devoid of all "human" signification, moreover. Thus the Moderns find themselves faced with the rather thankless task of restoring meaning, or direction, in spite of everything, to what they had nevertheless, through that Bifurcation, *deprived of direction*, that is, of reason: in other words, of prepositions.

The solution, painless at first but calamitous later on, was to create a world of "symbolic" realities charged with collecting the bric-a-brac of everything that had no place in "Nature" or the "real world." The Bifurcators put themselves in a position of continuous contradiction that obliged them to recognize *a little* (symbolic) reality in what had *no* (material) reality. Like those patients who have been deprived of half their field of vision by a brain lesion and who do not even notice that it is missing, the Moderns were driven out of the reality that is real but, alas, devoid of meaning, and that has become the private preserve of the (idealized) hard sciences; they have had to take refuge in the preserve of falsities, fortunately full of meaning, of the sciences of the mind! Who could be resigned to living in that sort of rump State, this Liechtenstein of thought? No, it's quite clear: after the obstacle of "Nature," which was blocking the way to any anthropology of the Moderns, we now have to remove the obstacle of "Language." As long as we remain faithful to our methodological principles, we should be capable of bringing this off. If we have succeeded in shedding our intoxication with "matter," it shouldn't be impossible to throw off the armature of the "symbolic," which is its inevitable counterpart.

TO AVOID CONFUSING SENSE WITH SIGNS ⊙ The solution might lie in a distinction between SENSE and SIGN. Let us recall that radical EMPIRICISM, the version that inspired William James and that this entire inquiry aspires to extend in a more systematic way, reconnects the thread of experience by attaching PREPOSITIONS to what follows them, to what they merely announce, utter, dispatch. To follow experience, for second-wave empiricism, is thus to follow—by a leap, a HIATUS, a mini-transcendence—the movement that goes from a preposition to what it indicates, prepares for, or *designates*. The "sense" is consequently the DIRECTION OT TRAJECTORY that is traced by a mode and that defines both the predecessors and the successors of any course of action whatsoever, as well as the path that has to be navigated in order for something to persist in being. The expression *ad augusta per angusta*, "to the heights by narrow roads" applies very aptly to the modes.

If we accept this point of departure, each preposition thus defines a way to *make sense* that differs from the others. If smoke follows fire, it is not because smoke is the "indication" of fire in the eyes of a human subject but because, for the beings of reproduction, such are the lines of force that dry wood struck by lightning is bound to follow. Smoke is indeed the sense, the direction, the movement in which fire hurls itself— yes, fire itself. We would have to take things quite differently, and thus *in another sense*, in another *direction*, with another *sensibility*, if we wanted to follow what goes before and what comes after in the manner of beings of metamorphosis, reference, or technology. To avoid misinterpretation, to avoid what we have called category mistakes, is precisely to identify the tonality in which we must take what follows: how to direct our attention, how to find out "what to do next," to borrow Austin's expression. For every mode, there is a distinct theory of meaning, a particular semiotics. If there is an even slightly general metalanguage, it has to be entrusted to the mode of prepositions [PRE]. To define the sense of an existent is to identify what is lacking, what must be added in order to translate it, to take it up again, to grasp it anew, to interpret it. Consequently, in this inquiry, *trajectory*, *being*, and *direction*, *sense*, or *meaning*, are synonyms.

But *if everything is meaningful*, if everything *makes sense*, this does not mean that *everything makes signs*. What can we say about signs? Let us say that a sign emanates from a mode of existence in due form. Whereas the sense comes ahead of the sign, a long way ahead, since trajectories are consubstantial with all modes, a sign would be a particular mode of meaning or sense that would form a sort of *regional* semiology and ontology proper to a particular mode. To verify this hypothesis, we shall proceed as we did with equipped and rectified knowledge, preventing it from floating off untethered, accompanying it in its networks, learning to feed it with a specially prepared ontological diet. The operation is delicate, but it is the only way to benefit from the plurality of genres without confusing them—and, of course, without coming back to the Bifurcation between signs (multiple) and being (unique). If pluralism does not succeed in getting us past this obstacle, it will never be anything but a pretense.

⊙ WE HAVE TO COME BACK
TO THE EXPERIENCE OF THE
BEINGS OF FICTION [FIC].
Our anthropologist is used to it now; she has
understood that every time her informants insist
on the importance of a domain—first the "external
world," then the "internal world," then "technolog-
ical infrastructure," now "the symbolic"—it is because there is something fishy going on, something she needs to ferret out so as to extract
its mode, which the institutions of modernism do not always allow us
to recognize. Just as matter confused two unrelated modes of being, the
symbolic has been confused with another type of being, one perfectly
worthy of interest and recognition, but one whose specifications have
not always been respected.

We have already referred to these characters more than once, these
entities encountered everywhere that weigh on us with a quite particular
weight of reality: for brevity's sake we shall call them *beings of fiction*
(noted [FIC]). As we shall see, this term does not direct our attention
toward illusion, toward falsity, but toward what is fabricated, consistent,
real. Trained as we now are to make room for invisibles as diverse as
metamorphoses [MET], means [LAW], references [REF], the Circle of representation [POL], or risky reproductions [REP], it shouldn't be impossible
for us to credit the beings of fiction with their proper consistency.

We would then understand why the Moderns thought they were
making the symbolic a separate world: just as "Society" derives from
the amalgamation of all the associations whose threads we have given
up trying to disentangle and that only networks [NET] allow us to trace,
the "symbolic world" would be an artifact produced by the superimposition of all the invisibles necessary to meaning whose interpretive keys
have been intermingled and which, by dint of being piled up on top of
one another, would give the vague impression of possessing a "certain
reality"—just as other modes, also piled up, gave the impression of
forming a "material external world."

BEINGS OVERVALUED
BY THE INSTITUTION
OF WORKS OF ART ⊙
Let us note first of all that the situation in
which the Moderns have placed the beings of fiction
is entirely different from that of the other modes we
have managed to identify up to now. If it has taken
us so long to realize that equipped and rectified knowledge possessed
its own mode [REF], if it has seemed so risky to restore weight to the

beings of metamorphosis [MET], if we hesitate (as we shall see) to give too much or too little weight to the beings that bring salvation [REL], if the political phantom appears so evanescent [POL], everyone will agree that the beings of fiction—in whatever medium—possess a particular type of reality that it is appropriate to cherish and respect. We have never ceased, at least in our own tradition, to develop, recognize, celebrate, and analyze the specific character of that reality. These beings seem to have enjoyed the same privilege as those of law [LAW], whose institutional rendering correlates fairly well with a mode of veridiction recognized as peculiar to it.

Consequently, our investigator does not have to twist herself out of shape to convince herself that the beings of fiction, like those of law, indeed possess full and complete reality in their genre, with their own type of veridiction, transcendence, and being. She has no trouble understanding her informants when they say that the adverb "fictitiously" immediately engages *all that follows* in a certain form of reality that cannot be confused with any other. In this case, at least, the *pre-positioning* of the preposition seems to pose no problem of detection. If she hesitates on one of these points, a hundred treatises on the "worlds of fiction" are available to help her in her analysis. Even if, up to now, we have observed only larger or smaller *maladjustments*, more or less aggravated by recent history, between contrasts and their institutions, at first glance the beings of fiction seem to benefit from a rather favorable situation: they seem to be appreciated for what they are.

And yet, with the Moderns, nothing is ever simple. If it is true that the beings of fiction have been swamped by honors, they have paid a big price for the central place they have been given in the collective: they have had to agree to be assimilated to the same type of reality as the much more recent institution of "works of art." They have been valued to an extreme, while too hastily denied any objectivity.

⊙ AND YET DEPRIVED OF THEIR ONTOLOGICAL WEIGHT.

The danger, here again, here as always, lies in using Double Click's benchmark and treating them with a touched or amused condescendence, as if they were "of course" incapable of truth since they are, precisely, "fictional" [FIC · DC]. (This is the way Austin treats them, moreover: as "etiolations of language"!) They are viewed with just as little respect when they are taken to be "imaginary creatures." We have already found out where these phantoms originate: in the "human mind," in that famous interiority, that artifact of Bifurcation, the bookend paired with exteriority. Our ethnologist rejoiced too soon: it isn't so easy to define the felicity and infelicity conditions of the beings of fiction.

To agree to treat them as "indifferent to truth or falsity," equally distant from reality and unreality, as a form of "truthful lying," or as beings authorized to live thanks to the "suspension of disbelief" is to have already accepted on their behalf the state of debasement in which too many of their supporters have settled complacently. Just as politicians too quickly take the absence of truth and falsity for granted so they can indulge in praising trickery and violence in the name of their highly distinctive type of right to lie [POL], it is much too easy for "artists" to take the unreality of their creations for granted and indulge in "real fakery" in the name of the "enduring rights of the imagination and creativity." "Poetic license": how much self-indulgence we risk allowing ourselves in your name ...

Notwithstanding the homage rendered to the beings of fiction, our ethnologist would thus be making a manifest category mistake if she took them to be mere products of the imagination.

NOW, THE EXPERIENCE OF THE BEINGS OF [FIC] INVITES US TO ACKNOWLEDGE THEIR PROPER CONSISTENCY ⊙ Moreover, the most ordinary experience, if we were to agree to follow it, would quickly dissuade us from making that mistake. Without any doubt, there is some *exteriority* among the beings of fiction: they impose themselves on us after imposing themselves on those responsible for their instauration, for the latter are more like constituents than "creators." They come to our imagination—no, they *offer* us an imagination that we would not have had without them. Don Juan exists as surely as the characters in *Friends*; President Bartlett occupied the White House for some time with more reality than his pale double George W. Bush; as for *The Magic Flute*, while there are hundreds of ways to perform it, the opera itself is what authorizes and incites us to perform it in all those ways and more. In any case, we discuss these beings among ourselves with as much passion, precision, and taste for proof as we discuss interpretations of a piece by John Cage or the restoration of a painting by Veronese.

A work of art *engages* us, and if it is quite true that it has to be interpreted, at no point do we have the feeling that we are free to do "whatever we want" with it. If the work needs a *subjective* interpretation, it is in a very special sense of the adjective: we are *subject* to it, or rather we *win* our SUBJECTIVITY through it. Someone who says "I love Bach" becomes in part a *subject* capable of loving that music; he receives from Bach, we might almost say that he "downloads" from Bach, the wherewithal to appreciate him. Emitted by the work, such downloads allow the recipient to be moved while gradually becoming a "friend of interpretable objects." If listeners are gripped by a piece, it is not at all because they are projecting their own pathetic subjectivity on it; it is because the work demands that they, insignificant amateurs, brilliant interpreters, or passionate critics, become part of its *journey of instauration*—but without dictating what they must do to show themselves worthy of it.

If the interpretations of a work diverge so much, it is not at all because the constraints of reality and truth have been "suspended" but because the work must possess many folds, engender many partial subjectivities, and because the more we interpret it the more we unfold the multiplicity of *those who love it* as well as the multiplicity of *what they love in it*. Someone who does not feel *held* and *engendered* by the requirements of the work will never be inhabited by it. The fact that we have to learn how to *make ourselves sensitive* to works of art proves nothing about their degree of objectivity (which takes a special form, to be sure). Such works populate the world, but in their own way.

"Populate the world? But that's impossible—   ⊙ AN ORIGINAL TRAJECTORY ⊙
where would the beings of fiction *go*? Into limbo?
Here's still another mystification! You still have this immoderate taste for invisibles!" Before accusing the analyst of a new excess of ontological realism, the reader should remember that the world the Moderns are going to be able to penetrate is now spacious, full of folds and niches, or, more precisely, that we are attempting to empty it of all the unwarranted fillers—Nature, Society, Language—that were keeping it from giving the various modes of existence their proper weight of being. As in Japanese paintings, the lines and brushstrokes stand out now against a background of paper left blank—and if we add a few ideograms, these will be placed neither above nor below the painted area but in the blank

spaces, deposited in the vertiginous void. The reader should remember, too, that our informants are no longer asking all these "tribes of beings" to live in the same way with the same obstinacy and the same persistence as stones [REP] or tables [TEC], which have their own modes. Finally, the reader should agree to assess yet again the unrealism of the post-Bifurcation landscape, with its invisible primary qualities (invisible yet known by sciences that have remained out of play) on which, by "psychic addition" (in Whitehead's words), the phantoms of the secondary qualities would float. If he finds that world "reasonable" and "concrete," he is ready to lose his reason for real…

The invisibles that we have to reintroduce to drive out the representations are perfectly *attributable*, and if they disappear for a moment, it is because they are *leaping* into existence in a move whose interpretive keys are no more mysterious, no more irrational, than those of metamorphosis or technology. To say that beings of fiction populate the world is to say that they come to us and impose themselves, but with a particular wrinkle: as Souriau pointed out so rightly, they need our *solicitude*. According to him, we form their "equilibrium polygon"! Their proper status is that of a "composite" that "has to hold together on its own," as Deleuze and Guattari would say. But if we don't take in these beings, if we don't appreciate them, they risk disappearing altogether. They have this peculiarity, then: their objectivity depends on their being *reprised*, taken up again by subjectivities that would not exist themselves if these beings had not given them to us. It's weird, yes, but it is not up to the ethnologist to determine the art and the manner of what exists or not.

⊙ AS WELL AS A PARTICULAR SET OF SPECIFICATIONS.

Someone will object that this weirdness is nevertheless proof that we only "imagine" these beings. Not necessarily: perhaps their specifications include this particular clause according to which we have to keep them going even though we cannot invent them. They are sufficiently asymmetrical, unstable, and, as it were, *inclined*, that they come to us and require that we *prolong* them, but in their own way, which is never stated but simply indicated. We find ourselves on their trajectory; we are part of their trajectory, but their *continuous creation* is distributed all along their path of life, so much so that we can never really tell whether it is the artist or the audience that is creating the work. In other words, they too make networks.

To reconstruct these networks [FIC · NET], the inquiry can find inspiration, fortunately, in the magnificent literature on the various "worlds of art." History and sociology have made themselves capable of deploying the trajectories of a work without skipping a single segment of these arrangements, as always heterogeneous, in which one has to take the whims of princes and sponsors into account as well as the quality of a keystroke on a piano, the critical fortune of a score, the reactions of a public to an opening night performance, the scratches on a vinyl recording, or the heartaches of a diva. In following these networks, it is impossible to separate out what belongs to the work "properly speaking" from its reception, the material conditions of its production, or its "social context." Even more than the anthropology of the sciences and technologies, that of art has succeeded, by stunning erudition, in continuing to *add* segments without ever taking any away: in the work, truly, to sustain it, everything seems to count. All the details count, and the details, all together, pixel by pixel, are what sketch out the composite trajectory of the work. For Beauty, even more than for Truth, we know in what small change we have to pay its ransom.

As with the sciences, technologies, and law, once the multiform arrangements that keep us from mistaking the beings of fiction for distinct domains have been traced, we need to specify the particular way they have of extending themselves. For each mode of existence is in fact a mode of *extension*. How are we to determine the alteration proper to beings of fiction that gives them their allure, their status, their identity, or rather their singular avidity? I suggest situating it, quite classically, in a new way of folding existents so as to make them the blueprint for a kind of expression that nevertheless cannot be detached from them, a mystery that the hackneyed theme of form and content signals but does not analyze. The *raw materials*—unrelated, let us recall, to the idealism of "matter"—seem capable of *also* producing FORMS or, better, FIGURES (if we are careful not to connect this term too quickly to the question, proper to art history, of mimetic figuration). Such is the new deduction that beings of fiction are going to extract from existents.

THESE BEINGS ARISE FROM A NEW ALTERATION: THE VACILLATION BETWEEN RAW MATERIAL AND FIGURES, ⊙

This astonishing discovery obviously precedes by far the very recent, very Western institution of Art, although that institution has never stopped exploring its depths and amplifying its contrast.

It is through this new potential debited against technologies that we can first spot their presence [TEC · FIC]. This happens every time a little cluster of words makes a character *stand out*; every time someone *also* makes a sound from skin stretched over a drum; every time a figure is *in addition* extracted from a line drawn on canvas; every time a gesture on stage engenders a character *as a bonus*; every time a lump of clay gives rise *by addition* to the rough form of a statue. But it is a vacillating presence. If we attach ourselves to the raw material alone, the figure disappears, the sound becomes noise, the statue becomes clay, the painting is no more than a scribble, the words are reduced to flyspecks. The sense has disappeared, or rather *this particular sense*, that of fiction, has disappeared. But—and this is its essential feature—the figure can never actually *detach* itself, either, from the raw material. It always remains held there. Since the dawn of time, no one has ever managed to *summarize* a work without making it vanish at once. Summarize *La Recherche du temps perdu*? Simplify Rembrandt's *Night Watch*? Shorten *Les Troyens*? And why? To discover "what they express" *apart from* and *alongside* their "expression"? Impossible, unless we imagine Ideas embodying themselves in things. This impossibility is the work itself.

By dint of profiting from it, we forget the stunning originality of fiction. Here we have a mode of existence like no other, defined by hesitation, vacillation, back-and-forth movements, the establishment of resonance between the successive layers of raw material from which are drawn, provisionally, figurations that nevertheless cannot separate themselves from this material. Just as technology, as we have seen, manages to extract metamorphoses [MET] and persistences [REP], new and totally unforeseen folds, so the vibration of fiction will *once again* fold those folds, renew them in a renewal that will engender something unforeseen, something still more unforeseen, as it were! For hundreds of thousands of years, clay lay on the floor of that cave before it found itself folded into an earthenware pot baked over a fire, but it finds itself transformed, transported, a second time when, from this earthenware pot held at someone's fingertips, some surprising anthropomorphic figure is extracted [TEC · FIC]. Will we ever be able to represent the amazement of the one who first found himself *made capable* of encountering such a being? What powers of transformation in this branching of beings of metamorphosis onto beings of fiction [MET · FIC]!

And it is from this necessarily fragile vibration—the figures are first of all disturbed materials, and they hold together only as long as the disturbance continues—that we draw the words used to speak of

⊙ WHICH GIVES THEM AN
ESPECIALLY DEMANDING
MODE OF VERIDICTION.

the requirements of this type of veridiction. For it is an incontestably and terribly demanding mode. "It fell flat." "It leaves me cold." Or, conversely, "It's come together." "It works." "That's it!" Through this discrimination between good and bad, true and false, a path is traced between two chasms: getting too far away from the materials, or remaining cold as ice before blocks of material that don't make sense. This path of truth and falsity is all the more demanding in that, as is the case with the other modes, there is no external judge, no arbiter of elegance and taste, no transcendence apart from that of the work, which is itself momentarily tested. "Guess, or you will be devoured": this is the question that Souriau attributes to the being he calls the "Sphinx of the work of art." It is about the work rather than about geometry that we should say *verum index sui*: what is true verifies itself.

Provided that it learns to make itself sensitive to its own truth by a quite particular path of verification. We have all experienced this situation: how unpleasant it is to hear, coming out of a movie or a play, blasé voices suspending all discussion, all evaluation, with the cliché: "I liked it" or "I didn't like it"; "there's no arguing about taste." A discomfort symmetrical to the unease experienced by all those ready to open themselves up, make themselves sensitive to a work, when they hear a critic decree, without discussion, without exploration, what everyone is supposed to think about the beauty or the worthlessness of a play or a film. We realize then that a key has been lost, the one that would have made it possible to discuss tastes to the point where a common taste is formed, while avoiding both the pretentions of a misguided objectivity and the abandonment of all criteria. What powerful normativity a work of art has! Anyone who complains that artists, artisans, creators don't know how to speak well about what they do is surely mistaken. They do better than talk about it: they indicate with a gesture the narrow gate through which one has to slip to take up the work again, a bit further on. How many times a day do we make judgments about what is beautiful or ugly, well or badly made? And this knowledge is supposed to be illegitimate because it is not judged by the yardstick of that big lunkhead Double Click [**FIC · DC**]? The fact that Double Click stands idly by doesn't mean that there is not, in the connection between the work, its constituents, its critics, its admirers, and its audiences, an implicit knowledge, trenchant as a scalpel, which is made explicit only in its own mode, prolonging the work *by another work*, often tiny and clumsy, that will extend its disturbance to another bearer of art. Judgment on and through the work is part of the work.

WE ARE THE OFFSPRING OF OUR WORKS.

It is quite true that the work depends on its receiver, but the notion of imagination does not account very well for this dependency. Like technologies, let us say that works of art are always *anthropomorphic* or, better, *anthropogenic*. Which does not mean that the artisan or artist has given a particular work the "form" of a human, but rather that the work has *gained* the form of a human in a rebound effect. Imagination is never the source but rather the *receptacle* of beings of fiction. Just as one becomes objective by connecting oneself to chains of reference, just as one becomes ingenious upon receiving the gift of technological beings, just as one receives what is needed to get a grip on oneself thanks to the beings of metamorphosis, in the same way one becomes imaginative when one is gathered in by works of fiction. "We are the offspring of our works": nothing more accurate has ever been said about the ontological unsettling caused by works of art. It is the *anthropos* stunned by the offerings made by his own hands who is made to draw back in surprise in the face of what is *morphing* him.

DISPATCHING A WORK OF ART IMPLIES A SHIFTING ⊙

How can we qualify with more precision such displacements, such behaviors, such transits? The experience is so common that we risk losing our sensitivity to it. Music begins, a text is read, a drawing sketched out and "there we go." Where? Elsewhere, into another space, another time, another figure or character or atmosphere or reality, depending on the degrees of verisimilitude, figuration, or mimeticism of the work. In any case, we go onto another *level*, in a triple shifting that is spatial, temporal, and "actantial" (to borrow from the jargon of semiotics). Will we come back? Perhaps, though that is not the question; for now, we are exploring all the forms of ALTERATION. Unquestionably, if we have gone somewhere, if we "have gone along with it," it's because we have been dispatched. But who dispatched us?

This is the most intriguing part: we have surely not been dispatched thanks to the flesh-and-blood author, who doesn't know very well what she has done and who, as a good artist, may lie like a rug about her own identity. And to whom is she addressing herself? Certainly not to "me," here, now, but to someone, a function, a position, that varies with each work, with each detail of the work, and that in no way preexists her—a function or position that I agree to fill and to occupy, or not. Here is a second level, situated *beneath* the work, that begins to shape both a virtual sender and a virtual receiver—speakers and addressees inscribed within the folds of the work. It is not for nothing that works "make up whole worlds": they even produce their authors and their admirers. Nothing precedes them, because they can make anything exist, as it were, "from scratch." Put a placard on stage saying "Asia begins here"—and there you have it, Asia begins. This is rather odd way to *make existence*.

It is a big mistake to accuse the SEMIOTICS of works of fiction of isolating the work from its "social context" or from the "interlocutory situation." Semiotics has done something better: it has discovered the originality of works of fiction, their proper ontological dignity, which no other mode of existence can replace. It was a stroke of genius on the part of A. J. Greimas to have understood this, despite the "good sense" objections that had represented the work of art, up to that point, as a kind of ersatz "communicative situation." A speaker, a medium, a message, and an interlocutor; nothing more had been seen in works of fiction. We recognize the familiar territory of good old Double Click, who believes that something is communicated to someone "through the intermediary of a message," as if those "fictional characters" of "communication" could *precede* the agencies of fiction in time! All these positions beneath and beyond originate in the work and because of the work; they emanate from the work and from it alone. It is fiction that has *made* us; the word "fiction" itself says as much.

⊙ DIFFERENT FROM
THAT OF THE BEINGS OF
TECHNOLOGY [TEC · FIC].

How could we be produced by what we produce? By the same effect of SHIFTING we encountered in the previous chapter. As we have seen, as soon as the raw materials begin to vibrate toward forms or figures that cannot, however, be detached from them, and toward whose peculiarities they never cease to refer, two new levels are immediately generated, the one ahead of, beyond, what is expressed, level n+1, and the other, beneath, behind, but also ahead, level n-1, that of the virtual addressee. It is through this double movement of sending ahead and pulling back that the world populates itself with other stories, other places, other actors, and that the possible positions of actor, creator, and subject appear. This is how a being of fiction is altered and folded. Alteration in alteration. Fold within fold. Reprise within reprise. To this dislocation, decisively recognized by semiotics, the work of fiction is going to add something to the technological object that will never stop serving it as a point of departure or launch pad. And to which it will be reduced as soon as it tips into failure, obsolescence, or oblivion—abandoned stage sets, rolled-up canvases, now-useless accessories, incrusted palettes, moth-eaten tutus.

If *Homo faber* and *Homo fabulator* actually derive from the same source, they differentiate themselves at once [TEC · FIC]. The technological dislocation folds materials that *stay in place* once the resistance gradient has been explored; the guardrail above a precipice that keeps you from jumping into the void keeps on protecting you with its steel uprights, whether you want it to or not. It is not like the story that has made your heart race when the hero threatened to throw himself into the precipice and was held back, at the last minute, by a guardrail of words. The latter is a story you have to read, you have to hold it together, charging it with your own knowledge and your own emotions. You have to have become, owing to the quality of the writing, capable of vibrating by distinguishing in your own body—without ever quite bringing it off—between material (that beating heart) and figure: figure for this material, material for the figure. The requirement of continuity is at once less strong than for the steel guardrail (you don't have to forge it) and stronger (you have to keep on holding it so that it will hold you!).

The difference is both obvious and subtle. Being "carried away" by a subway train and "carried away" by the beauty of a narrative are two different things. Two transports, two dislocations, but the two do not rely on the same linkages and do not conclude in the same way. It would be a mistake, however, to say that the first one transports "for real" while the second displacement is fake. Fiction is not fictional in opposition to "reality" (which in any case possesses as many versions as there are modes), but because as soon as those who are being displaced lose their solicitude, the work disappears entirely. This is indeed objectivity [FIC], but in its own mode, which requires being taken up again, accompanied, interpreted. There are still occultations, but their frequency, rhythm, and pulsations differ from those of all other modes.

No one takes that path through the mountains any longer, yet the steel guardrail still stands, even if the blacksmith who made it, the person who ordered it, and the walkers it used to protect are long gone. A terrible obligation, Péguy called it, the one that makes every reader responsible for Homer: the absence of Homer's reader swallows up the word; the negligence of Homer's reader reduces it to rubble. If we call the beings of fiction fictive or fictional, it is not because they are false, unreliable, or imaginary; it is, on the contrary, because they ask so very much from us and from those to whom we have the obligation to pass them along so they can prolong their existence. No other type of being imposes such fragility, such responsibility; no other is as eager to be able to continue to exist through the "we" whom they help to figure. In a sense, the beings of reproduction may be the ones they resemble the most [REP · FIC].

All the more so because they are everywhere. They have a kind of ubiquity that allows all the other modes to *figure* their own reality for themselves. What fiction does for technology and metamor-

THE BEINGS OF FICTION [FIC] REIGN WELL BEYOND THE WORK OF ART; ⊙

phoses—it folds and reprises them—will be done by all the other modes with the help of fiction. Without figurations, no politics is possible—how would we tell ourselves that we belong to any particular group? [FIC · POL]; no religion is possible—what face would we put on God, his thrones, his dominions, his angels and his saints? [FIC · REL]; no law is possible—*fictio legis* being indispensable to the daring passage of means [FIC · LAW]. Still, this doesn't mean that we live in a "symbolic world"; it means, rather, that the modes lend one another certain of their virtues.

Conversely, all the other modes vibrate differently if we grasp them the way we grasp beings of fiction, thus offering an excellent counterproof to the definition given above. If you find this mountain landscape "spectacular," it is because you are grasping the beings of reproduction [REP · FIC] as if their arrangement were a work projecting around itself a virtual arranger who has set up for you, a virtual spectator, a series of planes each of which plays the role of material for a form that cannot be detached from it. Taken according to this interpretive key, everything can be aestheticized, as the saying goes: machines, crimes, sciences ("What a beautiful theorem!"), even law ("What a splendid ruling!"). This is what gives meaning to the term AESTHETIC, which designates both the infinitude of the work and that of its creatures. The impression of infinitude does not come from the exact number of possible interpretations but from the intensity of the vibration between the successive planes, which seem at once detachable—chosen, arranged, selected—and undetachable—embodied, material, rooted. Suspend the vibration and there is no more infinitude at all.

⊙ THEY POPULATE A PARTICULAR CROSSING, [FIC · REF], ⊙

To get a good sense of the ubiquity of the beings of fiction, it suffices to consider the branching, now quite clearly visible, that they form with the beings of reference. The [FIC · REF] crossing is actually a very fertile one, for it is from the collaboration between these two worlds that we get a major part of our idea of the "WORLD" and its beauty. There is no other world "beyond," no other world "beneath," except the double dispatch from fiction and reference.

On one side, of course, no chain of reference can be established without a *narrative* populated by beings who can come only from fiction. How can we speak about remote galaxies, particles of matter, upheavals of mountains, valleys, viruses, DNA, or ribosomes without having at our disposal *characters* apt to *undergo* such adventures? They are all beings of paper and words, which have to be launched through the world like so many carrier pigeons. Every scientific article, every story of an expedition, every investigation is populated with stories experienced by these beings who always seem to have sprung from the unbridled imagination of their authors, and who go through tests alongside which so-called adventure movies seem entirely lacking in suspense. As Deleuze and

Guattari saw, no science is possible, and especially no abstract science, unless the world is populated by these little beings capable of going everywhere, of seeing and submitting to the most terrible trials, *in place of* the researcher trapped in her body and immobilized in her laboratory. It is these *delegates* that we have trusted, since the seventeenth century, to go off and travel everywhere.

And yet these little emissaries are distinct from the beings of fiction in one way: they have to *bring something back*. We know from Chapters 3 and 4 what they bring back: it is REFERENCE, that aptitude for maintaining a constant across the often very lengthy and very trying cascade of INSCRIPTIONS, ideographs, each of which differs from the previous one and from the one that follows. Factual narratives do not differ from fictional ones as objectivity differs from imagination. They are made of the same material, the same figures. What would scholars be without imagination? With what would they think? What stories would they be able to tell? What would they talk about? In what world would they move around? And yet, starting from the same basic raw materials, the two modes differ through the treatment to which we subject them: if we authorize beings of fiction to travel far and wide, to "carry us away," as we say, into another world, the same domesticated beings, disciplined by chains of reference, have to come back home to bring back, "report on," remote states of affairs, which they also are responsible for unifying in a verifiable common narrative. Multiple and partial narratives that are then summarized, simplified, unified, in great stories to be presented in CinemaScope, projected stereophonically and in 3D, under the name of "scientific vision of the world."

If the word "reference" has a meaning, it is because we have learned to *charge* the delegated, partial little observers, which proliferate with every instrument, not only to go off but also to *come back*. Think about any scientific apparatus at all: you will see walking around in it beings that have to be deemed "of fiction" (what else would they be made of?), but with one difference, which is decisive: these little beings have been repatriated, and they can be sent out again to start over; their comings and goings alone ensure the objective quality of access to remote states of affairs. The characters of fiction, like those of reference, take off

⊙ WHERE THEY UNDERGO A SMALL DIFFERENCE IN THE DISCIPLINE OF FIGURES ⊙

(they *shift* toward other spaces, other times, other actants), but whereas we readily allow the former to leave without thinking about their return, we need a whole delicate procedure to bring back the latter (to *shift them back in*). It is because it returns to the dovecote with its message rolled around its leg that the pigeon is said to be a messenger. It is because of that nuance that people have begun to insist so much on the distinction between the "figurative" and the "literal." And yet the LITERAL is nothing but the *disciplined and domesticated* replaying of the flight of figures. The literal is to the figurative what the dog in the fable is to the wolf. The sciences may be fictions, but they are domesticated enough to report, refer, inform; yes, they in-form, if we remember the weight in fiction and technology of that little word "form."

What is designated very awkwardly by the opposition between the "treasures of the imagination" and "cold, hard objective truths" is the fact that the level of enunciation n - 1 never allows us to validate the presence of beings of fiction—we shall never find Madame Bovary's birth certificate in Flaubert's study, or, if we do, it will be a fake—whereas we require that beings of reference allow us to attach in an uninterrupted succession—in fact interrupted at each inscription by the dissimilarity between two successive stages—the narrative published along with what the enunciator can guarantee in its favor. It is in the laboratory, in the researcher's drawers, even, that we demand to see (especially in the case of alleged fraud) the inscription at the n - 1 level that serves as head of the network following all the delegates dispatched throughout the world. Shifting out on one side; shifting in on the other. There is indeed a difference, but it really doesn't give us anything on which to make "the" distinction between the objective and the imaginary, the true and the false, the True and the Beautiful.

⊙ THAT CAUSES THE CORRESPONDENCE TO BE MISUNDERSTOOD.

Making room for the beings of fiction amounts, paradoxically, to authorizing ourselves to be materialists at last. If the reader is beginning to be familiar with the crossings between the modes uncovered one after another in the inquiry, he will be able to understand not only how one mode straddles another—for example, the fictional characters that have been domesticated to bring back reference—but how two modes can either collaborate or multiply their mutual misunderstandings. This is surely true of "mimetic figuration," that great moment in the history of art and the sciences when the same tools—calculation of perspective, projection onto a canvas or a blank sheet of paper, the establishment of conventions for reading distances and shadows, the unfolding of cartography, and later the acquisition of descriptive geometry—end up giving both artists and scholars the feeling that they are exploring "the same world," the one *right in front of them*, which they take to be a *spectacle* seen through a window. What happened in that decisive moment when the notion of correspondence—the one we finally uncovered at the end of Chapter 4, that of [REP · REF]—transformed itself into an idea of correspondence as the *mimetic resemblance* of a model and its copy? How has Science become the mirror of the world?

It is as though the very success of a whole series of "arts of describing"—to borrow Svetlana Alpers's lovely title—had produced a 90° shift in the relation between chains of reference and the remote beings to which they achieve access through a pavement of transformations generative of constants. The logic of paintings contemplated from the outside by a viewer who sees them as copies of an equally external world *wins out* over the length, the cumbersomeness, the cost, the complexity of chains of reference, even though the latter are being set up at the same time. As if the worlds of art were imposing their epistemology, or rather their aesthetics, on those of the sciences. Scholars begin to think of the known world according to the model that is being offered them at the same moment by paintings said to be "realistic," produced by artists who are themselves imbued with the new sciences. The real world, the one described by the sciences, then appears to bear an uncanny resemblance to the world depicted in paintings, in the sense that there is an original to describe and a copy that must be faithful to it. As if there were two elements, *and only two*, linked by resemblance, by mimetic figuration. The sciences start to forget how much their forms of inscription differ from paintings—and through how many dizzying stages chains of reference have to pass—and they begin to believe that their terminal objects really reside *in* that painted world.

We probably have here one of the origins of the RES EXTENSA, which would be in the last analysis only a fairly innocent misunderstanding about the crossing between fiction and reference, a scholarly enthusiasm originating in the arts. The "known world" would proceed from an aestheticization of the sciences! A vast construction site that our inquiry will soon have to confront, when we retrace the origin of what Philippe Descola calls NATURALISM, something that has deprived the Moderns of the possibility of understanding those who had never combined fiction and reference in the same way. Those about whom it is rightly said that, deprived of access to an "external world," they could not have grasped reality except through the mists of their "symbolic representations."

WE CAN THEN REVISIT THE
DIFFERENCE BETWEEN
SENSE AND SIGN ⊙
Aren't we now in a position to dissipate these mists? Not to discover the "external world" that they have purportedly been hiding for so long but to find the mechanism, the "fog machine," that has spread that world everywhere. As we saw earlier, a sign does not cover all the meanings of sense or direction but only one of the varieties of sense that derives from fiction.

Let us note first of all that the canonical definition—a sign is something that stands *in place of* something else, something that is necessary to its interpretation—remains a very general property that could define all types of senses or meanings, even all the invisible beings that we have to learn to capture in order to sketch the trajectories of being. To discover the sense is not in the first place to seek the connection between one word and another, but the connection between a word, a speech act, a course of action, and what must be put *in its place* if the latter are going to continue to have meaning, to *make sense*—that is, for them to continue to exist. Interpreting meaning is thus not to set aside all ontological questions while isolating the symbolic domain but on the contrary to take up the stray thread of ontology again. The idea that a sign, to be comprehensible, must be linked to another sign is only a particular case of the much more general situation, proper to the ontology of being-as-other: there has to be something other *in the place* of the same in order to persevere in being through the hiatus, the mini-TRANSCENDENCE of alteration.

Let us note, next, that the famous distinction between the "signifier" and the "signified"—the anode and the cathode, they say, of all our symbolic energies—amounts to repeating with respect to signs what we have already said about the beings of fiction. We find the same vibration between raw material and figure, and the same impossibility of detaching them from one another. The distinction is important, to be sure, but because it designates the particular case of the being of fiction whose vibration allows it to be always graspable—or rather ungraspable, by definition—either as raw material twisted toward form—the signifier—or as form inseparable from the material—the signified—without direct contact with the referent, which is not the "real world" but the result, merely glimpsed, of the proliferation of the *other* modes of existence—which fiction indeed always takes on obliquely.

Following this logic, the obsession with signs derives from an exaggeration of the place of the fictions that have been asked to define *all the modes of meaning*. Etymologically, the word "SYMBOL" designates one half of a token that travelers used to break in two as they were separating, with the other half destined to serve as a sign of recognition when they met again; yet there are many ways to separate, to come back together and reconnect the two pieces. Nothing in particular obliges us to link a given symbol to another symbol. To be interested only in the relations of signs among themselves would be to take as the point of departure something that would be literally devoid of meaning, and even at bottom senseless, since the sense would *already* have been lost—the *sense*, that is, what it anticipates and what follows and is necessary to continued existence.

Just as it is possible to see to what temptation the notion of matter responded—it sufficed to yield to the slippery slope and explain the success of knowledge as if the form necessary to reference were also the real and invisible foundation of the entities to which one had access [REP · REF]—we can understand in the same way on what temptation the notion of a symbolic world may hinge. Being-as-other, in fact, alters itself and renews itself; it is never *in itself* but always *in and through others*. Every existent thus turns out to be in part veiled, torn like the *symbolon* between what dispatches it and what is to follow. It is thus rather tempting to *replace* this hiatus between the preposition and what follows by a distance between a sign *and what it signifies*. Especially if one can link signs to one another as if, after all, they formed a world, a system, or a STRUCTURE.

This can only be a second-best solution, a matter of making do, but we can understand that it appears credible: "Having lost the word and meaning, let us act in spite of everything as though these *insignificant* signs formed a world *of their own*, connected not by some reality but by their own *rules* of association and transformation." Thus, to the "material" world, a first artifact, we are now adding a second, the "symbolic," a second artifact. And the more we insist on the expansion of matter on one side of the Bifurcation, the greater the temptation, on the other side, to give verisimilitude to this artifice of language.

Can we go back up this slope? Yes, because we are now seasoned enough in the recognition of category mistakes to avoid deceiving ourselves as to what is put "in the place" of something else, and to know that we always have to be suspicious when we are asked to take one thing *for another*. The sign is not necessarily there "for something else"—and still less for another sign with which it would form a "minimal pair," a somewhat desperate endeavor to make sense with non-sense. The sign is there for and through its predecessors and its successors. This time, to go back to Magritte's example, neither the painted pipe nor the pipe in our narrative nor the briar pipe set before the painting resides simply *in itself*, but always also *in the others* that precede and follow.

⊙ AND FIND ANOTHER WAY OF ACCESSING THE ARTICULATED WORLD.

In other words, the sign is "arbitrary" only for those who, having agreed to lose the experience of relations, try to reinject relations on the basis of the "human mind" into a "material world" that has been emptied in advance of all articulations. Now, as we are beginning to understand more clearly, it is *the world itself that is articulated*. If living beings manage to make out an "index" in the link between smoke and fire, it is because since the dawn of time fire has leaped, launched itself, announced, uttered, expressed, exhausted itself in smoke. Give existents back their ins and outs, what goes before and what comes after, and you will find that they are full of meaning, that they collect many differences besides that of the "minimal pair" dear to advocates of structure, that they register the world's alterations admirably well. Yes, of course, *cheval* in French is "horse" in English! What conclusion are we to draw from this, except that there are many ways for a large number of horses galloping on the plains to enter into relation with many tribes garbling French and English? Why draw from this rich fabric made of multiple intersections only the lesson of the "arbitrariness of signs"? Why remain so indifferent to the *other* differences?

That the world is articulated and that this is why we sometimes manage to take up certain of its articulations through the intermediary of expressions, only an infinitesimal number of which are produced through the channel in which air currents slip past the glottis—is this not a more realistic, more economic, more elegant hypothesis than imagining a human projecting from his head signs lacking any purchase on an inarticulated material world? Everything flows, everything creeps in the same *sense*, in the same direction: the world and words alike. In short, beings UTTER THEMSELVES, and this is why, from time to time, we are capable of speaking truthfully about something, provided that we go at it over and over. If natural language takes itself in hand to take up the world, it is because the world has taken itself in hand and is still doing so, time after time, to persist in being. A linguist should never circumscribe the isolated domain of "Language," unless it is to interrupt this movement of articulation for a moment, to make the analysis easier. Language becomes an isolated domain only through the desperate effort to make the continuity of beings hold up despite the drift of beings-as-other: then, indeed, we find ourselves with a sign emptied of sense that seeks to catch what is fleeing from it and that, unable to do so, resigns itself to clinging to another sign to try to "make world" in spite of everything; but it is a poor world, a world that has lost the world. Those who are going to have to decipher Gaia's injunctions very quickly would do well to learn to speak that language at last, without opposing their "articulated language" to what they perceive as an unarticulated world.

LEARNING TO RESPECT APPEARANCES

···

To remain sensitive to the moment as well as to the dosage of modes ⊙ the anthropologist has to resist the temptations of Occidentalism.

Is there a mode of existence proper to essence?

The most widespread mode of all, the one that starts from the prepositions while omitting them, ⊙ habit [HAB], too, is a mode of existence ⊙ with a paradoxical hiatus that produces immanence.

By following the experience of an attentive habit ⊙ we see how this mode of existence manages to trace continuities ⊙ owing to its particular felicity conditions.

Habit has its own ontological dignity, ⊙ which stems from the fact that it veils but does not hide.

We understand quite differently, then, the distance between theory and practice, ⊙ which allows us to define Double Click more charitably [HAB · DC].

Each mode has its own way of playing with habits.

This mode of existence can help define institutions positively, ⊙ provided that we take into account the generation to which the speaker belongs ⊙ and avoid the temptation of fundamentalism.

W E ARE BEGINNING TO UNDERSTAND WHAT ACROBATICS ARE REQUIRED TO HOLD ONTO ALL THE MODES AT ONCE. Every time they manage to extract a new contrast, the Moderns have a tendency to *weaken* or, on the contrary, to *exaggerate* another one to which they are equally attached. This is not necessarily irrational behavior, but rather what would be called flawed staging or lighting in the theater: the director or the lighting designer has brought out some nuance in the acting while plunging another into the dark. This is the kind of setup error we are trying to remedy. Each mode of existence can be wrong about all the others, and no single one can serve definitively as an unchallengeable standard for all the others—this is what provides the framework for the Pɪᴠᴏᴛ Tᴀʙʟᴇ. And yet we have promised to give each mode of being its own template, and to address each one in its own language.

But what complicates things even further is the fact that, depending on where in history we place the cursor, the same category mistake can be found in all possible states. At first a simple, unfortunate consequence of the extraction of a contrast, it is only much later that it can become dangerous, and then, perhaps, fatal, before it disappears altogether; or, on the contrary, it may find itself comfortably instituted in institutions adapted to it. By denouncing the deleterious effects of *res extensa*, for example, we don't mean that we wouldn't have been excited about it in the mid-seventeenth century: we would surely have seen

Cartesianism as the ideal solution for the simultaneous development of matter, the sciences, thought, and God. It is only gradually, and through the shock waves that reverberate in each of the histories proper to each mode, that we find ourselves lamenting, three centuries later, the simultaneous loss of the sciences, subjects, and gods. A few decades ago, we would have been excited about the power of critical thought, for it was finally making it possible to overturn institutions that were unable to shelter the values they were claiming hypocritically to defend. It is only today, owing to completely different circumstances, that we are obliged to renounce critique and learn to respect institutions again—perhaps even to cherish them.

This inquiry thus does not consist simply in highlighting the modes but also in identifying for each one the inflections that come up throughout what it would be appropriate to call their *ontological history*— with apologies to the real historians. Ivan Illich called these moments MALIGN INVERSIONS, taking as examples the threshold above which expenditures on health, useful up to that point, cause more illnesses than they cure, or the moment when, by dint of multiplying automobiles, we end up, on average, going more slowly than on foot. Each contrast is like a *pharmakon* that slowly builds up: over the long run, and at high doses, the remedy becomes a poison. We can never avoid all poisons, but we could balance out certain of their effects by carefully administered counterpoisons. There would then be a whole system of dosages and dietary advice, a whole pharmacopeia of modes of existence with which we would have to familiarize ourselves in order to avoid speaking too harshly about category mistakes—while running the risk of being mistaken about the moments when these errors become truly toxic.

Here is where our anthropologist begins to have doubts about her work. Of course, she is rather pleased to see that one can go from discovery to discovery, after all, while staying put at home, in Europe. Her colleagues on assignment in faraway lands may bring back extraordinary stories, but she holds to her conviction: no anthropological enigma is more exciting than the one offered by the Moderns. How could anyone have expected them to succeed in housing monster Transformers in some interiority? How could anyone have imagined that

⊙ THE ANTHROPOLOGIST HAS TO RESIST THE TEMPTATIONS OF OCCIDENTALISM.

the decisive ploy of technology could be transformed by them into simple objects, as obtuse as they are massive? That solid, stubborn "matters of fact" arise from wild-eyed idealism? That the Moderns would invent a symbolic world alongside the real one, to house beings of fiction? And yet, from another standpoint, she senses that she has committed more than one sin against method and given in too often to the temptation of exoticism—particularly when she treats her informants as people who deceive themselves and don't understand what they are doing. She feels that she is in grave danger of yielding to the delights of OCCIDENTALISM. If she has supposed, up to this point, that it was possible to avoid category mistakes simply by paying closer attention, she is well aware that this was just a ploy for bringing out the contrasts between the modes of existence. She now has to find a different response that no longer consists in accusing the Moderns of irrationality.

To help her pull herself together, we need to familiarize ourselves with a new mode, one that will make it possible to account for the *apparent continuity* of action. It will also allow us to give a more charitable version of Double Click [DC] and to provide a more precise definition of IMMANENCE, as well as of the notion of INSTITUTION we have used and abused up to now without really specifying its meaning.

IS THERE A MODE OF EXISTENCE PROPER TO ESSENCE? Fortunately, the anthropologist has noticed that misunderstandings pile up every time the question of ESSENCE is raised. Socrates, during his own inquiry into the modes of existence, annoyed all the tradesmen of Athens by claiming he could get back to the "essence" of cooking, beauty, horse training, even delousing, by making them spit out their little "ti esti?"—"what is ... ?" And he was disappointed every time by the practitioners' inability to express what they were doing. Whence the scorn he chose to adopt toward those who could not speak of essences in the right way, those who were limited to *doxa* alone, who had lost the pathway to the Idea and had perhaps "forgotten Being." However, while a question that disqualifies those to whom one is speaking may be of polemical use, it does not correspond to the empirical philosophy we claim to be following in this inquiry. To speak well in the agora with practitioners is to hope that they will nod their heads in approval when we propose a version of their practice that may be totally different from theirs but at least commensurate with their experience and, if possible, shareable. And above all a version that will allow them to respect, in turn, other modes that they had learned to scorn, through a sort of positive contamination that would be the exact contrary of the negative contamination introduced by the Socratic question. We cannot say that someone who populates the world with irrational people through his questioning is expressing himself rationally.

This amounts to detecting a new category mistake involving the very question as it has been posed in Socratic fashion. As if it were not at all Being, Idea, essence that had been forgotten, but BEINGS. The mistake would stem from the fact that the question of the essence of a practice, any practice at all, was raised in a single mode, the mode of equipped and rectified knowledge. As if it were impossible for the Moderns to propose as many arrangements for instauration as there are modes or prepositions. And that mistake would be all the more troubling, as we saw in Chapters 2 and 3, in that it would not even succeed in capturing the essence of knowledge [REF]! We would have sought to define the essence of every practice on the basis of an idea of knowledge already deprived of its mediations [REF · DC]. Here is where the ethnologist is happy to have chosen, as a metalanguage allowing her to speak rationally at last about all the modes, not knowledge as Double Click understands it (an emasculated form of [REF]) but the mode that protects all modes, that of prepositions [PRE].

She is well aware, however, that to posit such a diagnosis would plunge her into a dreadful contradiction. She would start to accuse Socrates of having been mistaken about knowledge as well as about all the other modes, because he had chosen the wrong touchstone. She might even reach the point of accusing philosophers of having *forgotten to forget* Being-as-being! She would still not have left critical thought behind, and moreover she would have made mistakes, and the Socrates-style detection of category mistakes, the only horizon of her inquiry. As if it sufficed to detect a mistake to put an end to it! She concludes from this that the question of essence cannot be entrusted either to a single mode—a shaky one at that—or eliminated as a simple methodological error. As always, she has to take the practices of her informants seriously, including those of Socrates and his descendants. Underneath this question of essence, this age-old obsession, another question, another manner of being, must be hidden. In other words, we now have to ask ourselves the question *of the mode of existence of essence.*

THE MOST WIDESPREAD MODE OF ALL, THE ONE THAT STARTS FROM THE PREPOSITIONS WHILE OMITTING THEM, ☉
The question may look like hair-splitting. But if the reader has something to complain about, it's not that we have reached the point of splitting these particular hairs, but rather that it has taken so long to introduce the most important, the most widespread, the most indispensable of the modes of existence, the one that takes up 99 percent of our lives, the one without which we could not exist, obsessed as we would be with avoiding category mistakes. The one that allows us to define the courses of action that we have learned to follow through the notion of association networks [**NET**].

There was of course a good reason for the delay: this new mode would have concealed those that we wanted to relearn to detect first of all, since it has the particular feature of *veiling* the prepositions. It doesn't forget them, it doesn't deny them, it doesn't reject them; no, let us say only that it dissimulates them, or, still more accurately, that it *omits* them, that it *must* omit them (we shall attribute a technical sense to the difference between "forgetting" and "omitting").

Let us remember, after all, that the **PREPOSITIONS** indicate the direction of a trajectory, but that they never propose anything further; and in particular they never serve as a foundation, as potential or as possibility conditions for what is to follow; they never do anything but announce it, signal it, prepare us to take what comes next in the right way. If you were to imagine any existent at all constantly nagged by the choice of the right preposition, it would never start existing! No action would follow. It would die of hunger and thirst, like Buridan's donkey, or would remain frozen in place like one of the hikers we met in Chapter 2, anxiously staring at the signposts and never deciding on a path. To go back to the example of the indication "novel," "report," or "document" at the beginning of a text, what sense would it make for a reader to contemplate these three words indefinitely without ever looking through the book? These notices give a sense of what is to follow, of course, but *provided that something follows,* provided that the reader "turns the page" and doesn't remain *stuck* on this single indication. And yet—this is the key point—he will do this *without ever* *completely forgetting* the indication. The prepositions that we have followed up to now thus find their real meaning only thanks to a new branching, the one that *makes it possible to add a continuation to what the prepositions merely indicated*; or, to put it in yet another way, the branching that gives the *position* of which they are precisely only the *pre-position*.

Someone will surely object: "This isn't, this ⊙ HABIT [**HAB**], TOO, IS A
can't be a real mode of existence!" But it is! And the MODE OF EXISTENCE ⊙
most common, the most familiar of all, the one that
William James—here he is again—designated with the only word that
fits it perfectly: habit, blessed habit (noted [**HAB**]).

Look around. Existents are not constantly preoccupied with their descendance; most of the time, they go about their business enjoying existence [**REP · HAB**]. The beings that produce psyches do not always make us vibrate in the anguish of surfing on metamorphoses; we simply feel "comfortable in our own skin" [**MET · HAB**]. As long as I am unskilled at putting up cinder-block walls, I feel the rapid passage of the technological upsurge, but once the subtle arrangements of muscle and nerve reflexes in relation to each tool and material have been established, I line up the sequence of works and days without even being aware of it, as if I were totally adjusted to my task [**HAB · TEC**]. A priest who is converted at every Mass at the moment of transubstantiation would remain like Saint Gregory, so stunned by what he is celebrating that he could never get beyond the first words of the Canon [**HAB · REL**]. A researcher who is exclusively concerned with understanding by what miracle of correspondence she manages to maintain a constant across the dizzying transformations of distinct inscriptions would never succeed in reaching remote beings [**HAB · REF**].

Habit is the patron saint of laid-out routes, pathways, and trails. Every lost hiker who has to "hew out his own path," at the price of scratched hands and sore feet, hesitating at every step, understands very well, when at last an opening in the brush signals the presence, however minimal, of a trail already used by others, the extraordinary blessedness of habit: he no longer has to choose, he can finally follow, he can finally put himself "in the hands" of others, he knows what to do next, and he knows this without reflecting, even as he verifies with an attention that is both casual and lively that there are indeed, here and there, indications that this is the right track. Without habit, in other words, we would make new mistakes, no longer through *ignorance* of the various prepositions, but because, this time, we would be *limiting ourselves to them* without heading toward what they designate, that toward which they propel us. The action would no longer follow any course. No trajectory would ensue. We would constantly hesitate as to the path we should take. We would be a little like Narcissus mistaking the contemplation of his navel—an incontrovertible signature left by the most initial of the prepositions [REP]—for life itself.

⊙ WITH A PARADOXICAL HIATUS THAT PRODUCES IMMANENCE.

So we now have to recognize two different senses in the notion of category mistake: being mistaken about the mode on the one hand and on the other limiting ourselves to the search for the right mode *without advancing* toward what it indicates. But would this not mean abandoning our own definitions, since each mode has been identified up to now thanks to a particular form of hiatus, of discontinuity, of TRANSCENDENCE? Habit, in fact, seems to have the characteristic of *no longer* needing transcendence at all, of leaping over obstacles so well that there is no more threshold, no leap, no discontinuity of any kind. True; but this proves that *even immanence* needs to be engendered by a mode of existence that is proper to it. If it is true that mini-transcendence is the default position, that it is thus *without a contrary*, immanence is not going to be introduced in this study as what is opposed to transcendence but only as *one of its effects*, as one of its ways—a particularly elegant one, to be sure—of adjusting the junction points *without splices* and without any *visible* break in continuity. Habit has the peculiar feature of *smoothing over*, through what must be called an *effect of immanence*, all the little transcendences that BEING-AS-OTHER explores.

There is nothing troubling to common sense here. We find nothing paradoxical in watching an animated film, even though we know perfectly well (but we forget it even more perfectly) that it is made up of a sequence of fixed images. So there does have to be some *special effect* to engender continuity: the effect that is outlined by acquired habits, but provided that the film is run at a certain speed, and only after each image has been painted with great care. Immanence is there, but it is never anything but an *impression*, and even a retinal impression, left by something else that passes by. A paradox? Yes, according to the only touchstone usually acknowledged, that of Double Click [HAB · DC]. But let us not forget that *all* the modes of existence are paradoxical, each in its own way, as perceived by all the others. It is precisely this feature that obliges the investigator to draw up the Pivot Table, and that prohibits her from taking a single mode to use as a metalanguage for the others (with the exception of [PRE]).

Whence the feeling, as old as thought, that phenomena are "hiding something from us." And it is true, they really are hiding something, yet there is no mystery to worry about: continuity is always the By FOLLOWING THE EXPERIENCE OF AN ATTENTIVE HABIT ⊙ effect of a leap across discontinuities; immanence is always obtained by a paving of minuscule transcendences. The big challenge is not to make a category mistake here. Especially because philosophers of habit are even less numerous than those of technology: we too often see habits as proof of irrationality, for want of being able to follow the thread of that particular reason. Now, the thread really does exist, even if it becomes very thin. And it is precisely the role of this mode to *make it thin*; otherwise we could never pass into the networks and could never deploy their surprising associations [HAB · NET]! Through habit, indeed, the discontinuities are not forgotten, but they are temporarily omitted, which means that we remember them perfectly well, but obscurely (clearly) in a very particular sort of memory that we risk losing at any time.

Here, too, the experience is a common one. You rent a car in England and have to drive on the left, having driven only on the right up to that point; well, in a few minutes, all your reflexes turn out to be redistributed. You wake up sleepy in the morning and discover a leak under the bathroom sink, and at once you change your routine and shift from making coffee to dealing with rags and plumbing. You're sitting down peacefully in your armchair to read the newspaper, but as soon as you notice the pained expression on your loved one's face, you put the paper down and try to take care of him or her. Proofs that, *underneath* forgetful and reflexive habits, *something has remained awake* throughout your long existence of driving on the right, waking up in the morning, enjoying your bourgeois comforts, something, as James has shown so well, that can "take things in hand" and redirect the flow of attention (it's up to the neurobiologists to show us how this works). Habit thus does much better than *losing* the preposition; it *presupposes* it even while *preserving* it carefully. Let us say that habit is the mode of existence that *veils* all modes of existence—*including its own.*

⊙ WE SEE HOW THIS MODE
OF EXISTENCE MANAGES TO
TRACE CONTINUITIES ⊙
It is this veiling, this omission, that we have to inscribe in its specifications. If attention had disappeared for good, we would be automatons, robots (stubbornly driving on the right in England; preparing our coffee while water kept on burbling out under the sink; invariably behaving like insensitive louts). But we would then be committing a double category mistake, regarding both machines and their human operators [HAB · TEC] AUTOMATONS are never wholly automatic. Every robot manufacturer knows quite well that in case there is a breakdown he must always anticipate, in addition to the automatic mechanism, what is called in the trade a "MANUAL RESTART" (he is obliged to do this, moreover, by his insurance contracts). A flesh-and-blood pilot has to be able to do, manually, everything the automatic pilot was doing before the breakdown. The expression itself can be our guide: if there is a restart, it is because the being in question has to pass once again through the intermediary of another, because there is a discontinuity, a leap, and thus a mode of existence, a type of alteration.

The special contribution of habit is that it is very good at defining essences, continuities that appear to be durable and stable because breaks in continuity are omitted even though they remain "highlightable" and "retrievable" at every moment. It is not that "existence precedes essence" but that behaving like an essence is a mode of existence, a way of being that cannot be substituted for any other and that no other can replace. Without habit, we would never have dealings with essences, but always with discontinuities. The world would be unbearable. It is as if habit produced what stays in place on the basis of what does not stay in place. As if it managed to extract Parmenides's world on the basis of Heraclitus's. We can say of habit that in effect it makes the world *habitable*, that is, susceptible to an *ethos*, to an ETHOLOGY.

⊙ OWING TO ITS PARTICULAR
FELICITY CONDITIONS.
If each mode is defined both as a particular "right to draw on" being-as-other and as a type of articulation, it must also obey particular felicity and infelicity conditions. Now, while habit has great qualities, doesn't it

remain indifferent to lies and to truth alike? Isn't this precisely what is shocking in habit, and why so many philosophers have taught us to speak of it as a matter of mere opinion, as *doxa*, a sin against enlightened knowledge? How can we reply to the objection that one cannot speak, with regard to habit, of a particular type of veridiction? And yet it suffices to slip from omitting prepositions to *forgetting* them to pass from the truth of habit to its *falsity*. If we were to forget habit, we would tip from rationality into irrationality: not by critiquing habit but by no longer being able to distinguish "blessed" habit from its exact opposite. Habit, too, habit especially, can lie or tell the truth. Might we not have been mistaken in dismissing *doxa* too hastily?

It is the most common experience. No touchstone is more discriminating than this one: there are habits that make us more and more obtuse; there are habits that make us more and more skillful. There are those that degenerate into mechanical gestures and routines, and those that increase attention. Either habit knows how to find the path of alteration by going back to the preposition that initially "dispatched" it, or else it has lost all traces of that path and begins to float without signposts. To follow a course of action because we have understood in what ethology we were operating is not at all the same thing as ceasing to follow any indications as to what we should do next time. Harold Garfinkel, one of the very few analysts of habit, has proposed this admirable characterization of a course of action: "for another first next time." Here is a fine felicity condition: next time we shall do what we did last time, yes, but it will also be the first time. Everything is the same, smooth and well known, but difference is standing by, ready for a "manual restart." Paradoxically, there is no inertia in habit—except when it tips into its opposite, automatism or routine. But there, no doubt about it, habit will be *lost*.

To repeat is not at all the same thing as to keep harping on something, to keep flogging a dead horse. Even the most exhausted hiker doesn't follow his well-marked trail "robotically," otherwise he would get lost at the first badly marked turn. Ethologists know how to distinguish, in animal behavior, between what depends on their own observational routines—a rat will "always act like a rat"—and what induces change in the most predictable rat as soon as one changes the conditions to which it is being subjected. As Vinciane Despret has shown, there are scientists capable of making a rat much more interesting, in the laboratory, by making it a little more *interested* in what it is doing. Ethology is always a habit, and one that can indeed *change* because it is always watchful, always *on standby*. The vultures that circle above the hiker in the canyons of Aragon, vultures that used to be considered strictly carrion feeders, may have learned to eat fresh meat since the European Commission passed a decree forbidding them to feed on sheep carcasses that shepherds no longer have the right to leave in the fields for them. The animal most faithful to its habits is always watchful enough to get a grip and change them.

But to be able to change we have to be able to go backward and thus always keep the tonality of the action in sight, as if under a veil. Let's say that bad habits are to good ones what spam is to electronic messages. Shreds of existence floating around without an author, with no responsible party, no receiver, polluting the world, offering on our screens an image of what the world would be like if we had really lost the direction given by the prepositions. Struggling against *doxa* seems legitimate if we designate by that word the loss of address, the rootless utterances that retain no scars indicating their regime of enunciation, that make their way blindly and, as it were, out of network before ending up in the trash. Here we recognize the "hearsay" philosophy has been inveighing against since it began. Whatever you may say or do, specify at least the preposition, or, to extend the metaphor, the IP address from which you are sending the message. Philosophy has always conceived of itself, and rightly so, as an antispam apparatus.

HABIT HAS ITS OWN   But we also see that it would be dangerous to
ONTOLOGICAL DIGNITY, ☉   make a category mistake about this mode of existence, by confusing the rejection of spam—a legitimate and necessary antipollution operation—with the rejection of all omissions and all veiling on the other side. Without omission and veiling, it would be impossible to engender the existents, these crossings between habits and prepositions [HAB · PRE]. It is here that "appearances are deceiving" for real, and we treat them badly if we conflate the struggle against unattributable beliefs with a totally different exercise: the search, behind utterances, for a SUBSTANCE that would really explain the continuity of essences. Here we find the philosophy of being-as-being

sidestepping evasively, but this time we can respond more tactfully without adding any accusation: the whole problem arises from the fact that the philosophers of being have not seen how they could actually *respect appearances.*

This doesn't mean that we have to be suspicious of depth and applaud the modesty of Nature, which "likes to veil itself," as Nietzsche puts it so nicely (with a good dose of Orientalism and machismo in his predilection for the dance of the seven veils). No, veiling has a function, an ontological dignity, that we can miss in two different ways. First, by seeking *direct* access to "unveiled" things: at best we would simply come upon association networks stripped of their differences [NET], or find differences only in tonalities, prepositions lacking trajectories, follow-ups, and networks [PRE]; second, by resigning ourselves definitively to dealing solely with appearances, without ever again seeking "that of which" they would be the appearances.

Here is the category mistake proper to this mode: appearance does not stand in front of "what it hides," like a cloth covering a precious casket; nor is it, as in Japanese gifts, a series of envelopes embedded within envelopes with no content other than the beauty of the folds and the successive embeddings—which would amount to aestheticizing it [HAB · FIC]. "Behind" appearance there is not "reality," but only the key that allows us to understand how reality is to be grasped—and this key does not lie underneath, but *alongside* and *ahead.* Appearance allows itself to be seen in the *direction* given by the preposition, like the path followed by a hiker who is reassured but nevertheless careful not to make a mistake. To follow this direction really amounts to leaving the placard behind, heading in the direction it has indicated, without there being in this forgetting the slightest denial of the direction it has indeed *given* you. No one will say that the term "novel," "provisional report," or "documentary fiction" on the first page (appropriately called in French the *page de garde,* the "warning" page) "founds" the reality of the volume that follows, but no one will say, either, that such notices "conceal" its contents. While no one would think of saying that a signpost obscures, contradicts, denies the direction it designates, no one can claim, either, that it would be much more rational to do without any signs at all. In other words, we must seek neither to get rid of appearances nor to "save appearances"—to save

face—nor to traverse appearances. We must simply head in the direction
indicated by the preposition, without forgetting it. Appearances are not
shams. They are simply true or false depending on whether they veil or
lose what has launched them.

⊙ WHICH STEMS FROM
THE FACT THAT IT VEILS
BUT DOES NOT HIDE.

It is understandable that, having arrived at
such a branching point, philosophy has hesitated—
a hesitation whose effects we have seen time and
time again—when it has chosen to "speak straight"
(imitating what it thought it had understood about knowledge) rather
than to speak well. Indeed, the smallest shock suffices for things to
be taken in the wrong way. We owe this ambiguity between being as
SUBSTANCE and being as SUBSISTENCE to what habit leaves behind in its
wake, since habit—this is its virtue but also its danger—obtains *effects of
substance on the basis of* subsistence. By forgetting the effect, one would
of course be making a mistake—the existents remain and they are really
there, with all their habits, their ethology, and their habitat—but the
mistake is just as serious when we forget the price they pay in discontinuities in order to succeed in subsisting. The contrast between being-
as-being and being-as-other arises from this slight tremor, this hesitation. In fact, since the presence of other prepositions is lightly veiled
by the effect of habit, it is not so surprising that this hesitation and this
veil have ended up provoking the suspicion that something else had to
be sought "underneath" and "behind" appearances. Rather like a viewer
who, unable to determine how a sequence of images fixed on film can
produce continuity of movement, seeks the source of this movement
outside of the film and outside of the projection booth. For we have to
admit that there is only the subtlest nuance between on the one hand
what lies underneath, explaining the continuity of essences, and on the
other the simple smoothing operation carried out by habit. The latter in
no way explains continuity, but it defines another type of discontinuity,
a special one to be sure (but every mode is special!), through which the
phenomenon has to make the risky passage in order to subsist. A moment
of inattention, and we tip from mini-transcendence into the wrong transcendence, the one that requires a *salto mortale* to reach the substance
"behind" and "beyond" appearances.

And this is why philosophers have always felt that there was in continuity, in sameness, something undecided, veiled, incomplete, and, let's be blunt, something "not quite right," and that, as a result, it would be somehow lazy to stop there. The mistake certainly did not lie in this legitimate feeling of unease. We begin to go wrong (and especially to get the mistake wrong!) only if we claim that we can get out of our discomfort by backing up continuity with a genuine, solid substratum, that of substance *causa sui*. As if sameness had to be guaranteed by sameness.

It is here that, to "save appearances," philosophers began to invent the scenography of phenomena and reality, the world and the world beyond, immanence and transcendence. From a legitimate hesitation between "upstream" and "downstream" with respect to the same flow of beings, they created an incomprehensible and sterile scenography of a world that would collapse if it were not held together by an *other* world. Yes, this is one saying we can't argue with: "Appearances are deceitful."

By restoring a little of its ontological dignity to habit, the anthropologist can now revisit an opposition that has probably perturbed the reader. We have too often claimed that the Moderns did in practice the opposite of what they said. The trope was quite awkward, implying as it did that it was impossible for the actors— owing to false consciousness?—to say what they were doing. To be sure, the inquiry has explained why they tended to lose the thread of experience, on the one hand because of the confusion between knowledge and the known [REP · REF] and on the other because of the crack introduced by constructivism. We are now discovering a more charitable explanation: habit has the effect of rendering IMPLICIT the vast majority of courses of action, though the adjective EXPLICIT does not mean "formal" or "theoretical."

WE UNDERSTAND QUITE DIFFERENTLY, THEN, THE DISTANCE BETWEEN THEORY AND PRACTICE, ⊙

We no longer have to confuse making something explicit with imposing a difference between those who don't know what they're doing because they have "forgotten" the essence of Beauty, Truth, and Goodness, and those who know these things by way of "formal" knowledge. For habit, making explicit is simply to specify the key to reading that it veils while maintaining its presence through vigilant attention. This doesn't mean that we have to grasp every course of action according to the mode of reference alone, as Socrates requires of his interlocutors, while unduly exaggerating the empire of that mode. This false dichotomy between practical knowledge and formal knowledge is imposed by the Socratic question itself; this is what empties PRACTICE of all explicit knowledge. In fact, if you interrogate all the modes as if they necessarily had to produce a FORM—in the third sense of the term defined in Chapter 4—all modes will fail the test, including chains of reference [REF], an irony that ought to have struck that great master of irony! In itself, *the implicit lacks nothing.* It is not the mark of a defect that a philosophy seeking foundations would have to repair in order to keep practitioners from remaining in ignorance. The practitioners *would know* perfectly well what they are doing, if only we adjusted each of their samples to the principles of judgment, of veridiction, that are appropriate to them. Once again, the metalanguage of equipped and rectified knowledge abused by Double Click [REF · DC] is very ill suited to respecting all the other categories. For habit, it suffices that the key be at once specified and delicately omitted for a practice to have all the explicitation it is capable of having. Consequently, when we complain that the Moderns do not know how to account for their own riches, we are not trying to extend the critical question, the Socratic question, to their entire anthropology: we are asking, proposing, suggesting that they *no longer* raise that question, so that all the other keys can be made *explicit,* each according to its mode.

⊙ WHICH ALLOWS US TO DEFINE DOUBLE CLICK MORE CHARITABLY [HAB · DC].

Such a distinction between two definitions of the explicit may allow us also to rehabilitate the double-click information that we have been denigrating so insistently from the start, and that our ethnologist has set up as the Evil Genius of the Moderns, the one who has polluted all their sources of truth by inventing a single shibboleth borrowed from knowledge (and which doesn't even manage to understand knowledge!). She needed a template so manifestly false that it couldn't help but make the felicity conditions of each mode stand out, by contrast. But obviously, whether objective knowledge, technology, psyches, gods, even, or law, politics, or fiction were at issue, it was the search for a displacement without deformation that had to appear less well adapted and more ludicrous every time. By dint of pretending to torture all the modes on that narrow Procrustean bed, she had to end up abandoning the idea of forcing them to lie down on the same mattress.

But now we discover that Double Click, too, can be justified: this is what happens when habit has so well *aligned* the discontinuities that everything takes place *as if* we were seeing transports without deformation, simple DISPLACEMENTS. This is what we are saying without thinking about it when we say "all things being equal": that is *never* the case and it is *almost always* the case. It is indeed according to this double mode that things *happen*, that courses of action unfold smoothly at first—until the next crisis. Our heartbeat is regular; our household trash is picked up by the trash collectors; we follow the path without thinking about it any longer; when we press the switch, the light comes on; conversations flow easily; and when we double-click on the icon for a program, it opens. "It's working." "Everything's cool." The error is not that we trust Double Click—it's our whole life—but that we slip unwittingly from omission to forgetting. For if a crisis arrives—our heart beats too fast, the trash collectors go on strike, the trash-burning factory upsets its neighbors, we've strayed off the path, the fuses have blown, the computer is crashing—then we're really lost, unable to repair, start up again, find the branching points we missed. What was only a slight, legitimate veiling, a necessary omission, has been transformed into oblivion. There is no "manual restart." Without a restart, it's a catastrophe; there's only an automatic pilot in the plane now. Will we succeed in saving Double Click from himself? Can we make him aware of the dizzying quantity of mediations required for a mouse click to produce any effect at all? Can we reconcile him with his real ethology, that of the thousands of lines of code that had to be written at great expense so that a double click could actually produce an effect?

To treat Double Click and prevent him from leaving irrationality everywhere in his wake, it would be necessary to recognize the particular way each mode has of unfolding and folding back up, of making itself explicit and of "implicitating" itself. In fact, we have been a little negligent in claiming that mediations *always had to be deployed in the same way* for us to follow the networks, detecting the appropriate pathway every time [NET · PRE]. That would be too simple: each mode of existence has its own

EACH MODE HAS ITS OWN WAY OF PLAYING WITH HABITS.

way of unfolding and *refolding* itself. It is rather as though, on the pretext that the general category "arranging" exists, in the sense of "tidying up," we would begin to confuse the various ways of folding a fan, sheathing a sword, putting away a picnic table, or rolling up a tent.

For example, chains of reference obviously need to deploy each of their links in order to reach remote beings [REF]. This is true, but as long as a planetologist is in the process of setting up these links one after another, he sees nothing of the planets; it is only when he can finally omit all the intermediaries and retain just the two extremities—his computer screen and the image of the impact of a robot on Mars millions of kilometers away—that he begins to work for real. There is thus a mode of veiling that is particular to reference, so that one can *first* study the mediations and *then* bracket them because they are aligned thanks to the play of constants maintained from one form to the next. It is in the nature of a scientific instrument *not* to make visible the thousands of indispensable components that permit visibility. If a single one of these components fails, since the chain is worth no more than its smallest link, the instrument becomes worthless from the standpoint of reference. It is completely opaque.

But this mode of presence and absence differs entirely, to take another example, from what is also called an "instrument," for example in the arts [FIC]. A composer of electronic music who distorts the voice of a soprano by an ingenious treatment would see all his efforts lost if the listener were no longer sensitive to the set of subtly intermingled harmonics of the computer and the glottis. If there were only the two extremes [FIC · REF], the pleasure would have disappeared; one would have shifted from art to science. The way an artistic instrument [FIC] makes its components resonate is thus entirely different from the way a scientific instrument functions [REF]. And yet the composer would be very annoyed if his computer's motherboard had given out, or if a bad cold had left his soprano hoarse. The sudden visibility of some of the mediations would surely not be part of the "effects sought." On the other hand, it is what would allow the repairman, the engineer, the otolaryngologist to designate their points of intervention with certainty, for technology unquestionably has a third, entirely distinct mode for making its own mediations present or absent [TEC · FIC]. And, moreover, the same technicians would know how to repair the bugs in the scientific instrumentation [TEC · REF].

They would also know how to fix the broken-down microphone that prevents the faithful from hearing the pastor's sermon or the voices of the choir singing Bach chorales [TEC · REL]. But if you drew the conclusion that a Lutheran service is a "spectacle" on the pretext that the voices are admirable and the church magnificently lit, you would have made a new category mistake, this time by mistaking the pleasure procured by the staging of the mediations for the itinerary of the spirit of conversion [FIC · REL]. You would have "aestheticized" the worship service. This happened to Bach himself: his music was "too beautiful"; it made the shocked congregants shout out "Blasphemy!" But the blasphemy would be greater still, as we shall soon discover, if the faithful had profited from the music no longer in order to enjoy the arrangement of instruments and voices, but in order to pretend to reach the other world by behaving as if the music had transported them "far away," inducing them to commit the far more serious sacrilege of abandoning their less fortunate neighbors [REF · REL]. Go that route and you are no more than a "clanging cymbal," as Saint Paul says (1 Cor. 12:31). As we can see, "*forgetting being*" *is not a general category mistake:* each mode of existence requires that it be forgotten *in its own way.* Here we have a whole ethology of modes of existence that the Moderns have learned to recognize and that may be as subtle as that of an Amazonian ecosystem. These are manners that we too have to learn to respect.

It will be said that if habit is so important, it must have received a particularly careful treatment in modernism. Yet the opposite is true. Volumes have been written about the importance of science, technology, law, art, and religion for civilization, but habit has had only a few champions. Although it contributes so much to the *maintenance* of institutions, it has benefited, paradoxically, from extremely poor institutional returns. We might even say that the ever-so-subtle contrast of habit has not been instituted—except negatively. This is not going to help the Moderns come to grips with it.

THIS MODE OF EXISTENCE CAN HELP DEFINE INSTITUTIONS POSITIVELY, ⊙

For those who have really grasped the type of veridiction proper to habit, even routine actions can be taken either in the mode of (necessary) omission or in that of forgetting. Now, this distinction, internal to this particular mode of existence, must not be confused with another that has recently been substituted for it, one that is wholly external to it and even parasitical, one that *opposes* institutions—which it accuses of being routinized, artificial, bureaucratic, repetitive, and soulless—to the initiative, autonomy, enthusiasm, vivacity, inventivity, and naturalness of existence. Here again, we can recognize a slippage from good to bad transcendence. In the latter case, indeed, there is life only on condition of *getting out* of institutions, even destroying them, or, short of that, getting as far away from them as possible in order to subsist on the periphery and, as it were, "on their margins." A great iconoclastic temptation: before an institution that one can no longer mend, there is no solution but to raise up against it the vital forces of "spontaneity." Habit with all its appurtenances would have gone over to the forces of death; as for life, it would now be purely a matter of initiative, autonomy, freedom, and invention.

⊙ PROVIDED THAT WE TAKE INTO ACCOUNT THE GENERATION TO WHICH THE SPEAKER BELONGS ⊙

More than any of the other modes, habit offers a contrast whose tonality depends crucially on the historical moment. We have to approach this "MALIGN INVERSION" with some trepidation, for it depends on the generation to which one belongs. Taking generations into account may appear shocking in an inquiry that puts so much emphasis on the predecessors and successors of every course of action. On the topic at hand, we have no choice: readers and investigators along with the author are going to have to specify their pedigree while agreeing to speak in the first person.

I myself belong to the generation designated as baby boomers, at least until age has earned us the dreadful replacement moniker "golden agers." Without this indispensable reference point, it won't be possible to tell whether it is reactionary or not to propose, as I did in the introduction, that we should "learn to respect institutions." Unless we know the genealogical cluster in which you are located, it will be impossible to know, given that habit has so many enemies, whether you want to protect a value by instituting it or, on the contrary, whether you want to betray it, stifle it, break it down, ossify it. Now we baby boomers have drained that bitter cup to the dregs. Confronting the ruins of the institutions that we are beginning to bequeath to our descendants, am I the only one to feel the same embarrassment as asbestos manufacturers targeted by the criminal charges brought by workers suffering from lung cancer? In the beginning, the struggle against institutions seemed to be risk-free; it was modernizing and liberating—and even fun; like asbestos, it had only good qualities. But, like asbestos, alas, it also had disastrous consequences that no one had anticipated and that we have been far too slow to recognize.

In particular, it took me a long time to understand what effect such an attitude was going to have on the subsequent generations from whom we were threatening to *conceal* the secret of institutions owing to our own congestion (and also owing to our numbers and our appetites for living lavishly and for a long time). We expected these generations to *continue* (as we had?), through the vigor of their critical spirit, *to hold onto* the originality of their initiatives, their spontaneity, their enthusiasm, everything that institutions were no longer able (and no longer knew how) to keep going. This was to sin against blessed habit; it was to claim to be continuing institutions without offering any way to ensure continuity. We thought we were protecting values and contrasts by extracting them from institutions—from which we had profited before we destroyed them—like fishermen who claim to be saving fish from asphyxiation by bringing them out into the air. One little hypocrisy too many; we have to hope it won't be stamped on our foreheads on Judgment Day...

And here is the "malign inversion": by losing the thread of the means that could have ensured subsistence—habit being no longer able to ensure the relay—we have involuntarily pointed in the direction of a return to substance without specifying to the next generation that this return would be truly fatal, precisely for want of defining its *means of subsistence*. In Pierre Legendre's words (provided that we extend them to all the modes and not just to psyches), we have broken the "genealogical principle," that is, the search for antecedents and consequents. Being-as-other can gain its subsistence through the exploration of alterity, through multiplicity, through relations; it cannot ensure continuity by entrusting it to a substance. But without the scaffolding of habits, it cannot subsist at all! Here is where the trap closes, where the miracle product called asbestos begins to make the employees who breathe its microfibers cough their lungs out.

⊙ AND AVOID THE TEMPTATION
OF FUNDAMENTALISM. I may be overdramatizing the situation, but I cannot help thinking that if those who are starting to succeed us inadvertently sought to keep speaking of what is true or false, they would have no choice but to plunge headlong into a search for foundations, since institutions can no longer guarantee continuity. In other words, to those who, tired of spontaneity, are nevertheless still searching for truth, we have left no recourse but FUNDA-MENTALISM. Now all the contrasts I have talked about up to this point are lost forever if we set out in search of their "incontrovertible foundation": God, of course, as we shall see, but also law, science, the psychogenics, the frenzied world itself, in short, the *multiverse*. If the reader has grasped the weight, or rather the lightness, of habit, he has also understood that *there is nothing true except what is instituted*, thus what is *relative*: relative to the weight, the thickness, the complexity, the layering, the multiplicity, the heterogeneity of institutions; but relative especially to the always delicate detection of the leap, the threshold, the step, the pass necessary for its extension. Exactly what Double Click teaches us to miss. By confusing the rejuvenation of institutions with their dismantling, hasn't the baby-boomer generation made it possible to slip, almost unwittingly, from the critical spirit to fundamentalism? As if a first category mistake about blessed habit had triggered a second, infinitely more calamitous, concerning the radical distinction between what is true and what is instituted. The late modernism that thought it was digging the grave of its predecessors would thus have been digging its own grave!

I am well aware that we would be committing a new injustice, however, if we were to go on flagellating ourselves too long. If it is hard for our children to inherit our muddled passions, how could we have inherited the whole history of Modernism without difficulty? If it has seemed impossible for us to utter the words "truth" and "instituted" in the same breath, it is surely because of the lamentable state in which we had found the aforementioned institutions. If we have criticized them, it is surely because they had not been functioning for a long time—or at least because there was no longer a recipe adapted to their various regimes. If there were just one way to take habits, there would have been just one way to stand guard over institutions while keeping them from degenerating and tipping unnoticed from omission into forgetting. But as each mode has its own particular way of *letting itself be omitted* by habit, these are the differences that have made it so difficult for a civilization to provide the care that would have been required to maintain all the contrasts extracted by the ontological history of the Moderns.

If our predecessors had spent even a fraction of the energy devoted to the critique of institutions on differentiating all these cares, all these attentions, all these precautions, our generation would never have found itself before empty shells. But the very idea of care and precaution had become foreign to them, since they had hurled themselves blindly into this modernizing furor for which the time for care and attachments, as they saw it, had definitively passed. As if that archaic time were henceforth behind them and they had before them only the radiant future, defined precisely by a single emancipation, by the absence of precautions to be taken, this reign of irrational Reason whose cruel strangeness we have come to understand. I grant that it is hard for the young people born after us to inherit from the so-called May '68 generation; but can someone tell me what we were supposed to do with the legacies left behind by the generations of "August '14," "October '17," and "June '40"? Not an easy task, to inherit from the twentieth century! When will we be done with it? But we must try to be patient: once we have deployed all the modes, we shall know what we are to inherit and what we can, with a little luck, pass on to our descendants. In any case, in the face of what is coming, are not all generations, like all civilizations, equal in their ignorance?

ARRANGING THE MODES OF EXISTENCE

....

Wherein we encounter an unexpected problem of arrangement.

In the first group, neither Objects nor Subjects are involved.

Lines of force and lineages [REP] emphasize continuity, ☉ while the beings of metamorphosis [MET] emphasize difference ☉ and those of habit [HAB] emphasize dispatch.

A second group revolves around the quasi objects ☉ [TEC], [FIG], and [REF], originally levels n + 1 of enunciation, ☉ produced by a rebound effect at level n - 1.

This arrangement offers a conciliatory version of the old Subject/Object relation ☉ and thus another possible position for anthropogenesis.

As she begins to deploy the plurality of modes of existence, the anthropologist of the Moderns comes across an unexpected problem. Let us recall that she is trying to reconstruct their value system just as her colleagues have always aspired to do for more exotic terrains: by finally managing to reconstitute, all at once, the totality of these collectives' experience of the world. Thus she cannot avoid seeking to be systematic. At the same time, for situating courses of action, she can no longer rely on the system of coordinates between Object and Subject proposed by the Bifurcation, a system that had at least offered the convenience of defining the one through contradiction with the other. So she now finds herself obliged to propose a different system of coordinates that corresponds to a double constraint: it has to regroup all the modes at once, but without confusing them. In other words, she is going to have to invent a principle of *arrangement*. A daunting operation, and moreover completely outdated: who still believes in the possibility of a systematic philosophy, a systematic anthropology?

The investigator would have recoiled before the scope and even the ludicrousness of the task, if she didn't know that the goal of her classification principle would be to facilitate future negotiations. Since what is at stake in the first place is comparative anthropology, the principle hardly matters as long as it will allow her to designate the modes that lend themselves best, or worst, to confrontation and then to negotiation. Anyway, she has little choice: any alternative system of coordinates will be preferable to the current one—and preferable to the absence of all reference points. Moreover, a pitiless critique of Master Narratives will never keep readers from surreptitiously making one up for themselves. All things considered, she may as well propose a version that is at least compatible with the experiences collected here.

Let us first try to group the three modes of repro- duction [REP], metamorphosis [MET], and habit [HAB]. They have in common the fact that they explore, in being-as-other, three specific and complementary

IN THE FIRST GROUP, NEITHER OBJECTS NOR SUBJECTS ARE INVOLVED.

forms of alteration. Multiply persistences; multiply transformations; throw oneself headlong into existence: three ways of exploring being as alterity. Now here is the crucial point: even though they never pass through the Object box or the Subject box, they have DIRECTION. Direction, for us, is what precedes and what follows any entity whatsoever—its vector, its TRAJECTORY—as well as the PREPOSITION that spells out how we are to take what is to come. This first group is neither mute nor without sense. Even if it precedes the human infinitely, it is well articulated without in any way resembling an "external world" to which the world of the symbolic would be opposed. An entity is ARTICULATED, let us recall, if it must obtain conti- nuity through discontinuities, each of which is separated from the others by a juncture, a branching point, a risk to be taken—something called, precisely, an articulation.

Let us come at these three modes from another angle by restoring the movement that makes them mesh with one another. What is the status, for example, of the beings that persist via the risky abyss

LINES OF FORCE AND LINEAGES [REP] EMPHASIZE CONTINUITY, ⊙

of reproduction [REP]? They are mute, that goes without saying, since they precede articulated *language*. But it would be absurd to say that they are not articulated on that account. They unquestionably *enunciate them- selves* (in the etymological sense), since they thrust themselves into exis- tence across the ever-so-perilous hiatus of maintenance in persistence. Among those that are dismissively called inert and mute, what activity of ENUNCIATION! The LINES OF FORCE, inert beings, are wholly marks of enun- ciation, as it were, since the passage into another that is almost the same is defined by the insistence and the transformation of forces—which the sciences, much later on, will learn to define as energy [REP · REF]. Mute, persistent, obstinate, if you like, but surely not dunces just stubbornly "there." To exercise a force, unmistakably, is to *be determined to pass.* There is thus, in this mode of being, a particular *"pass"* through which Earthlings may slip, but on which they will never settle for "projecting" relations of identity and difference on the basis of the "categories of the human mind." If you become capable of speaking, you have to be able to insinuate yourself into itineraries of force that already resonate and enunciate themselves in you, overwhelming you on all sides.

While there may still be some hesitation about endowing inert beings with such a capacity for articulation, doubt is no longer an option with the LINEAGES of the living, since the proliferation that extracts what is almost the same—something we have learned to respect thanks to Darwin, this hiatus of reproduction, this miraculous continual disequilibrium of ethology—is never obtained by a sempiternal maintenance of sameness but by a repetition that is riskier each time. What a confounding "combination of circumstances" it takes for an organism to reproduce! Understanding the organism prepares us to grasp the *logos.* Through this process we are inevitably attached by those from whom we come and of whom we are, in the strict sense, the risky enunciation across the abyss of reproduction. They persist in us, but only if we ourselves manage to persist in others (almost) like us (and even to persist as ourselves, a little while longer, by dint of aging).

The word "enunciation" might appear exaggerated, unless we emphasize that we are dealing with a particular case in which the enunciator dispatches itself, sends itself off, persists, in *another enunciator* without ever being able to turn back, to retrace its steps. What matters is knowing how to determine the hiatus, the little transcendence, the articulation that allows us to locate the specificity of this mode of existence. At all events, it is not "Nature," it is not "the world," it is not "the PRELINGUISTIC."

⊙ WHILE THE BEINGS OF
METAMORPHOSIS [MET]
EMPHASIZE DIFFERENCE ⊙

The power of metamorphosis is another form of enunciation—a very strange one, to be sure. If the previous form explores a maximum of identities through the risks of reproduction, this one explores a maximum of transformations. These beings indubitably possess an ontological dignity that explains both their intimidating objectivity (they can change us into another at any time), their fecundity (they can change us into another at any time!), and, finally, their invisibility (they are as impossible to pin down as Proteus). Their institution under the auspices of the psyche defines them only vaguely, from afar; their institution by divinities seems less improbable; yet collectives other than our own have been able to invent hundreds of ways of welcoming them and profiting from their initiative. Moreover, haven't living beings themselves already explored this power of metamorphosis in the form of *mutations* [REP · MET]?

To use a linguistic metaphor, if the beings of reproduction define some kinds of syntagmas (lines of force for inert beings, lineages for the living), might we not say that the beings of metamorphosis define *paradigms*, possible series of transformations, vertiginous *trances*? We would then be sketching a matrix made of the crossings between horizontal lines—reproductions—and vertical lines—metamorphoses or substitutions. They would form the warp and the woof of which all the rest is woven. If, much later on, humans begin to speak, it is because they slip into these horizontal and vertical series that they could not have invented. If humans act and speak, it is because the worlds are already articulated in at least these two ways: *they reproduce, they metamorphose.*

Habit, too, may seem heterodox with respect to the canons of the theory of enunciation. And yet it is as if habit directed attention, for its part, toward

⊙ AND THOSE OF HABIT [HAB] EMPHASIZE DISPATCH.

utterances without insisting on their attachment to what has thrust them forward, what enunciates them. Habit consists, as we have just seen, in thrusting oneself into a course of action while veiling the dispatcher, as it were, but without completely omitting it. Habit would predispose us, in a way, to detach an utterance from its enunciation (to cut it out in advance). As if we had already prepared what would later become a SHIFTING. Here, too, there would be something like a prefiguration of the *logos*.

There is no point placing too much emphasis on this regrouping, but it nevertheless has the advantage of sheltering these modes against any accusation of being unarticulated, immanent, "thingified," external, natural, or—especially—prematurely unified. While this group is first in the order of our categorization, it is not "primitive" or "primary." These initial explorations of the alterations of beings-as-other are simply going to allow many other modes to hazard other alterations. These three, in any case, for want of speech, pass through one another other, let us say that they "parlother," a verb to be invented somewhere between palaver, parley, jabber, pass, speak, exist. "You're speaking from where?" "*From here*, necessarily from here."

It is astounding to comparative anthropology that these modes, despite their importance, are the ones that have been at once most elaborated by the other collectives and most ignored by our own. It is not so surprising that misunderstandings have piled up between the Moderns and those they purport to be modernizing. Would it be possible, today, to take up history differently by respecting these modes from here on? "Ah! *You too*, you have something to do with these beings. This is what is going to allow us to make contact with you. Here, yes, really, at last, we can begin to negotiate a bit more seriously." Whereas if we continue to think of them as prelinguistic, we shall never get out of the impasses of modernism, we shall never be civilized: we shall remain barbarians besieged by inhumans—and before Gaia we shall remain without a voice.

A SECOND GROUP REVOLVES AROUND THE QUASI OBJECTS ⊙

That takes care of the FIRST GROUP. Let us try to define a SECOND GROUP, the one that would revolve around objects. After all the trouble we have taken to avoid the difference between Objects and Subjects, the expression may be startling. But we are actually dealing here with QUASI OBJECTS, to borrow a term from Michel Serres. Up to now, our ethnologist has criticized the BIFURCATION as if it were a matter of a category mistake, a congenital malady afflicting the Moderns. This is a serious distortion of the deontology of a work that cannot settle for treating its informants as totally delirious! The time has come to situate more charitably the origin of a distinction that our investigator was right to call badly instituted, but she was wrong to act as though it were a matter of a mistake that could be corrected by paying a little more attention. There is indeed a distinction, but it has to do with quasi objects and QUASI SUBJECTS, the THIRD GROUP that we shall set forth below.

As we saw in Chapter 9, the beings of fiction [FIC] occupy an interesting position between those of technology [TEC], encountered in Chapter 8, and those of reference [REF], from Chapters 3 and 4. Without the first, it would be impossible to form or to figure anything at all. The beings of fiction have lent powers of delegation to the beings of technology, powers that have allowed the sciences, starting from a limited viewpoint that condemned them to blindness, to traverse the whole world and cover it with chains of reference paved from end to end with instruments [TEC · REF] and with delegated and domesticated virtual observers [FIC · REF]. Hence the idea of grouping these three modes together.

It is in this group that the most orthodox agencies, as elaborated by semiotics, are developed. With technological folds [TEC], we obtain what semioticians have taken for granted in narratives but ⊙ [TEC], [FIG], AND [REF], ORIGINALLY LEVELS n+1 OF ENUNCIATION, ⊙ what first has to be engendered by a very particular mode of existence: a shifting outward of the utterance, the speaker, and the addressee, a dislocation that would be completely impossible without the invention of technology. With this mode, renewal, and thus the proliferation of spaces, times, and actors, can truly begin. The baked earthenware pot *remains* in place even though the being that instituted it has disappeared, and it *addresses itself* to many uses and users besides its craftsman.

But what the beings of fiction alone allow, what no other mode could anticipate, is the fact that the figures dispatched in forms by shifting into other times, other spaces, and other actants are also capable, by retroaction, of *figuring* the speaker as well as the addressee. Works of art, it is true, sketch out *other worlds*—the only other worlds worthy of the name, perhaps—inhabited by characters that are visible only as long as the raw material of which they are made vibrates with forms. And if the work of fiction gives one additional fold to beings of technology, chains of reference [REF] are going to fold, twist, and translate them a *third time*, as we saw at the end of Chapter 9, in order to domesticate them and make them serve a novel purpose, one that is also totalizing: providing access to remote beings.

It is all the more important to stress the "quasi" in these expressions, in that each of the modes grouped here results in engendering, by a rebound ⊙ PRODUCED BY A REBOUND EFFECT AT LEVEL n-1. or recoil effect, particular forms of *subjectivities*. If this second group revolves entirely *around* fabricated things [TEC], dispatched things [FIC], or known things [REF], it is as though the quasi objects, by dint of turning, designated by default the places that potential subjects could come to fill later on. We saw this in Chapters 3 and 4: one *becomes* an objective mind little by little as chains of reference grow, since the mind is only one of the extremities, the telomere, to use a biological metaphor, of which the known object forms the other extremity [REF]. We encountered the same rebound effect in Chapter 8: competence, know-how, skill *come* to those who have to do with technological beings [TEC]. It is because these three modes turn around quasi objects that they produce, by a sort of centrifugal movement, original forms of subjectification: skills, creations, objectivities. Shifting produces both the n + 1 levels, ahead, and the n - 1 levels, behind.

Here we have a new example of subjects consisting of distinct layers that have been engendered by each mode of existence. We have seen that one could gain *interiority* through the attention finally paid to the beings of metamorphosis [MET]; and that we *persisted* thanks to the formidable leap of reproduction [REP]. We are going to see that we are *attached* to our utterances by the chains of law [LAW]; that we *give our opinions* thanks to the renewal of the Circle of representation [POL]; that we become PERSONS owing to the *present* of salvation-bearing angels [REL]. The reader will understand without difficulty that the ancient distinction between Subject and Object surely could not register the diversity of the successive layers necessary to the production of subjectivities thanks to the successive passages of the modes.

THIS ARRANGEMENT OFFERS A CONCILIATORY VERSION OF THE OLD SUBJECT/ OBJECT RELATION ☉

As we see, the Subject/Object opposition is troublesome only if we take these two terms as distinct ontological regions, whereas it is really only a matter of a slight difference between two groups, themselves composite, moreover—and *both are different from the first*, whose fully articulated character modernism had no way to grasp. Thus it ought to be possible to relocalize and, as a result, to mitigate this major issue of subjectivity and objectivity, before learning to reinstitute it in nonmodern institutions that are at last better adapted. For want of an appropriate metaphysics, perhaps the Moderns merely exaggerated, to the point of making an incontrovertible foundation out of something that should always have remained just a *convenience of organization:* some modes are more centripetal with respect to objects, others revolve more around subjects. Nothing to make a scene about; nothing that would make Nature begin to bifurcate!

However cobbled-together it may be, this arrangement makes it possible to shift the emergence of Earthlings slightly. In a multiverse repopulated by beings each of which goes its own way according to its own type of trajectory, it becomes less implausible to conceive of the birth of humans through a crossing of these beings, by interpolation, by amalgamation. It is not impossible to imagine animate beings becoming humans little by little, because they welcome these invisibles just as plants draw from the sun possibilities that that star didn't know it had. It is less astonishing that these Earthlings should have dispersed over the surface of the planet, according to their different ways of entering into contact with these beings, of giving them bread and salt, and, especially, of extracting contrasts from them. In other words, the diversity of "CULTURES" can no longer testify against the truth of their access to *what is real.* Reality and plurality are no longer necessarily opposed—something that is not without consequences for future diplomacy, for the invention of an alternative way to produce universals and for the various ways of finding oneself "on Earth."

⊙ AND THUS ANOTHER POSSIBLE POSITION FOR ANTHROPOGENESIS.

In any case, the hypothesis is worth exploring. In the long, tiresome quarrel over the chicken and the egg—which came first, the Subject or the Object?—perhaps something was left out: the beings that make us exist! Why not suppose that humans are the ones who made encounters and proceeded to instaurations? By joining forces and addressing ourselves in common to these beings, we may be able to open up with the other collectives a negotiation that the strange idea of a civilization that had "discovered" objectivity and subjectivity could not inaugurate. Most important, we can prepare ourselves to make other encounters together.

Especially because by listing the modes recognized up to now and by agreeing to grant them an order of precedence we see that the "modern human" (this is the canonical term used to describe Homo sapiens) perhaps begins only with the beings of fiction (this is what the paleontologists say, at least) and that it is probably only at the time of the overinvestment in equipped and rectified knowledge that one can begin to distinguish the originality of the true Moderns. Thus, by deploying the modes of existence according to this classification, we can sketch out an already more universalizable world: we share four of these modes with nonhumans, five with all the other collectives. Is this not a more engaging way to take the inventory of our own inheritance? And, above all, a less provincial way to prepare us to inhabit a world that has become common at last?

HOW TO REDEFINE THE COLLECTIVES

Chapter 11

WELCOMING THE BEINGS SENSITIVE TO THE WORD

....

If it is impossible not to speak of a religious mode, ⊙ we must not rely on the limits of the domain of Religion ⊙ but instead return to the experience of the love crisis ⊙ that allows us to discover angels bearing tumults of the soul, ⊙ provided that we distinguish between care and salvation as we explore their crossing [MET · REL].

We then discover a specific hiatus ⊙ that makes it possible to resume Speech ⊙ but without leaving the pathways of the rational.

The beings of religion [REL] have special specifications— ⊙ they appear and disappear— ⊙ and they have particularly discriminating felicity conditions ⊙ since they define a form of subsistence that is not based on any substance ⊙ but that is characterized by an alteration peculiar to it: "the time has come" ⊙ and by its own form of veridiction.

A powerful but fragile institution to be protected ⊙ as much against the misunderstandings of the [REL · PRE] crossing ⊙ as against those of the [MET · REL] crossing ⊙ and the [REF · REL] crossing, which produces unwarranted rationalizations.

Rationalization is what produces belief in belief ⊙ and causes the loss of both knowledge and faith, ⊙ leading to the loss of neighboring beings and remote ones alike ⊙ as well as to the superfluous invention of the supernatural.

Hence the importance of always specifying the terms of the metalanguage.

In PART TWO, WE LOOKED AT A NUMBER OF MODES THAT MAY HAVE APPEARED SOMEWHAT EXOTIC. THE GOAL WAS TO EXTRICATE OURSELVES AS THOROUGHLY AS POSSIBLE FROM THE NOTIONS OF Nature, Matter, Object, and Subject, so we could let the experience of the various modes be our guide. We are still only halfway there. In Part Three, we are going to approach modes that are closer at hand, ones that have been addressed more directly by the "human" and "social" sciences. For we still have to get around two major obstacles: the prevailing notions of SOCIETY and especially of ECONOMY, the most recalcitrant of all. Only then will we be able to make the collectives comparable to one another, without using any system of coordinates except that of modes of existence.

By using the term COLLECTIVE, as we have done up to now, instead of speaking of "culture," "society," or "civilization," we have already been able to emphasize the operation of gathering or composing, while simultaneously stressing the heterogeneity of the beings thus assembled. Let us recall that what allows networks [NET] to unfold is precisely the fact that they follow *associations*, whatever these may be. "The social," in ACTOR-NETWORK THEORY, does not define a material different from the rest, but rather a *weaving* of threads whose origins are necessarily varied. Thus, in this inquiry, "the social" is the concatenation of all the modes. But the inventory of these modes still remains to be completed. It is hard to imagine an ethnography that would not speak of religion or politics or law or the economy. These are the topics we have to tackle next.

As our anthropologist continues to inventory the inherited values to which her informants
cling, she knows she will have to face the question of religion sooner or later. She may put it off as
long as possible, but she will not be able to avoid it. Her lack of enthusiasm is understandable. On the one hand, there are so many passions,
so many splendid elaborations: traces of the passage of religious beings
still occupy all the old space of the old Europe; our languages, our arts,
everything is full of them. But on the other hand, where will she find the
patience to untangle the knots in which these beings have been jumbled
up and perhaps lost forever? A cascade of category mistakes seems to
have made them unspeakable, unpronounceable, the very same beings
that made their fathers speak, got their ancestors excited, led them to
move mountains—and commit more than one crime. How is she to get
her bearings here? And yet her method requires her to "speak well" about
these things—even to those for whom "religious matters" have become
incomprehensible. And also to those who believe, alas, that they understand these matters, although they seem to have lost their interpretive
key long ago. So many misunderstandings!

But at the same time, what a test for her method, if she were to
manage, in spite of everything, to make what these beings have to say
audible once again, in their own language; if she could offer them their
exact ontological tenor, alongside the others! After all, is it really more
difficult to rearticulate religion than to restore objective knowledge
[REF] to its rightful place, along with the multiple existents to which
this knowledge sometimes allows access [REP]? In any case, we have no
choice: how could we approach the tasks of diplomacy while requiring
the others, all the others, to renounce religion before having the right
to sit at the negotiating table? Religion presents itself as too universal
to be dismissed with an assertion that it is "behind us" forever. Clearly,
there's no way around it: for this contrast more than for all the others, the
diplomat has to rely on a precise inventory of what has happened to those
who gave her their mandate. What are we presumed to inherit? What are
they really clinging to, those who say they hold to religion and those who
say they don't?

In fact, we have already met these beings: they allowed us, in Chapter 1, to begin to define INSTITUTION by noting the meticulous care the Churches have always invested in the distinction between truth and falsity. We ran into them again in Chapter 6, in the serious charge brought against fetishes and idolators. It was in the surprising idea of a God "not made by human hands" that we situated one of the two sources of the speech impediments characteristic of the Moderns: to extract a contrast of decided novelty, they attacked another contrast, a wholly different one quite innocent of the crimes of which it was accused, namely, rituals generative of psyches [MET · REL]. And it was through the intersection with a very different problem, that of objective knowledge (deprived of its networks), that the Moderns, according to this analysis, reached the point where they couldn't declare in the same breath that what was well constructed could also be true [REF · REL]. Hence an astonishing consequence: they had made it impossible for theory and practice to mix. Always this embarrassment of riches that makes it so hard for the Moderns to maintain all the contrasts that they have extracted for want of a metaphysics and an anthropology adapted to their ambitions.

⊙ WE MUST NOT RELY ON THE LIMITS OF THE DOMAIN OF RELIGION ⊙ "At least," the ethnologist says to herself with satisfaction, "in the case of religion, I don't lack for an institution; it's the oldest, most meticulous, most widespread, most fastidious of all." However, she quickly has to change her tune. With the Moderns, nothing is simple; she ought to know this by now. Precisely because institutional religion was hegemonic for so long; because it took on responsibility for all domains— politics, morality, art, the cosmos, law, even the economy; because it believed it could extend itself to the entire planet as a universal form of "the religious": for all these reasons, it has never been able to make concessions at the right moment to preserve the contrast, the only contrast, that it should have been intent on instituting. It is even more astonishing that the institution has been demolished more forcibly by those who call themselves religious than by those who call themselves secular. Rotten luck for our investigator here: if there is one indication that she must not follow to define religious beings, it is the well-demarcated domain of Religion.

She must be all the more prepared to look elsewhere inasmuch as she has to set aside another parasitical phenomenon: "the return of the religious." Nothing would lead her further astray in her investigation than to go along with those who accept her project by saying: "Yes, of course, you're right, you can't inquire into contemporary values without looking into this universal phenomenon." The ethnologist has to reject this good-sense advice. It seems to her, indeed, that the beings of the Word are buried more deeply still by all the talk about the return of religion, the need to maintain it, or the need to see it disappear.

How can she be so sure of this? Quite simply—assuming one has grasped the principles of this inquiry—because this phenomenon is not defined by an original type of subsistence, of risk, but, on the contrary, by an often desperate quest for substance, guarantees, some substratum. For those who use the term "religion" are really appealing to another world! And this is exactly the opposite of what we are trying to identify. There is no other world—but there are worlds differently altered by each mode. The fact that people speak tremulously of "respecting transcendence" hardly encourages the ethnologist to take this phenomenon seriously, since she sees quite clearly here the wrong TRANSCENDENCE, the one that has IMMANENCE as its opposite rather than its synonym. What is so disagreeable in the appeal to the "supernatural" is that the "natural" is accepted in the same breath. And if someone speaks, in hushed tones, of "spirituality," we are warned that a peculiar idea of "materiality" has just been swallowed whole. Why should our investigator be concerned with those who raise their eyes toward Heaven to speak ill of the things of the Earth, of "rampant materialism," of "humanism": what do they know about matter, reason, the human?

What passes for religion today can offer only a particularly discouraging avatar of the quest for immobility, for the incontrovertible, the supreme, the ideal. Some have gone so far as to take religion as a quest for the absolute, and even as a nostalgic portal to the beyond! Religion turned into a "rampart against relativism" and a "supplement of soul" against the "secularization" and the "materialism" of "the world here below"! No targeting mistake is more spectacular than this one. Really?! All those treasures of intelligence and piety only to end up with this? Thousands of years of uninterrupted translations, continual variations, prodigious innovations, to end in a quest for foundations? How can anyone be so mistaken as to worship these false gods?

Even if she steps completely out of her role, the indignant investigator no longer even dares to call such a perversion a category mistake. "Heresy" would be a euphemism; should she speak of category horror? How puerile they seem to her, the ancient confusions between Yahweh and Baal or Moloch! The idolators would never have dared confuse their God with an undistorted transport, an immobile motor, an uncreated substance, a foundation: at least they knew that one could not institute Him without a path of alterations, interpretations, mediations. Fetishism is only a peccadillo alongside the idolatry in question here: the replacement of the religious by its exact opposite, the confusion of the relatively holy with the impious absolute. And this blasphemy is uttered in the temples themselves, at the heart of the churches, before the tabernacle, from the pulpit, under the wings of the Holy Spirit! Where are the prophets who could have spewed forth their anathemas against these pollutions, these ignominies, these abominations? Where are Jeremiah's tears, Isaiah's lamentations?

No, if the investigator wants to hold onto her sanity, she has to look for the religious outside the domain of religion. She has to hypothesize that what is called "the return of the religious" manifests only the return of FUNDAMENTALISM. And we can understand why. Incapable of situating multiform values in institutions made for them, reactionaries of various stripes fall back on an ersatz solution that seems superficially to "defend the values"—by placing them out of reach! Between this search for an ultimate foundation and the beings of religion, there is nothing more than a relation of synonymy. Here is where the investigation has to begin, even if such a decision augurs nothing good where future diplomacy is concerned: how to make our informants renounce religion (as well as BELIEF) so as to restore the beings of religion to their rightful place? How to convince them that learning to redirect attention is religion itself?

"Where your treasure is, there will your heart be also."

⊙ BUT INSTEAD RETURN TO THE EXPERIENCE OF THE LOVE CRISIS ⊙ But then what thread can we rely on to locate the presence of religion-bearing entities? The inquiry must return, as always, to experience itself, even if this seems quite remote from the domain officially recognized as religious. Let us recall that in Chapter 6 we were surprised at how hard it was to place metamorphoses [MET] at the heart of the arrangements and apparatuses capable of producing psyches; we had identified the enigma in which a character so obviously dependent on her *owners* (the "possessors" of those who were "possessed") viewed himself, as a good modernist, after the fashion of an original "self": native, primordial, autochthonous, and autonomous. How was it that a character who could access her own interiority only by entering into transactions with the beings of metamorphosis that had formed her dreamed only of naturalness, plenitude, and authenticity? The attitude struck us as almost suicidal, for the more the "self" wants to be full, well rounded, and complete, the *less* it can defend itself against transformations. If there is anything guaranteed to produce insanity, it is an autonomous "self," without attachments and without an owner; it will be left without care, without defense against attacks; it will encounter the beings of metamorphosis only as entities that threaten or betray it.

Now, if there is such a huge gap between two requirements as contradictory as welcoming beings capable of metamorphosing you on the one hand and searching for primordial authenticity on the other, we can diagnose a new category mistake for certain. There must have been some confusion between the character visited by the blind invisibles and an additional layer of subjectification, the one that produces PERSONS. There must be new beings that have the "soul" as their target and make no mistake in their aim—and with which it is impossible to compromise. This would explain why it is so hard for the Moderns to speak with respect about the beings that produce psyches: were they to do so, they would be taking those beings to be *others*. Here again, here as always, the ontological famine of their theory prevents the Moderns from restoring the varieties of their own experience.

Our ethnologist has decided to take her chances by approaching the most banal of interactions among friends, among intimates, between lovers. Here is where she hopes to detect the distinction between profiting from metamorphoses in order to *manufacture composite interiorities* on the one hand and feeling the passage of words that *make the subject exist as a unified person* on the other. Whereas the first metamorphoses are not aimed at us, even though they arouse and transport us if we don't manage to avoid feeling targeted, the second ones gather us in and straighten us up by *addressing* us unmistakably: "It's you, it's me, it's us." The nuance may be subtle, but it doesn't escape those who are involved in intimate interactions and who have to sort out this crossing in the tumult of crisis [MET · REL].

There are tests that make it possible either to avoid a curse or, on the contrary, to curse and condemn, with a word, a gesture, a ritual. To exit from a crisis, as we have seen, one has to set up a procedure—often a treatment apparatus, a therapy, whether invented by amateurs in the course of conversation or entrusted to specialists. In this sense, we are all sorcerers or spell-breakers to some extent, each of us coping as best we can; the result is always either a form of detachment between *aliens* and subjects—white magic—or, conversely, a sort of possession of *alienated* subjects by the forces that have been unleashed—black magic.

Now this is not at all what happens in love crises, when other words, other gestures, other rituals, although they seem almost the same on the surface, result in "bringing back together" the "persons" who said of themselves before the crisis that they were "not close"—even if people sometimes go from one form of crisis to another in tears and outbursts without even noticing it, given that the two passes look so much alike. Words of love have the particular feature of endowing the person to whom they are addressed with the existence and unity that person has lacked. "I felt far away"; "I didn't care about anything"; "I was as good as dead"; "Time was standing still; now I'm really here, present, beside you"; "Here we are, together"; "We've grown much closer." In this situation, the mistake concerning "address" would lie in believing that these words are not aimed at us, that they must be avoided, as if other words were in question, the ones that allow us to address the beings of metamorphosis. Here, on the contrary, to close one's ears to these words— or never to pronounce them for others—is to disappear for good, or to make the others disappear for good. What could be more miserable than never being the intended recipient of a loving word: how could anyone who had never received such gifts feel like a person? Who could feel like *someone* without having been addressed in this way? What wretchedness, never to have aroused anything but indifference! For we don't draw the certainty of existing and being close, of being unified and complete, from our own resources but from elsewhere: we receive it as an always unmerited gift that circulates through the narrow channel of these salutary words. Our experience as recipients of such gifts is what gives us the confidence to start over, again and again.

Very special words: words *that bear beings capable of renewing those to whom they are addressed.*

At the beginning of this inquiry, it would have been difficult to identify a layer of subjectivity that would be added in this way to all the others: the anchoring in an aboriginal subject would have been ⊙ THAT ALLOWS US TO DISCOVER ANGELS BEARING TUMULTS OF THE SOUL, ⊙ too strong. But now that subject has been completely unmoored, dislocated, distributed, divided up. It no longer serves as a hook. Never again as a point of departure. On the contrary, it is approached from all sides by those who seek to grasp it, seat it, or ensure it. If it holds up, this is because it has found the path of instauration for everything that is to come not in its inner sanctum but in its *outer sanctum*. Nothing prevents us now from recognizing other beings bearing other layers of subjectivity that are no longer characterized as capable of *fabricating* or possessing, but as capable of *saving* characters by transforming them into *persons*. These beings have the peculiar characteristic of bringing persons from remoteness to *proximity*, from death to *life*. Let us say, to use more direct language, that these words *resuscitate* those to whom they are addressed—in the etymological sense, that is, they arouse them anew, get them moving again.

It may appear astonishing to establish such a rapid bypass between the vocabulary of interactions between lovers and that of religion, quite rightly called "revealed"; what matters, however, is giving beings their true names. Throughout the tradition, those who bear not messages but tumults of the soul have been called *angels*. Double Click doesn't get this at all, of course, since he is after nothing but displacements, transfers of information [REL · DC]. "Messengers that transport messages with no content but transformations of persons? You must be out of your mind!" "Out of our minds," no. "Beside ourselves" would be more accurate. We can only believe that the Evil Genius has never received proofs of love! And that he is completely unaware of the prime example, painted and sculpted tens of thousands of times: that of the angel Gabriel whose address comes not only to overwhelm young Mary's soul but to make her give birth to life itself. There is no better way to define beings linked to a particular type of word capable of converting those to whom they are speaking. By greeting them, they save them and impregnate them. No one has ever been able to define the soul or even decide whether or not it exists, but no one can deny that this "whatever-it-is" lurches and vacillates in response to such words, to such mutual overwhelming. Who among us does not have locked in his heart the treasure of injunctions like these that he has given or received—without possessing them for all that?

⊙ PROVIDED THAT WE DISTINGUISH BETWEEN CARE AND SALVATION AS WE EXPLORE THEIR CROSSING [MET · REL].

If this hypothesis is correct, we can understand why it is as hard to institute angels as to institute psyches. But there is a constant risk of interpolating, confusing the two, failing to respect the contrasts. To care for is not to save. To initiate the circulation of psychogenics is not at all the same thing as letting oneself be overwhelmed by angels—and not at all the same thing, either, of course, as obtaining verified knowledge [REF]. We may want to discover the truth about subjects, but, since there are numerous beings productive of subjectivities, there are necessarily numerous such truths. Anyone who is not prepared to control the traffic among all these beings *without confusing them* should not be talking about interiority or subjectivity—should not claim to be a psychologist.

Which does not mean that the same places, the same priests, the same rituals, the same passes cannot serve both purposes. It suffices to enter a sanctuary: we sense that people often come there in search of both cure and salvation. A single statue of the Black Madonna can offer both; a single pilgrimage to Lourdes rarely cures, but it often converts. When we have drawn up the specifications of the various modes more systematically, we shall often encounter troubling resemblances between such varied requirements; these will allow us to become sensitive not only to INTERPOLATIONS but also to AMALGAMATIONS, syncretism, and, finally, HARMONICS. But it is not because subjects are multilayered, as it were, that we have to blend their components. There is invisibility and invisibility. Learning to encounter *aliens* does not necessarily mean expecting salvation from them.

The investigator would like someone to help her identify the angels bearing salvation. She is a little annoyed with herself for having to limit them for the moment to the intimacy of love crises. But she has to resign herself: she cannot count on getting any assistance. Just as epistemology could not help her define objective knowledge, theology cannot be relied on to help her speak correctly about salvation-bearing beings. We shall soon see why: Double Click has struck both down with a single blow, obliging theologians to escape into belief through an erroneous conception of knowledge. It would be hard to do worse! And yet our investigator clings to her method, knowing that she has to take this new HIATUS as proof of a mode of existence that is truly sui generis. In all the sermons about the beyond, about the "afterlife," there must be something that can sometimes ring true. How can anyone overlook this distinction between the far and the near, the dead and the living, the disappearance and the appearance of persons, salvation and loss? Here is surely a contrast that must not be attenuated. If the word VALUE has any meaning, surely it lies here.

WE THEN DISCOVER A
SPECIFIC HIATUS ⊙

And this is the point at which the astonishing miracle always occurs; she absolutely has to record it in the SPECIFICATIONS of these beings. No sooner has she reformulated in these banal terms the little difference that identifies the words capable of arousing persons anew, no sooner has she set aside the aid of theology, than she suddenly finds herself at a surprising branching point with the most technical vocabulary of religion—at least of the Christian religion, the only one she is trying to capture for the moment. And yet how she quaked, initially, at the idea of looking into love crises for traces of religious beings! She was doubly afraid: that this audacious branching would sound like an irreverent move destined to reduce the "great ideas" of religion to sentimental crises; or, conversely, that she would seem to be seeking, through a sort of apologetic perversion, to capitalize on secular love so as spout nonsense about sacred love. But no, she has no reason to tremble, for the most explicit words of the tradition have never done anything but emphasize the very contrast she is trying to specify. In other words, at the very moment when she despaired of theology, she finds herself perfectly *aligned* with the tradition. Yes, *faithful*. No ethnology can rival in explicitness the know-how of believers. "There is only one love," says the Portuguese nun. Our investigator no longer has a choice, she's been nailed: she has to reconsider everything all over again, certainties and doubts both, all equally ill-placed.

⊙ THAT MAKES IT POSSIBLE TO RESUME Speech ⊙   And this is fitting, because religion is REPRISE par excellence, the ceaseless renewal of speech by speech itself. This is its own enlightenment: it starts over, it begins again, it goes back to the starting point time after time, it repeats itself, it improvises, it innovates: moreover, it never stops describing itself, self-reflexively, as Word.

If there is one mode of existence that ought to be at home in natural language, it is the religious mode; they share the same fluidity, the same simplicity, the same flux, the same flow—as Péguy, "saint Charles" Péguy, grasped with great precision—as long as it starts over, repeats itself. This is what makes it comprehensible to the simpleminded and to children, and what conceals it right away from the wise and the scholarly, for the latter do not want speech to flow, they want it to transport the literal with no distortion, they want it to speak of something *that does not speak*, or to spout out words that do not save those who hear them. Here it is, flowing again, overflowing, passing by way of parables, plunging into rituals, getting a grip on itself in a sermon, snaking around in a prayer, circulating indifferently at first and then suddenly converting, just like that. It follows the thread of the Word itself. The Thread that is said to be "the Way, the Truth, and the Life" (John 14:6).

Why does speech always have to *be renewed*? Well, because this Logos cannot rely on *any substance* to ensure continuity in being. It too is then a mode of existence, a particular form of alteration of being, and that is why it is so far removed from "the religious phenomenon" or from the institution of religion, alas, as it has belatedly consented to define itself. The very institution that for such a long time had understood this linking between reprise, invention, and expansion of its content— before brutally interrupting its course. With no other institution do we find ourselves facing this contradiction between a loss that seems total and definitive and, simultaneously, the overwhelming ease with which one can rediscover what religion has always sought to transmit and bear. As if the relation between the proximate and the remote were part of its very definition: it has lost the world, yet what it says is within arm's reach.

With religious beings as with psychic beings   ⊕ BUT WITHOUT LEAVING THE
or the beings of fiction, the danger lies in losing the   PATHWAYS OF THE RATIONAL.
thread of reasons by agreeing too quickly to view
them as "irrational." The fact that Double Click doesn't manage to follow
them doesn't mean that they are illogical or mysterious. Our investigator
knows that she has to be suspicious of those who want to set "limits" to
the rational when religion is under discussion. As if one had the choice
of mode of reasoning! If it is quite true that there are several modes of
existence, we still have just one and the same reason for following them.
No one has an extra brain. I hope the reader will do me justice on this
point: not once in this inquiry have I required anyone to give up the most
ordinary logic; I have only asked that, with the *same* ordinary reasoning,
the same natural language, they follow *other* threads. As we shall see, it
is precisely because the religious institution has used and abused the
"limits of the rational" that the ethnologist must not bat an eye when she
has to define the nature of these beings. They are rational through and
through. Like psyches. Like fictions. Like references.

There is a risk, obviously, that this requirement to treat religion
rationally will be mistaken for a return to the critical spirit, that is, to the
good old "good sense" of the social sciences. But it should be clear by now
that we can expect nothing at all from the "social explanation" of religion, which would amount to losing the thread of the salvation-bearers
by breaking it and *replacing it with another*, while seeking to prove that
"behind" religion there is, for example, "society," "carefully concealed"
but "reversed" and "disguised." Such an "explanation" would amount
to losing religion, to be sure, but also to betraying the very notion of the
rational—not to mention that we would not understand anything about
"the social," either. There is nothing "behind" religion—no more than
there is anything at all interesting "behind" fiction, law, science, and so
on, for that matter, since each mode is its own explanation, complete in
its kind. The social consists of all of them together.

To remain rational, we just have to spell out as   THE BEINGS OF RELIGION
precisely as possible the specifications as well as the   [REL] HAVE SPECIAL
type of veridiction one can expect of such beings   SPECIFICATIONS— ⊙
without losing the *direction* they give to what is to
follow or the clarification of the conditions through which they are to be

taken. It would not be of much use to say that religious beings [REL] are "only words," since the words in question transport beings that convert, resuscitate, and save persons. Thus they are truly beings; there's really no reason to doubt this. They come from outside, they grip us, dwell in us, talk to us, invite us; we address them, pray to them, beseech them.

By granting them their own ontological status, we can already advance quite far in our respect for experience. We shall no longer have to deny thousands of years of testimony; we shall no longer need to assert sanctimoniously that all the prophets, all the martyrs, all the exegetes, all the faithful have "deceived themselves" in "mistaking" for real beings what were "in fact nothing but" words or brain waves—representations, in any case. Fortunately, investigators no longer have to commit such reductions (not to say such sins!), since we finally benefit from a sufficiently emptied-out universe to make room not only for the invisible bearers of psyches but also for the pathways of alteration—we can even call them networks—that allow the *processions of angels*, the conversion-bearers, to proceed on their way. Where a psychophor has managed to pass, another soul-bearer ought to be able to slip in . . .

And we also know that there is no ontological laxity in this attitude, but simple respect for the plurality of experience. It appears infinitely simpler, more economical, more elegant, too, to stick to the testimony of the saints, the mystics, the confessors, and the faithful, in order to direct our attention toward *that toward which* they direct theirs: beings come to them and demand that they be instituted by them. But these beings have the peculiar feature of *appearing* to those whose souls they overwhelm in saving them, in resuscitating them. If we are to be empirical, then, these are the ones we must follow. What was impossible with interiority becomes respectable with the modes of existence. Thus no one will be surprised that angels can follow different paths from technologies, demons, or figures. The world has become vast enough to hold them all.

⊙ THEY APPEAR AND DISAPPEAR—⊙ Especially because their demands are very specific. Indeed, nowhere in the specifications of religious beings do we find an obligation to imitate the type of persistence manifested by tables [TEC] or cats [REP]. Like the beings of metamorphosis, religious beings belong to a genre "susceptible to being turned on and off." With one difference: if they appear—and our cities and countrysides are still dotted with sanctuaries erected to harbor the emotions these apparitions have aroused—they *disappear* even more surely. Moreover, this intermittence has provided the basis for mockery, and has been taken as proof of their lack of being, of their phantasmatic and illusory character; the critical spirit has not held back in this regard. But the big advantage of an inquiry into modes of existence is that it can, on the contrary, *include* this feature in the specifications: one of the characteristics of religious beings is that *neither their appearance nor their disappearance can be controlled.*

In this they differ radically from metamorphoses: one can neither deceive them nor deflect them nor enter into any sort of transaction with them [MET · REL]. What matters to them, apparently, is that no one ever be exactly *assured* of their presence; one must go through the process again and again to be quite confident that one has seen them, sensed them, prayed to them. Another difference from metamorphoses: with religious beings, the initiative *comes from them*, and *we* are indeed the ones they are targeting. They are never mistaken about us, even if we constantly risk being mistaken about them; they never take us "for another," but they invite us to live in *another*—totally different—way. This is what is called, accurately enough, a "conversion."

Clearly, then, even if it is hard to distinguish with certainty between beings of religion and beings of metamorphosis, their ontology cannot be the same. Moreover, this is why it would be prudent to use the name DIVINITIES for the powers of transaction through which one addresses the beings that bear psyches, reserving the name GODS for the beings with whom no transaction is possible, the ones that come not to treat but to save. If there is reason to worry, as we shall see, about "taking the name of God in vain," there is no reason to abstain from using a word that has nothing irrational about it, that does not refer to anything supernatural, that is not there to solve any sort of problem of foundations, but simply to designate the trajectory left in its wake by a particular type of being that can be recognized by a particular mark, by this hiatus of conversion through the word that saves and resuscitates.

The word "God" cannot designate a substance; it designates, rather, the renewal of a subsistence that is constantly at risk, and even, as it were, the pathway of this reprise, at once word and being, *logos*. It can only be said with fear and trembling, for the expression ought to be given its full weight of realism: these entities have the peculiar feature of being *ways of speaking*. If you fail to find the right manner of speaking them, of speaking well of them, if you do not express them in the right tone, the right tonality, you strip them of all content. Merely ways of speaking? Doesn't this deprive them of any ontological basis? On the contrary, it is a terrifying requirement that ought to silence hundreds of thousands of sermons, doxologies, and other preachings: if you speak without converting, you *say* nothing. Worse, you sin against the Spirit.

⊙ AND THEY HAVE PARTICULARLY DISCRIMINATING FELICITY CONDITIONS ⊙   All the testimony agrees on this point: the appearance of such beings depends on an *interpretation* so delicate that one lives constantly at risk and in fear of lying about them; and, in lying, *mistaking them for another*—for a demon, a sensory illusion, an emotion, a foundation. Fear of committing a category mistake is what keeps the faithful in suspense. Not once, in the Scriptures, do we find traces of someone who was called who could say he was sure, really sure, that the beings of the Word were there and that he had really understood what they wanted of him. Except for the sinners. This is even the criterion of truth, the most decisive shibboleth: the faithful tremble at the idea of being mistaken, while infidels do not. Exactly the chiasmus that the transmigration of religion into fundamentalism has lost, replacing it by a differentiation—as impossible as it is absolute—between those who believe and those who do not!

It is because the nuance is so subtle that we have to be especially careful not to confuse the felicity and infelicity conditions owing to which we are going to be able to understand what is coming. "Watch out! Be careful! He *is no longer* here. See the place where they laid him," the angel says to the women who have come to seek Him in a tomb: "What? You haven't yet understood what it is all about? Why do you look for the living among the dead?" The Scriptures are only an immense hesitation about how to comprehend a message whose distinctive feature is that it transports no information and requires that it always be given a new direction in order to correct its interpretation. The Good News does not inform. Zero informational content. It's enough to make Double Click starve to death.

We see to what extent the point is being missed when someone asserts that in religious matters "you can say anything you like": no other regime of veridiction distinguishes the true from the false, speaking well from speaking *badly*, in such a radical way. The paintings in our museums, the tympana of our churches ought to have made us familiar with that difference, the one that distinguishes Heaven from Hell, salvation from perdition. The angel Gabriel carries our souls away and weighs them. It would be hard to be clearer, more decisive, more radical. The image is naïve, but how could one speak more simply of the requirement of conversion? "He who has ears to hear, let him hear." No one is afraid of Hell today? Very well, but one must still tremble before the prospect of confusing true religion with false versions. "Thou shalt not take the Name of the Lord thy God in vain." And this is the regime about which the purported "rationalists" assert so loftily that it is "indifferent" to truth and falsity alike, that it proves that reason has been abandoned! Some of them, through bland condescendence, are even ready to attribute this abandonment to the "limits of reason" or the "necessity of defending values." How low they stoop, with this false respect!

If we want to understand the continual risk run by the Word that saves, we can turn, strangely enough, to the beings of reproduction [REP · REL] to find a powerful resemblance. Let us recall that lines ⊙ SINCE THEY DEFINE A FORM OF SUBSISTENCE THAT IS NOT BASED ON ANY SUBSTANCE ⊙ of force and lineages do not rely on any substance, either, to exist—stones no more than chairs, electrons, horses, obelisks, languages, or bodies. Continuity over the long term is obtained, here too, by a particular form of discontinuity that we have called persistence. It would thus be absurd to ask religious beings to rely on more durable, more assured, more immediate, more continuous substances while the beings of reproduction continue to risk their existence by stunning acrobatic discontinuities. The idea of something "supernatural" is clearly ludicrous—as if there were something "natural"! Instead, we have to use one of Occam's well-honed razors to specify the type of discontinuity through which the beings of the religious mode have to pass in order to win, to warrant and resume, their continuity, their subsistence.

⊙ BUT THAT IS
CHARACTERIZED BY AN
ALTERATION PECULIAR TO
IT: "THE TIME HAS COME" ⊙

In the preserve of being-as-other, if we may use these terms, let us ask the following question: what then is the ALTERATION on which the religious is going to *draw*? This will allow us to understand why it is such a special mode of existence, why it has been so overinvested, until the recent past, and why it has been so unsuccessful at resisting the category mistakes that have multiplied since the dawn of modern times with the emergence of other modes—and especially owing to the irruption of Science.

It is true that this way of grasping alteration is quite peculiar. We can understand how the extraction of this contrast must have over-whelmed our fathers and mothers, made them weep for joy, let them see the sky open up and legions of angels descend upon them: *alterity can be final*, it can come to an end. "The time has come." There is no substance, and yet subsistence also procures something like a goal, something defin-itive, in any case something of an *ending*, some sense, some promise of plenitude. Which is translated, very awkwardly to be sure, by "eternity" and "eternal life"—but in time, always taken up again in time. A stupe-fying contradiction—all those who have prayed have been dazzled by it, and they have given it the most fitting names they could find: Presence, Creation, Salvation, Grace. An astonishing innovation in alteration, difference in difference, reprise in reprise: one can have the passage by way of the other and the definitive acquisition of salvation at the same time—as long as one always *begins again*. Here is the crux, the junction point, the decisive innovation: it is in time and it escapes time without escaping it; above all, it does not abandon time. An astounding develop-ment worked under the name *Incarnation*.

How can we give the full measure of this drama? On the one hand it has nothing to do with substance—a trap into which the religious fall as soon as they begin to doubt the promises of this Word—and on the other the discontinuity leads to something of substance after all, and this explains the trap, for what subsists is the *same*, always renewed. If it is not renewed, it is lost, hidden; if it is renewed, it is indeed in fact the same. Isn't it the very name of Reprise that is given to "God": "The one that is, that was, and that is to come"? An overwhelming contrast indeed. Enough to provoke a flight into the desert. The sacrifice of one's life. Time gives everything, including salvation. Outside time? Of course not. In time *differently grasped*: "The Spirit will renew the face of the Earth."

What is most striking in the specifications of the beings of the religious mode is their fragility: they depend continually on the refreshment of the interpretation that makes it possible to restate exactly the same thing that had been said before ("Then you have not understood what our Fathers believed?"), yet if anyone claims to be "maintaining intact" the treasure of what has been said without transporting it a step further by a new discontinuity, he is lost. Manna poisons as soon as someone tries to hoard it. Very quickly, it threatens to authorize every crime: woe to those who do not believe!

Of all the modes we have identified, this one is the most durable but also the most open to misunderstandings, the most apt to turn into poison, even into rage. While it has unceasingly developed, innovated, spread across the Earth by maintaining constant vigilance over these misunderstandings, it has gradually become incomprehensible in its own eyes. Kilometers of theological elaborations; thousands of lives devoted to this discovery; hundreds of rituals; texts to bring tears to one's eyes every Sunday. Then, nothing more. No more picture. No more sound. Interruption of service. No more networks. Inanities. Platitudes. Preaching. Religions as war.

It is this instability that explains why people have never stopped interpreting the Scriptures, revisiting them, patching them up, commenting ⊙ AND BY ITS OWN FORM OF VERIDICTION. on them. One cannot simultaneously speak religiously and, as they say, "preserve the treasure of Faith." Groping, contradictory exegesis: this is religion itself. Etymology attests to this: religion is the relationship among or, better still, the *relativism of interpretations*; the certainty that one obtains truth only through a new path of alterations, inventions, deviations that make it possible to obtain, or not, against rote reiteration and wear and tear, the faithful renewal of what has been said—at the risk of losing one's soul. Betrayal by reiteration, betrayal by deviation. Between the two, the risk of reprise. There isn't a single prophet, saint, martyr, confessor, or reformer who is not defined by the renewal of this contrast, which had been lost and which finds itself (provisionally) revived at last.

It's enough to make one weep with pity: there was nothing in the religious mode to nourish the two fanaticisms, that of an uninterpreted and absolute word, and that of a total word that would be at once the underpinning and the superstructure of all the others. Can we forgive Modernism for this failing? How can one resign oneself to losing a difference as essential as the difference between life and death, salvation and damnation, the passage of time and its completion, risky subsistence and the plenitude of times? How numerous they are, the paths to perdition, the "ways to Hell" ... In extracting psyches, we wondered how one could care for the Moderns without an institution that conformed to their dealings with metamorphoses: but how can they be *saved* if they have managed to lose this message as well?

A POWERFUL BUT FRAGILE INSTITUTION TO BE PROTECTED ☉

If fragility with regard to the precise conditions of enunciation is such an intimate part of the definition of this contrast, we may wonder why those who held to it so strongly have not developed the most cautious and most subtle of institutions for it. But in fact they have done just that! The whole story of Judaism, Catholicism, Christianity, and all their variants attests quite strikingly to the means that an institution can give itself in order to keep on preserving, despite continual crises and over immensely long periods of time, a contrast so contradictory that one can say "the time has come" while time is still going on; "God is here" whereas he is coming; "God is going to come" whereas he has come; and so on, in an uninterrupted chain—constantly broken—of renewals, conversions, reinterpretations, innovations, and fruitful or fatal betrayals. A practical *polytheism*, in total contradiction (this should not surprise us) with nominal *monotheism*: the latter figure, frozen by so-called rational theology in the word "God," has become the keystone of a substance put there to resolve the metaphysical problems that themselves result from a cascade of category mistakes.

If that "God" is immediately *emptied of its substance*, it is precisely because it has been turned into a substance that is supposed to persist "underneath" the renewal of interpretation, and especially—the supreme sin—*whatever* the quality of that interpretation and that renewal may be. No matter *how one speaks of it*. Whereas this "substance" is extremely sensitive to speech; with each word, we lose or gain the sense of what we are saying depending on whether we are pronouncing it well or badly. If there is an error to avoid in speaking "about God," it lies in separating continuity from discontinuity, repetition from difference, tradition from invention, subsistence from its renewal, monotheism from polytheism, transcendence from holy immanence. And it is indeed that fatal error that religions were trying to avoid in sketching out, quite explicitly, by successive innovations, the admirable series, yes, let's venture the word, this *network* of transformations: "Holy nation," "God," "Son," "Spirit," "Church," this chain of renewals and wild inventions through which only the continuity of a message could be retained without any content but the reprise itself. It's all right there, before your eyes, and you see nothing? You continue to see nothing? As in those paintings of the boy Jesus among the teachers in the temple: before the same arcane texts that everyone was leafing through, He reads and understands; they look elsewhere and continue to understand nothing.

It will be said that this is too particular an ontology to resist the critics' teeth for long. But no, there is nothing extraordinary here, it's simply what is special about this mode *in its genre*: "God" is to this network what objectivity is to chains of reference, the law to legal passages [LAW], persistence to the beings of reproduction [REP]. "God" is the name given to what circulates within this procession, if, and *only if*, all the rest of the betrayals, translations, fidelities, inventions are in place. In this sense, it is true, "God" has no special privilege, is not located in addition to or beyond other beings [REL · NET]. "God" has nothing with which to judge them. And this is indeed indicated by the very movement of the Trinity, which has been turned into an unfathomable mystery, a quiz for mad metaphysicians, whereas it only indicates with admirable precision the trajectory of successive reinterpretations of beings sensitive to the Word, to the way of speaking: God, taken up again by Jesus, again by the Spirit, again by the Church—and perhaps lost by the Church. Hard to be more explicit. Hard to be simpler.

⊕ AS MUCH AGAINST THE MISUNDERSTANDINGS OF THE [REL · PRE] CROSSING ⊙

The error of religion, in the Western context, was probably to make the Church take in too many modes and to establish it as a meta-institution. The Roman Empire must have weighed rather heavily on its shoulders; this is still the case today, with the rather mad idea of making religion serve as a pillar for bioethics, morality, social doctrine, canon law, the education of children, the vice squad . . . Whereas we understand now that *no mode* can be singled out to contain all the others, to serve as a metalanguage for all modes. A new category mistake appears to have been made here, this time with the mode that collects prepositions [REL · PRE], as if the beings of religion could be expected to ensure the meticulous maintenance of all other beings. But it is true that each mode requires hegemony for itself, or at least will have a mistaken view of all the others, which it cannot help but misunderstand. As we shall see later on, if we can rely on prepositions to protect the multiplicity of modes, it will certainly not be because of a totalizing power like the power that knowledge sought to gain and that theology had coveted earlier, because prepositions offer no sort of foundation: they say nothing about what is to follow. This is their great virtue, if not the way they must be taken, the language in which they must be heard. We shall lean on this fragile hope in an attempt to protect *all* the modes.

Nothing is more fatal than requiring of religion that it be the "whole" world; just like every other mode of existence, this one can do no more than add a thread, leaving behind as many empty spaces as full ones. We lose it if it sets out in its turn to "fill in the blanks," supply "padding," in an effort to cover all realities on the pretext that it has succeeded (but always through a new reprise) in maintaining sameness and achieving the ultimate goal, reaching the end (except that there is no end!). Like all the other modes, it deploys its own particular interpretation of the totality. It has only this one distinctive feature, but one that is entirely *of its own making*: the time has come *in spite of* and *thanks to* the passage of time. No other mode can offer this; no institution but religion has extracted this contrast, cherished it, amplified it, preserved it. Can it be brought to light once again? In modernism, it could have been saved only if the Moderns had limited themselves to this contrast alone, and this is what they didn't know how to do. Their fatal error was to believe that either the contrast had to agree to disappear or it had to combat "secularization" (as if there had ever been a *saeculum*, a century, in which it was at risk of dissolution!).

⊙ AS AGAINST THOSE OF THE [MET · REL] CROSSING ⊙

Before we can understand how such an important contrast could have been shunted aside by another, we have to take another look at the category mistake that was responsible, as we saw in Chapter 6, for the impossibility of CONSTRUCTIVISM. The moment it appeared, the religious contrast defined a radical break in the order of time: "Yes, the time has come"; "Salvation is possible, and not just care"; "There are beings that bear Salvation"; "These beings differ from all the divinities." But, in an astonishing detour, this contrast was translated into a squabble over the *form* of statues and images: "Our God is not made by human hands." And the accusation was directed against idols! These were made out to be obstacles standing on the path of Truth and blocking access. The idol-wrecker's hammer began to strike indiscriminately—whereas the only difference that needed to be extracted was the one that reconciled two forms of risk: the risk of losing the powers of metamorphosis, and the risk of losing the end times.

This is the contrast that the "Mosaic division"—of which the Moderns find themselves the proscribed heirs—was attempting to extract, however awkwardly. No, in fact, God is not made by human hands, but this is not at all because He is opposed to idols, to divinities; it is simply because He comes to arouse anew the persons to whom He addresses Himself, and this addressing has to be constantly renewed because He escapes all substance. Not being a substance is not at all the same thing as not being fabricated by a craftsman's hands. The idols are innocent of the crimes with which they are charged. They've been knocked down by mistake!

Can this contrast be reinstituted? Is reinstauration possible for the God of mediations—the one so aptly named Revelation? Probably not—unless we note that the difficulties did not really arise until the moment an effort was made to connect, in the double Bifurcation identified in Chapter 6, the question of knowledge with question of Scriptural interpretation, the moment that is known as the "scientific revolution." A curious blend: science and religion, amalgamated in Reason and thrust into a general planet-wide clearing-away of all idols, all "false gods." The Moderns know something about false gods, indeed... The irony is bitter: by mistaking its target at the moment it appeared, this contrast underwent, and was unable to resist, many centuries later, an attack just as unfair as the one to which it was subjected by the emergence of objective

knowledge. A double *iconoclash*: once the idols have been destroyed, it
was the icons' turn to experience the same fate.

⊙ AND THE [REF · REL] Even though our ethnologist is accustomed
CROSSING, WHICH to the distance between the values of the Moderns
PRODUCES UNWARRANTED and their institutional rendering, she is afraid of
RATIONALIZATIONS. losing her way, so extreme is the difference between
what she is discovering about religion—which is, as
always, tradition itself, simply renewed, without any irrationality whatsoever—and what she is told about religion, whether good or bad. Isn't
there something unacceptably implausible here? This is why we have
often stressed the importance, in this inquiry, of the *moment* when one
approaches a conflict of values. It may have become politically incorrect to date our era "before and after Jesus Christ," but we cannot say
anything about the religious mode without specifying whether we are
talking about what happened *before* or *after* the scientific revolution.

Here we see at work a phenomenon we have already encountered, RATIONALIZATION, which is the exact opposite of the rational.
Rationalization is, if we may put it this way, a nonsensical response to a
question that is itself meaningless, a sort of headlong rush toward falsity
or, to put it more politely, toward a category mistake. It is the response
that one mode of existence makes to what another mode is demanding
of it, without being able to clarify the origin of the conflict of interpretation. Since the modes of existence accused of having no substance have
no philosophy at their disposal except that of being-as-being (or the skeptical critique of that philosophy, which amounts to the same thing), they
can no longer defend their originality except by the most devastating of
solutions: they *exaggerate* this deficit and affirm against all evidence that
they too achieve displacements without deformation, arrive at truths
that are *literally* true [REF · REL]. To try to save their treasure, religions
panic and entrust it to the care of Double Click—the least reliable of
saviors.

Rationalization is the response—"we, too; we most of all!"—
offered by the old modes trapped into failing to understand themselves
any longer, for want of a replacement metaphysics. We have already met
this sort of recoil in the case of reference: RES EXTENSA was the logical
(but mad) solution to an artifact created by the difficulty of reconciling

knowledge with the extent, the complexity, and the humbleness of its chains of instruments [REF]. We observed it again with the metamorphoses [MET]: as it is difficult for them to be granted their exact weight of being, in the face of accusations of unreality (and they are indeed "unreal" by Double Click's standard), they exaggerate this lack by a paradoxical rationalization and affirm that "they really do exist, but in a *hidden* fashion." The mysterious is never anything but a rationalization provoked by the loss of the interpretive key—and this, moreover, is the only mystery it holds. In an entirely different sense from Hegel's, we can say that the real is indeed rational, but only if we follow the diversity of keys: the rational degenerates into parasitic rationalization as soon as we lose or confuse the keys.

Now, no regime is harder hit by rationalization than religion. Astonishingly, it is the excess of logic that has driven it mad [REL · DC]. As long as it follows the thread of exegesis, it is of biblical simplicity. But it loses its head as soon as it is made to go *backward*, back up the slope it had previously been descending. Anything at all can make it go off the rails: one has only to ask it to transport a substance different from "what" is said. At that point, religion is transformed into a big lie—which doesn't even have the excuse of being pious! The letter has been separated from the spirit, the thing said from the way it is supposed to be said. The impetus of religion is lost every time someone asks: "But, finally, *what* does it say?" It is immediately transmuted into a primordial monstrosity. For the religious mode informs about nothing whatsoever. It does something better: it converts, it saves, it transports transformations, it arouses persons anew.

When the scientific revolution came along, the religious mode had sixteen centuries of scholasticism and rationalism behind it, and so it succumbed RATIONALIZATION IS WHAT PRODUCES BELIEF IN BELIEF ⊙ to the temptation of jumping into a competition over questions of subsistence and identity [REF · REL]. And yet, to reach the end times is not to go more deeply into the issue of substance, it is to convert. If there is one question it should not have answered, it was this one: "And what do *you* say about remote beings?" It takes a religion very sure of itself to consent *not to respond*. As soon as it starts to doubt its capacity to speak while saving, it is going to start justifying itself, in self-defense, against its competitors, in the very terms of its skeptical contradictors: "But I *too* offer you a pathway to remote beings that *resembles* reference, even if it doesn't have the vehicle of IMMUTABLE MOBILES or the help of FORMALISMS to get it to its destination." There you have it: religion becomes "belief in something"—but something that will remain inaccessible! Orpheus has looked back, Eurydice is returning to limbo...

Pious souls ought to stop here, discern the impasse clearly, go back the way they came, return to the spirit, to the flow of the Word itself. They should have easily detected the abomination: asking the religious to believe in something substantive is as absurd as confusing reference with reproduction [REP · REF], reproduction with metamorphosis [REP · MET], or therapy with law [MET · LAW]. But no, to prove the strength of a mode of veridiction that they are already ceasing to inhabit, they insist on rationalizing it. *Perseverare diabolicum est.* And here we are again, with devils by the hundreds. We are going to pile up monstrosities one after another through a succession of "proofs," all derived "logically." What exquisite scholasticism we have in store.

As belief "in" this immobile something does not resemble what produces access to remote beings, the religious will claim that there is something "beyond" Nature. They could not go further astray: given that NATURE is already an artifact, just think what the *supernatural* must be like! And to reach this supernatural world that has no existence whatsoever, they proceed as though religion were something like a ladder that allows access—a ladder without rungs or rails. And the worst part, the diabolical aspect, is that, once our eyes are turned toward the *beyond*, we can no longer lower them again to see what is close at hand, our neighbor, the present, the *here and now* of presence, the only promises that religious words can actually keep: that of incarnation in time, provided that we never exit from time, that we start over and over.

Those to whom this religious language has been addressed up to now are confused: what would they want with the supernatural? Can something that doesn't exist, something that is inaccessible, outside time, step in and save them? These listeners shout in vain: "Come back, there's still time. Get out of the impasse where you have taken the holy word. Give us back its true meaning, get back in your right mind, find the Spirit again." But no, on the contrary, no one is going to come to them any longer *on a mission*, start talking to them again. No, they are going to be asked, in a supplementary perversion, to *understand nothing!* They are going to be presented with a seven-headed monster, the Beast of the Apocalypse: religion is a "mystery" that *must not* be understood. Suspension of reason—and in consequence, suspension of religion *as well*, suspension of the proper movement of this very special thread that could have made it possible to continue the work of rationality if only it were followed. At this point, there is nothing left to do. Religion is over. Go in peace.

This is a much too rapid summary of a very long, very sad history. But it doesn't take much to understand at what point mistakes in signage made it impossible to follow religious veridiction and scientific veridiction at one and the same time, rigorously and exclusively, by ordinary reasoning. This conflict never involved a struggle in which Belief was gradually "replaced" by Reason—whatever illusion the Enlightenment may have entertained on this point. If the conflict took place, it is because, by being wrong about the sciences, religion was sure to make the intermittent thread of the holy Word even more ungraspable. Indeed, if one is wrong about reference and begins to imagine displacements *without* deformation [REF · DC], then the quite particular mode of reprise of religion seems incongruous, inept, inconsistent, dishonest. To move about within its own mode of veridiction, religion necessarily (from the standpoint of information) has to "lie"—as do the sciences, I must recall with some insistence [REF · REL]. The modernist tragedy is to have been mistaken at the same moment, through a ricochet, as it were, about Science *and* about Religion.

> ⊙ AND CAUSES THE LOSS OF BOTH KNOWLEDGE AND FAITH ⊙

If it is true, as we have seen, that objective knowledge has never recovered from its commitment to the search for substance that obliged it always to advance by backing up, systematically denying the peregrinations that allowed the sciences to progress, it is even more true that the commitment of religion to the frantic quest for a Substance mistakenly confused with the name of "God" could only lead it to wander in the desert, turning its back on the Promised Land.

⊕ LEADING TO THE LOSS OF
NEIGHBORING BEINGS AND
REMOTE ONES ALIKE ⊙

In Chapter 6, we saw the *hole* that that double denial of mediations had left in the Moderns' self-awareness, making constructivism impossible. What is infinitely more serious is that it reversed the direction of piety, which is henceforth obliged to be directed toward the "other world," in an unhealthy competition with access to remote beings. An ontological skid without remedy, since religion has had to abandon the only access it could offer: *access to one's neighbors.* And this is why the Moderns, believing they have been freed of religion by Reason, run the risk of *losing both the proximate and the remote.* By a stunning comedy of errors, when they speak of Science they point to the Earth, and when they speak of religion they point upward, indicating that one should direct one's gaze toward Heaven.

We can appreciate how hard it is to do an even slightly systematic anthropology of the Moderns when we see that, as soon as they start talking about the "conflict between Science and Religion," they act as though it were a matter of opposing (or "reconciling," which is worse) two types of approach: one that would give us MATTER, the "here below," the rational, the natural, and one that would offer us the spiritual, the beyond, the supernatural, the supreme values! As if there were a world here below to which Science would give access and a world beyond to which Religion claims to give even faster access. With only one choice, a yes-or-no question: must Science absorb everything, including the beyond, or must we still, in spite of everything, reserve "a little room" for the spirit, Sunday morning, for example? How could our anthropologist not remain speechless before such a "MALIGN INVERSION"? And it is truly *malign.* By dint of stringing category mistakes together, the Moderns have managed to reverse the relationship between these two modes with almost perfect precision. Isn't it the Evil One himself whose forked foot we are beginning to perceive?

Quite to the contrary, it is when we speak of Science that we should raise our eyes toward the heavens and when we speak of Religion that we should lower them toward the Earth. For it is quite obviously objective knowledge that gains access to remote entities, and that goes everywhere with no limits whatsoever as long as it is given the means, while it is religion that has some chance of allowing access to what is nearby, to our neighbors. Good sense, for its part, deprived of both science and religion, will not accede to either one. What it calls the "ordinary world" is hardly ordinary at all. To rediscover COMMON SENSE, we would have to be able to redirect our gaze twice: up, toward the *others* that would be inaccessible without reference, which grasps remote beings, and down, toward the *others* who are inaccessible without speech, which brings them closer together.

How have we arrived at such a reversal? Well, because once again there has been "padding." Here we find the flooding by the *res extensa* that gradually took over the beings of the "material world" and left no place for other beings except elsewhere, above, high up, far away, or else in the depths of interiority, in the inner sanctum.

⊙ AS WELL AS TO THE SUPERFLUOUS INVENTION OF THE SUPERNATURAL.

Iconography offers a magnificent emblem of that expulsion. Starting from the moment when the immutable mobiles begin to circulate, painters had to work harder and harder to move the Madonna, for example, not through the new isotropic space but from a "terrestrial" to a "celestial" state through the intermediary of a conversion. The theme gradually changed direction. It had begun with Byzantine icons in which transports *through radical conversions* were suggested thanks to quite obvious discontinuities of form and matter—for example, the gold at the top of the icon and the earthy brown at the bottom; the corpse lying in the tomb and the little white soul borne away by Christ; the mandorlas where cherubim are flying about, and so on. But in the wake of the scientific revolution, painters no longer knew what to do to convert the body of the Holy Virgin, since the reign of Double Click information was extending and they no longer knew how to move anything through space except *what was not transformed* (or which pretended not to be transformed). What to do with that corpse? How to get rid of it? A perfect metaphor for the fate of religion: one ought to "send it flying," expel it into the beyond of *res extensa*. And so the unfortunate Virgin ends up looking like a sort of Cape Canaveral rocket taking off, propelled by little booster angels, to shift *with no more conversions of any sort* by simple DISPLACEMENT into an undifferentiated sky. The sky has replaced Heaven. Thousands of overwrought pilgrims may bend their knees before the icon and feel themselves, too, resuscitated; but before the flight of an immutable mobile, conversion is impossible. It is as though the Madonna had been gradually submerged by the rise of "matter"! (Let's not forget that the engineers we met in Chapter 8 fared no better: they didn't succeed in producing technical drawings adequately depicting NASA, the institution within which their rockets flew while they were also flying in the sky.)

Those who are preoccupied by global warming must not forget that other Flood that has already drowned so many other beings, those of the sciences above all, but also those of technology and the economy, not to mention politics. Only once the multiverse has been emptied of the unwarranted forms of extension—Nature, matter, language, society, the symbolic, God, and so on—will we have enough room to let in all the beings that our informants collectively cherish. And we haven't seen an Ark floating on the waters lately, or even a dove bearing the olive branch of salvation. When will immanence be restored to us? When will we Earthlings come back to Earth?

HENCE THE IMPORTANCE OF ALWAYS SPECIFYING THE TERMS OF THE METALANGUAGE.

In the face of this malign inversion, our ethnologist is discouraged in advance about the future chances of her DIPLOMACY: how can one reverse a reversal on such a scale? What to do to make one's gaze turn toward one's neighbors—rather than toward Heaven—when one is talking about religion, and make it turn toward remote entities when one is talking about science—and not toward the solid good sense of "things as they are." Especially because, to convince the friends as well as the enemies of the "things of religion," she has at her disposal only the fragile ontology of "ways of speaking"? She can already hear the complaint of those who will cry out in indignation: "But then these are only stories?!" And it will do her no good to reply: "Yes, but they're sacred stories"; she knows perfectly well that that won't do. It's too much or too little. "There must be more to it than that," say the religious, "much more than stories, there have to be real things here, objective things." And the ethnologist agrees completely. That's just what she wanted to say, too, in her own way, like the whole tradition, with the whole tradition, betraying it with homemade inventions in order to renew it, to those whom she anticipates addressing in their own language, for a brief moment.

She doesn't see any possibility of getting out of this impasse unless she manages to make her interlocutors admit that, in these eminently delicate questions, it is always necessary to specify by a sign the interpretive key in which this requirement of objectivity must be taken. For, finally, the reader must have understood by now that [REL] objectivity differs completely from [REF] objectivity, which itself has a complex relation with [PRE] objectivity and with [FIC] or [TEC] objectivity. If we were to manage to specify the modes for all the terms of the metalanguage, could we not open a way to future negotiations? Learn at last to speak well? Even about religion, without taking the name of God in vain? A God of Incarnation, finally back on Earth—is that not what would prepare us better for what awaits us than the strange idea of a religion that would lift us up toward Heaven?

Chapter 12

INVOKING
THE PHANTOMS
OF THE POLITICAL

Can a contrast be lost? The case of the political.

An institution legitimately proud of its values ⊙ but with no grasp of practical description: ⊙ before it can be universalized, some self-examination is required.

To avoid giving up reason in politics [POL] too quickly ⊙ and to understand that there is no crisis of representation ⊙ we must not overestimate the unreason of [POL] ⊙ but rather follow the experience of political speech.

An object-oriented politics ⊙ allows us to discern the squaring of the political Circle, ⊙ provided that we distinguish accurately between speaking about politics and speaking politically.

We then discover a particular type of pass that traces the impossible Circle, ⊙ which includes or excludes depending on whether it is taken up again or not.

A first definition of the hiatus of the [POL] type: the curve ⊙ and a quite peculiar trajectory: autonomy.

A new definition of the hiatus: discontinuity ⊙ and a particularly demanding type of veridiction, ⊙ which the [REF · POL] crossing misunderstands.

[POL] practices a very distinctive extraction of alterity, ⊙ which defines a phantom public ⊙ in opposition to the figure of Society, ⊙ which would make the political even more monstrous than it is now.

Will we ever be able to relearn the language of speaking well while speaking "crooked"?

OUR ETHNOLOGIST IS SOMETIMES ANNOYED WITH HERSELF FOR FEELING SO MUCH RAGE OVER THE APPEARANCE AND DISAP-pearance of religion. She will certainly be criticized for losing her cool, and told that she is at bottom a "believer"; and it is true, of course, that she believes, even if she has proposed a form of agnosticism thanks to which one could get along entirely without belief in belief. But she has no illusions; she knows that there is nothing worse, in late modernism, than to display infinite sadness over the loss of religion. Especially if she is already being accused of "not believing" in the sciences and of having a culpable tolerance for phantoms, spirits, and succubi! When she has to put on diplomats' clothing, later on, who will agree to have her as a representative?

And yet it would be a mistake to deem her pious. In fact, what distresses her in religious matters are not so much the usual aftereffects that a contrast undergoes when other contrasts are extracted in their turn; nothing is more normal, since each mode interprets the others in its own key. What distresses her is the possibility that a contrast may disappear completely, suffocated by proliferating category mistakes. And if we can share her pain in sensing the disappearance of the religious mode, it is because of the possibility that another value may vanish, this one infinitely more important for common life, for common decency: the political (noted [POL]). Now, that value has a common feature with the religious: it mobilizes beings that are just as sensitive, although differently, to the tonality in which they are enunciated. These beings, too, are "manners of speaking," in a way; consequently, they too become endangered languages once there are no more speakers to *speak them well*. We can understand that it would be enough to ruin all hope for a possible civilization if political speech, following the same path as religious speech, were lost for good.

We have met these beings already, in Chapter 5, when I sketched out the detours through which the emerging sciences grasped the political question by operating a short circuit, as fundamental as it was contingent, that would henceforth make it possible to construct common life on the basis of knowledge—misinterpreting both the chains of reference [REF] and the truth proper to the political [POL]. This strange POLITICAL EPISTEMOLOGY was a rival of religion [REF · POL] before it took religion's place, through a sort of moral rationalism, by claiming to reign over all metalanguage in the name of the "scientific view of the world" [REF · DC]. But while our study of this crossing informed us about the traps of reference, it did not yet allow us to identify the requirements proper to the political mode. We must now define the SPECIFICATIONS of these beings more precisely, spelling out their particular elocutionary conditions, before reconnecting them, in Chapter 13, to those of religion [REL] and law [LAW], for once these three modes productive of QUASI SUBJECTS have been brought together, they will help us finally get back to the source of the quite peculiar scenography of Object and Subject by means of which the Moderns claimed they could describe everything, even as they were constantly breaking the threads of experience.

At first glance, approaching the question of the political is to take on a domain full of vitality, strongly valorized, largely raked over by the media and by a myriad of scholarly disciplines, overseen by numerous observatories, hemmed in by fastidious statistics. From this standpoint, it has nothing in common with the religious question. For our ethnographer, then, it is unlike the other modes: however skeptical she may have become about what her informants tell her, she can rejoice at seeing the pride with which they all point to the importance of politics. Even more than for science or art, but exactly the same as for law, what is at stake here is a common bond where an institution, a domain, and a

AN INSTITUTION
LEGITIMATELY PROUD
OF ITS VALUES ⊙

contrast to which the informants claim to hold avidly overlap in part—at least in their discourse. If there is a value they claim to be maintaining unchanged since the *logos* of the Greeks and right up to the contemporary blogosphere, it is the one that made the peoples of "free speech" stand out in contrast to the "Barbarians." If a collective is defined by the list of supreme goods for which its members are ready to give their lives, there is no doubt that "political autonomy," the "rule of law," "representative government," "public freedoms"—the precise term hardly matters—have pride of place. Enough people have died for this ideal—enough are still giving their lives—for it to be shameful to doubt the good faith of the Moderns, on this point at least. They can rightly say that they are the ones for whom freedom counts above all. Who would dare mock the proud challenge "Give me Liberty, or give me Death!"

Even if the recourse to the Greeks and Romans smacks a bit too much of the historical epic, the backbone of today's democracies still really stiffens in the name of that ancient pride: "We obey only laws that we have freely given ourselves. If you take that away from us, we take no more pride in being called human." Students today continue to read Aristotle, Augustine, Bodin, Locke, Rousseau, and Rawls; deputies swear to be faithful to their principals in hemicycles inspired by the ancient ones, under cupolas whose shape is borrowed from the Roman Pantheon, decorated with Corinthian pilasters, sometimes ornamented with statues of Solon, Cicero, Brutus, Montesquieu, or Washington; as for their sovereigns, they still seek to enter history in the form of statues, disguised as Roman emperors, adding their figures to the gallery of their ancestors. Here is at least one value, it will be said, with an uninterrupted tradition, an indisputable legacy, an inheritance free of any debt. SPQR, Senatus Populusque Romanus.

⊙ BUT WITH NO GRASP OF PRACTICAL DESCRIPTION: ⊙   And yet if we start from the ideal of democratic AUTONOMY that has constantly been advanced over the course of history, it is not easy to work our way back to the practical experience of political speech. The chasm that we have learned to spot between the requirements of epistemology and the establishment of chains of reference [REF] looks like a modest ravine compared to the virtually infinite distance between what we require of freedom and what we are ready to give it. Paradoxically, if no value is held in higher esteem than the autonomy permitted by democracy, no activity is held in greater scorn than politics. It is as though we wanted the end, once again, but not the *means* to reach it. A new paradox that the inquiry must address head on: how can these same Moderns simultaneously define themselves as "political animals" and reduce the veridiction that is proper to politics to a bare minimum?

To the point that the very idea of political truth and falsity appears absurd—as absurd, alas, as the idea of truth and falsity in technology [TEC], in psychogenesis [MET], in fiction [FIC], or in religion [REL], so commonly is it acknowledged that the political world is not, cannot be, cannot ever become, must not become the kingdom of any veridiction whatsoever. The case is closed: to go into politics, to take courses in political communication, to participate in an electoral campaign would be to *suspend* all requirements of truth. As if it were written at the entrance to those neoclassical portals through which only Sophists can pass: "Abandon all hope of truth, ye who enter this kingdom of false pretenses." And so our ethnologist has rejoiced too quickly once again. No contrast is more highly esteemed; no means of bringing it out in practice is more degraded. So much so that, once again, it is impossible to know whether the Moderns hold, or not, to that value. Here as elsewhere, here as always, it is up to the inquiry to trace an appropriate path between experience and the account the Moderns give of it.

The question is all the harder to avoid given that, with Law and Science—and of course The Economy, which we shall soon encounter—what is at stake is the value that Westerners have been most inclined to universalize, without asking themselves too many questions about the logistics necessary to allow it to be delivered everywhere on the planet. In seeking to export democracy all at once to the whole world, the Moderns presupposed that Earthlings dreamed of nothing but taking their place in a Parliament so as to become citizens by getting laws adopted according to the subtle mechanics put in place by the invention of the representative governments that grew out of the revolutions of the eighteenth century. The proposition was daring, perhaps generous: it does not appear that the rest of the world has rushed to put on Roman togas or Phrygian caps, or to sit in the seat of a deputy or a congressman.

⊙ BEFORE IT CAN BE UNIVERSALIZED, SOME SELF-EXAMINATION IS REQUIRED.

It does not appear that the members of the other cultures wish to become citizens of a free government—at least not as long as they have not redefined the words "citizens," "freedom," and "government" in a thoroughgoing way, in their own terms.

We must no longer be astonished by this: the democratic ideal cannot be extended more rapidly, with fewer instruments, fewer costly mediations, than the scientific ideal [REF]. Democracy can't be parachuted in from the bay of a U.S. Air Force plane in the form of an "instantly inflatable parliament," as Peter Sloterdijk has ironically imagined. Just as the establishment of chains of reference demands a proliferation of apparatuses that our epistemology in no way prepared us to finance, similarly the delicate ecology of freedom requires precious technologies and countless habits that enthusiasm for democracy alone does not prepare us in any way to set up or maintain. If the universalization of knowledge remains a hypocritical pretense as long as we fail to extend the networks of laboratories and colleagues that make it possible to bring knowledge into existence, the universalization of freedom is only a gratuitous injunction as long as we don't take pains to build the artificial enclosures, the "greenhouses," the air-conditioned equipment that would finally make the "atmosphere of politics" *breathable*. How could our anthropologist defend that value if she first had to require all the others at the negotiating table either to renounce truth when they talk about politics or to endorse the ideal of freedom without specifying how to obtain it in practice?

And, of course, in giving up the goal of extending democracy to the planet we would be guilty not only of cowardice but of an even more ethnocentric scorn than when we wanted to universalize it without fanfare—or rather *with* fanfare—since we would be denying to other peoples the autonomy that we have made the supreme value, but only for ourselves. If we must not despair of the universal, then, we have to suppose once again that it is the embarrassment of riches on the part of the Whites that has prevented them from bringing their own ideals to fruition and from maintaining their own virtues. As in the case of religion, as in the case of the sciences, a precise inventory of what has been inherited needs to precede any project of universalization. The inquiry will thus have to turn back to its principals at some point to ask: "When you say you hold to politics, what are you really *holding onto?*" We have to learn to work on ourselves first of all.

The reader is now familiar enough with the principles of this work to suspect that the solution is not to wallow in the necessary irrationality of public life but rather to search for the particular type of

To AVOID GIVING UP
REASON IN POLITICS
[POL] TOO QUICKLY ⊙

reason with which public life is charged. If the task seems insurmountable at first, it is because the Moderns always hesitate between two alternatives: losing all hope of being rational in politics or making politics "rational at last" by using as its guiding thread a form of reason as foreign to it as possible. As they do with religion [REL]? And with technology [TEC]? The interpolation seems much more complete, the abyss between theory and practice much deeper. It is better measured by the rationalists' efforts to make politics reasonable at last—according to their not very reasonable definitions of reason. By the scope of the padding operations we can measure the depth of the gap they have tried to fill.

Even though Westerners have never stopped taking freedom and autonomy as their supreme virtues, they have never stopped making the exercise of that freedom and the outlines of that autonomy more and more impracticable by borrowing their principles from other forms of life. As we saw in Chapter 6, political reasoning *never goes straight:* this is what scandalizes, and what the Moderns keep trying to rectify with prostheses. They want it to be straightforward, flat, clean; they want it to tell the truth according to the type of veridiction that they think they can ask of the Evil Genius, Double Click [POL · DC]. This began with Socrates and has never stopped, through Hobbes and Rousseau, Marx and Hayek, to Habermas. "If only we could finally *replace* the crookedness of the political with the path of right reason or science—law, history, economics, psychology, physics, biology, it hardly matters!" The hope is always the same; the only thing that varies is the type of speech therapy—orthophonic, or rather "orthologic"—with which they claim to be *straightening out* politics.

Now, the transport without transformation of IMMUTABLE MOBILES (itself misunderstood, as we must keep reminding ourselves) can do no better at capturing this way of being. If religion [REL] has never gotten over the competition from transports without transformation, we may say that the political mode has suffered at least as much. Since the fraudulent Double Click has been in charge, the only result is that, comparatively speaking, the political mode has started *to lie more and more*. And this is how, by dint of measuring it with an inappropriate yardstick, we risk gradually losing a contrast. For three centuries, indeed, the abyss hasn't stopped deepening between demands for information, transparency, representation, fidelity, exactitude, governance, accountability on the one hand and what the particular curves, meanderings, and twists and turns of the political have to offer on the other.

If, as we have just seen, the angels aren't immune to this demand—"Gabriel, in your message, there were how many units of information, how many bytes?"—the *logos* of political reason gives an even more pitiful answer when it is asked "How much information do you transport?" Answer: "None! I lie, I have to lie." Measured in megabytes, the opacity increases: the representations are less and less faithful. The reign of lies extends everywhere. "We're being manipulated!" "They're deceiving us!" "It's all smoke and mirrors!" "Just a political spectacle." "Everything has already been decided without us." And finally, with a shrug of the shoulders: "They're all corrupt!"

The scale of the gap will be all the greater in that it is no longer a question of bringing able-bodied males together on the square of Rousseau's beloved Swiss village, but of building artificial spheres capable of restoring common life to billions of beings, women and children included—without forgetting those nonhumans on which they depend, whose number is constantly increasing with the end of Nature. This growing distance between representatives and represented, this ineluctable rise of lying, has even been made official, as a "crisis of representation." There is unanimous agreement: "The elites are out of touch with the masses." "Politics is no longer equal to the stakes." "My kingdom for a Science!"

⊙ AND TO UNDERSTAND THAT THERE IS NO CRISIS OF REPRESENTATION ⊙ There is indeed a crisis of representation, but only if we understand by this expression that people are obstinately critiquing political representation for something that it can never procure: they are asking it to "express faithfully"—and thus mimetically—the "political opinions" of billions of beings, or asking those beings to politely obey the injunctions of their principals while strictly applying the rules sent down from on high. We might as well as ask religion to transport "belief in God" [REL · DC]; ask reference to produce objectivity without transformations [REF · DC]; ask technology to find a way to make things work without any detours [TEC · DC]; ask the beings of reproduction to behave as well as the inscriptions that allow us to know them [REP · DC]; ask the passage of law to reach good judgments without being given the "means" and without hesitation [LAW · DC]. The purported crisis of representation is only an artifact that comes from the application to the political of a principle of displacement that is no more adapted to it, to borrow a delightful feminist slogan, than a bicycle is to a fish.

It is as though political thought had practiced a vast operation of transfusion by which people keep trying to replace the overheated blood of the body politic with one of those frozen liquids that quickly change into solids and allow the "plastination" of corpses offered to the admiration of the gawkers, according to the modus operandi of the sinister doctor Gunther von Hagens. Monster metaphors are mixed in here on purpose, because one cannot do political anthropology without confronting questions of teratology. If "the sleep of reason brings forth monsters," no more frightful ones have ever been created than by the confusion of politics with information, with Science, with management, with power—or, worse still, with a "science of power struggles." Not to mention the aberration known as "political science."

"But," someone will exclaim, "no one would have the absurd idea of accusing politicians of lying, because no one even imagines any longer that politicians might be in the business of telling the truth. ⊙ WE MUST NOT OVERESTIMATE THE UNREASON OF [POL] ⊙ 'True' and 'false' have no meaning in that regime. It's all about power struggles; didn't you know that? It's high time you learned." A new problem of calibration that can no longer be attributed to the application of a bad model of reason, but that results rather from the acceptance, also premature, of a type of *unreason* just as ill-adapted as the other. We find the same problem here as with fiction [FIC], which people have sought to reduce too quickly to the suspension of all requirements of objectivity and truth. No longer able to see by what thread one could follow the reason of the political, they began to overestimate unreason, and to brandish lies, skill, power struggles, violence, no longer as defects but as qualities, the only ones that would remain to that form of life. Such is the temptation of MACHIAVELLIANISM. Now, if people misunderstand the political mode by requiring transparency and information from it, they misunderstand it just as much by propagating the belief that it has to abandon *all* rationality.

We have just measured the damage done, in the case of the religious mode, by this way of overestimating the accusations made by those who judge according to a different mode. If rationalization makes religious beings lose all common sense—"Yes, it's true, we don't 'know,' we 'believe,' and we're proud of it"—it is even harder on politicians obliged to confess "Yes, it's true, we never go in a straight line, we lie; sadly, that's what we have to be proud of." Two opacities, two mysteries, two arcane secrets, all the more frightful in that they can be cumulated in a rather distasteful political theology [POL · REL]. "Cover up that bosom, which I can't endure to look on": if we call prudishness the attractive horror that Dorine's breast arouses in Tartuffe, what must we call the fascinated indignation of reason before the curves of the political mode? The Moderns have changed the definition of the reasonable, yet they have never lost the hope of finally making the scandalous secrets of politics disappear behind the severe clothing of the reasonable, the just, the moral, the common, the learned, the clear and clean—something known today by the sanitized term "governance"—or, conversely, the hope of indulging in a sort of pornography of violence, conspiracy, force, and ruse.

⊙ BUT RATHER FOLLOW THE EXPERIENCE OF POLITICAL SPEECH.

Fortunately, as we now know, there is no mystery, no opacity, no irrationality (this is at least the hypothesis of this inquiry) other than the INTER-POLATION of forms of reason. Yes, it is true, the polit-ical mode moves crab-wise, it is the "Prince of twisted words," but that does not make it irrational. As always, our inquiry has to manage to approach experience in order to grasp the peculiar thread that no other yardstick would allow us to measure. The problem is how to isolate, from within the political domain as a whole, an interpretive key that is specific to it.

What must we follow? Demonstrators asleep in their bus on the way to the traditional march from Nation to the Bastille? Passers-by indignant at the way the police are treating some dark-skinned young men? Elite officials plotting subtle tactical moves outside the door of a cabinet minister? Small-town mayors overwhelmed by the arrival of new decrees from the Prefecture? Regulars at a bar loftily asserting how they would have reformed the government if they were in charge? Young militants distributing leaflets in a marketplace? Attentive readers of Debord's *Society of the Spectacle* meeting in secret to dynamite that society? "Greens" chaining themselves to hundred-year-old oak trees to keep them from being felled? In any era, in any country, at any moment, from one subject to another, experience is so multiform, diversity so great that the ethnologist doesn't see exactly what her informants mean when they say they are "talking about politics" or that they have been "politicized," or when they despair, on the contrary, that "the masses" or "the young" aren't "politicized" enough.

To get our bearings in such a shambles, we need to be able to embrace the diversity of the stakes as well as the specificity of the Circle, and this implies
AN OBJECT-ORIENTED POLITICS ⊙
following two distinct paths. If politics has to be "crooked," this is first of all because it encounters stakes that oblige it to turn away, to bend, to shift positions. Its path is curved because on each occasion it turns around questions, issues, stakes, things—in the sense of *res publica*, the public *thing*—whose surprising consequences leave those who would rather hear nothing about them all mixed up. So many issues, so much politics. Or, in the forceful slogan proposed by Noortje Marres: "No issue, no politics!" It is thus above all because politics is always object-oriented—to borrow a term from information science—that it always seems to elude us. As though the weight of each issue obliged a public to gather around it—with a different geometry and different procedures on every occasion. Moreover, the very etymology of this ancient word— *chose, cause, res*, or *thing*—signals in all the languages of Europe the weight of issues that must always be paid for with meetings. It is because we disagree that we are obliged to meet—we are *held* to that obligation and thus assembled. The political institution has to take into account the cosmology and the physics through which things—the former MATTERS OF FACT that have become MATTERS OF CONCERN—oblige the political to curve around it.

⊙ ALLOWS US TO DISCERN
THE SQUARING OF THE
POLITICAL CIRCLE, ⊙

For the time being, we need to be concerned not with the things that, by their weight, force beings to circle around them but with the Circle itself: its originality does not seem to have been recognized by the institution that purports to be sheltering it, as if the political were ashamed not to correspond to the template of the other modes. Strangely, this Circle is *impossible to trace*. And if we are to understand what we are dealing with here we had better underscore the adjective "impossible" more than once: it is a Circle; it is impossible to trace; it must be traced, however, and once it has been traced it disappears; and we have to all start over again at once . . . This is the only way, the only trajectory, the only vector for acquiring the freedom and the autonomy we say we cherish but whose course we hate to pursue: it is so contrary to our rigidities, our other certainties, our other values—it *hurts* so much, it so threatens to do bad things we don't want rather than the good things we would like to do. We can understand both the pride of those who have extracted such a contrast and made it the cornerstone of their civilization and also their fear of losing a variety that is so delicate to define and that, detached from the things around which it turns, can become dangerous.

Like religion [REL] and law [LAW], political discourse [POL] engages the entire collective, but in an even more particular way: one has to *pass* from one situation to another and then come back and start everything, *everything*, all over again in a different form. However numerous and diverse the various examples of political life may be, the question posed by political discourse is always how to connect beings to others so that the collective holds together while respecting a strange condition that for the time being appears contradictory: the political has to allow beings to pass through and come back while tracing an *envelope* that defines, for a time, the "we," the group in the process of self-production, before it is taken up again by another movement thanks to which the others, called "they," find themselves fewer in number—unless the movement goes in the other direction and they become more and more numerous. "Here is what we are"; "Here is what we want." What do they do, the investigator wonders, to produce such expressions?

She perceives clearly, in particular, that in the search for examples of political speech, she must not make the very common mistake of distinguishing between questions of *representation* (half the Circle) and questions of obedience (the other half). Coming or going, the same movement is involved, and by separating the different phases one would risk losing the pass that is the distinguishing feature of this form. Millions of people say they crave representation; hundreds of people say they crave obedience. Now, these are *the same people*, necessarily, at different moments of a single trajectory, of which the multiple examples are only stages, since they try to form a group, since they try to say "we" and "they," since they try to agree on a common will—whereas they agree on nothing and want nothing that the other wants. So it is hardly surprising that they should gradually slide, in one case, from *multitude* to unity, and, in the other, from *unity* to multitude. It is this slippage, this passage and return, that has to sketch out the totally original circular phenomenon that we need somehow to isolate. Not only is it a Circle, but in addition it agglomerates on one side while it disperses and redistributes on the other. Unquestionably a very bizarre beast; it is easy to see why its passage has aroused dread.

A second precaution consists in isolating the overly variable *content* of the countless position-takings called political, distinguishing these stances from a certain *manner* of grasping them only to pass them along. The ethnologist has to apply to this type of practice the same method she used to isolate the beings of fiction, technology, and religion. To grasp them, she has to concern herself not, initially, with the result—with the position-taking—but, as always, with what sends the course of action in a certain direction, something that we might call "preposition-taking." As with the other modes, our attention must shift from the contents to the containers, from what is happening—what is being passed along—to the gesture of making it happen—the "pass."

⊙ PROVIDED THAT WE DISTINGUISH ACCURATELY BETWEEN SPEAKING ABOUT POLITICS AND SPEAKING POLITICALLY.

In fact, as we have done for every mode from the start, we will do well to shift our attention from the adjective "political" to the adverb "politically." If it is quite difficult to specify what a technological "object" is, we saw nevertheless in Chapter 7 what it can mean to act *technologically*; and, as we now know, the difference is immense—even infinite—between speaking "about" religion and speaking *religiously*. Let us try to do the same thing here, the ethnologist suggests: let us set aside the enormous mass of statements that bear *on* politics in order to try to understand what it may mean to act or speak *politically*. Is there an experience of speaking politically that is unique?

The distinction strikes her as all the more important in that she senses how easy it is to express oneself "about" politics: her informants seem to have an opinion about everything, as if they had "settled positions" that cost them nothing. But when she asks them about these "settled positions" and in particular *where they might lead*, in the literal sense, these same informants start to splutter a bit. Talking politics seems to mean something to them: it means running into a particular type of theme supplied by the media (the men and women known as "politicians," elections, scandals, injustices, the State, and so on), themes that seem to be sharply distinguished from the others—private affairs, the topics of society, culture, and so on. But speaking *politically*, on any subject at all, in a way that entails a certain manner of moving forward? There, no, the same informants "don't have any idea what she is getting at."

The investigator thus has to learn to slip from one distinction to another: from the one that sets the domain of politics apart from the others to the one that isolates a mode of existence from all the others. The first would lead her to a hopelessly muddled holdall—the domain of official politics; the second obliges her to detect a frequency, a wavelength, and one only, for the recognition of which we all have—we all used to have—a receptor, an ear of stunning precision. Always this distinction between the search for substance and the search for subsistence, between good and bad TRANSCENDENCE.

WE THEN DISCOVER A PARTICULAR TYPE OF PASS THAT TRACES THE IMPOSSIBLE CIRCLE, ⊙

What is it that "rings false" in politics? How do the informants define failure? What has to "run" when we speak politically? What baton is used in this relay race? In other words, what is the *trajectory*, the particular PASS thanks to which we are going to be able to recognize what is or is not true or false, or, to rehabilitate an expression that is actually very fitting at last, what is the way to be *politically correct* or *incorrect*? Our analyses of the other modes offer a valuable clue by drawing our attention to the suspended *segments* of a movement that is meaningless if it does not continue to advance. Interruptions of this movement are infelicities that mark what the FELICITY CONDITIONS of political discourse would be.

Indeed, the principal infelicity condition of the political is to have its course *interrupted*, the relay broken off. "That's not going anywhere." "That's pointless." "That won't do any good." "They're forgetting about us." "They don't give a damn about us." "Nobody's doing a thing about it." Or, in a more scholarly fashion, "We are not represented." "We are not obeyed." In other words, something rings false in each example taken separately precisely because it is *taken separately*. The examples ring true when they are connected. But connected how, by what, by what particular thread? That is what we have to find out.

For, finally, what form of life can bring off the following feat? Start with a multitude that does not know what it wants but that is suffering and complaining; obtain, by a series of radical transformations, a unified representation of that multitude; then, by a dizzying translation/ betrayal, invent a version of its pain and grievances from whole cloth; make it a unified version that will be repeated by certain voices, which in turn—the return trip is as least as astonishing as the trip out—will bring it back to the multitude in the form of requirements imposed, orders given, laws passed; requirements, orders, and laws that are now exchanged, translated, transposed, transformed, opposed by the multitude in such diverse ways that they produce a new commotion: complaints defining new grievances, reviving and spelling out new indignation, new consent, new opinions.

Wait, that's not all! One turn on this merry-go-round is just the beginning. What is most magnificent in the political, what makes those who discover its movement shed tears of admiration, is that one has to *constantly start over*: beginning again with the multitude—perhaps this time more confident, more reassured, more protected—to take up the thread of representation again—perhaps more easily, and more faith- fully, too; then go through the unification phase (the millions become one: what a strangling bottleneck!); next—the operation may have become a little less risky thanks to the preceding turn—establish the prescribed order, which may be a little better obeyed (unity becomes millions: can you imagine the impossibility of this new translation/betrayal?).

The Circle cannot subsist through a substance, only through a quest for its own subsistence. But wait, we're still not finished, for this Circle, even taken up again, leaves no more durable traces than if you had drawn it on sand or in water. You have to start over again: if you stop, it *disappears*. In its place, you have nothing but the multitude, dispersed, grumbling, restless, disappointed, violent—or, worse, indifferent, dispersed, unaddressable; you have nothing but elites pressing uselessly on "joysticks" that no longer obey, that transmit nothing; you're left with a people of impotent groaners just good enough to be indignant without knowing what to be indignant about. Suddenly the "crisis of representation" is back: a yawning gap opens up again between the "top" and the "bottom"; dispersal is guaranteed; no agreement is possible; the enemies are waiting outside to attack "us." Why do we always have to start over? Because the Circle is impossible, of course! Coming and going alike. The multiple becomes one, the one becomes multiple; it *can't* work; it has to work; so we have to start all over again.

⊙ WHICH INCLUDES OR EXCLUDES DEPENDING ON WHETHER IT IS TAKEN UP AGAIN OR NOT.

Hold on, we're still not done, for the Circle can either *include* more, or *exclude* more, depending on the number of people it manages to represent faithfully (by translating/betraying them through and through) and on the number of those whose obedience it secures (and these are the ones, this time, who betray/translate what is expected of them). The same movement of enveloping, encircling, embracing, gathering in can thus serve, according to the rapidity with which it turns, either to fabricate inclusion—those who say "We" leave outside only a few "They"—or else to fabricate exclusion—those who say "We" find themselves surrounded by ever more numerous Barbarians who threaten their existence and whom they treat as enemies. And nothing in this movement ensures its duration; here is the source of all its hardness, all its terrible exigency, since it can at any moment grow larger by multiplying inclusions, or shrink by multiplying exclusions. Everything depends on its renewal, on the courage of those who, all along the chain, agree to behave in such a way that their behavior *leads* to the next part of the curve.

The best one can hope for is that, by dint of tracing the Circle and starting it over again, beings form habits that make it possible, little by little, to count on a reprise. Each segment feels obliged to act and able to speak in such a way as to avoid interrupting this paradoxical movement [HAB · POL], as if each were preparing to take a position in anticipation of the following stage. When this happens, a *political culture* begins to take shape and gradually makes the maintenance, renewal, and expansion of the Circle less and less painful. DEMOCRACY becomes a habit. Freedom becomes engrained. But things can also turn in the other direction: they can literally "take a turn for the worse," "turn out badly"; obstacles can accumulate and make it less and less possible, more and more painful, to renew the Circle, the defining exercise of the political. Then it's over: time indeed to abandon all hope of being represented, of being obeyed, of being safe. The "We" has been torn to shreds. Soon there won't be anyone who is *in agreement* or who wants the same thing as his neighbor. The habit of democracy has been lost. If only politics could be reduced to skill, to lying, to balances of power, it would be so much simpler than this requirement of speaking in curves. And this is the movement, so difficult to grasp, so hard, so *forced* (in the sense of dynamics), so counterintuitive, that we wanted to spread throughout the entire planet without striking a blow? Our ethnologist, for her part, remains dumbfounded before the beauty of this movement.

As she sees it, her informants ought to fall on their knees before the dignity of this gesture of envelopment that always has to be renewed. Instead of trying to replace it by a series of straight lines or, worse, to wallow in the "arcana of power," they ought to deploy all the talents of their culture, their art, their philosophy, in order to respect its origi-nality, to follow—to caress!—its admirable volume. If the Moderns have something to be proud of, it is that they have been capable of extracting this contrast against all the evidence supplied by other regimes of truth. But do they even know that they possess this treasure? The investigator can't help making a note: "Strange Moderns: they honor their sciences, sometimes their technologies, rarely their gods, never their divinities, but they scorn those who devote their lives to holding onto this impos-sible envelope with their fingertips, and they find no praise for the Circle of representation and obedience but the adjective 'Machiavellian.'

O splendors of politics, O beauty without equal, O courage of those who plunge without a net into these dizzying acrobatics!"

A FIRST DEFINITION OF THE HIATUS OF THE [POL] TYPE: THE CURVE ⊙
What makes it almost impossible to do justice to the distinctive quality of those who trace the political Circle is that they try to approximate it with segments of straight lines, tangents, by an application, very approximate in itself, of an infinitesimal calculus. Now this is the surest way to *lose* competence, the talent proper to those who, to speak politically, must carefully follow the curve and *not* go straight at all. The metaphor of the Circle has to be taken literally, because it allows us to give a first definition of the HIATUS peculiar to this mode of existence (as we have already seen in Chapter 5). Political beings are always accused of lying, whereas they begin *truly* to lie, to lie *politically*, only if they "go off on a tangent," as the familiar expression has it, by beginning to proffer straight talk, that is, wanting be "faithfully" represented or "faithfully" obeyed. Can one imagine a greater injustice than making a mistake of this magnitude about the difference between lies and truth? Here is a category mistake par excellence.

The hiatus is thus found in the small gap between the temptation to go straight and the ever-recurring obligation to turn away, to "bend over backward" to ensure, within the limits of one's means, the passage of the political baton, which will *never come back*, we can be certain of this, if we take it upon ourselves to interrupt it or to send it straight ahead, elsewhere, straightforwardly. Being indignant is fine, but preparing yourself to pass on to something else is better; deciding is fine, but preparing yourself to be betrayed is better; wanting to "blow it all up" is admirable, but preparing yourself to redesign it entirely will prove that you weren't lying when you used the fine word "revolution"; claiming to "assuage the passions" by "debating calmly with reasonable people" is a magnificent project, but it has no meaning—no political meaning— if you're not already preparing yourselves to arouse new oppositions and new passions. In other words, at each point, the proof that one is not lying is given by what *follows* in the curve and by the anticipation, the hope, of its necessary return, its renewal, and its future extension— return and renewal and extension, a reprise that depends entirely on *the followers*, all along this chain in which the lack of *a single one* would suffice to make it collapse. And make no mistake about it, there is no grouping other than this movement of *collection*, no reserve on which we could count, no identity, no root, no essence, no substance on which we could rest.

What gives some consistency to this perilous exercise, in spite of everything, is that the Circle reconnects with an eminently classical tradition.

⊙ AND A QUITE PECULIAR TRAJECTORY: AUTONOMY.

This Circle has been celebrated under the name of autonomy. But to be autonomous you need to focus on issues, affairs, topics that force you to circle around them. It is contradictory—a torture worthy of Tantalus—to expect politics to take this autonomous circular form and deprive it of issues around which to turn. But if the Circle is pursued obstinately enough, if it is constantly taken up again, if we pass time after time from multitude to unity and from unity to multitude, we gradually become, in effect, those who receive *from on high* the orders that they have whispered *from below* to their representatives. We are no longer heteronomous; we become proud of our autonomy.

We win our political freedom gradually, provided that we pursue the Circle, just as we win our objectivity gradually when we understand the price in transformations that chains of reference have to pay in order to accede to remote beings through the intermediary of constants [REF · POL]; just as we gradually win our salvation by dint of inventing—yes, fabricating or in any case translating—the venerable words of religion [REL]. And the idea that this envelope may extend from one nearby being to another instead of constantly shrinking—this is indeed a lovely ideal, but on condition that we measure the conditions and calculate in advance the exact radius of curvature. Here again, here as always, we have to mistrust dry, "straight" idealism.

How can we ignore the stupefying series of *transformations* and mediations—betrayals—by which it was necessary to pass, both coming and going, along with the things, the issues, the controversies that oblige us to come together in their regard by turning around them, different every time? Who can claim to draw the political Circle capable of making billions of citizens autonomous? And to do this *without* the billions in question? Do you appreciate what that would represent in reprises and representations? And it's this regime of truth that we wanted to extend to the whole planet, without effort, as if it were the unquestionable foundation for a humanity thirsting for democracy?

A NEW DEFINITION OF THE
HIATUS: DISCONTINUITY ⊙
We thus have to deal with not one but two discontinuities, two chiasmuses, two hiatuses: the one that separates one political moment from the next, and the one that must take the temptation to go straight ahead and make it veer instead toward this very special form of curve. As long as they have not been qualified with precision, political networks cannot be described empirically. They will always be taken for *something else*. We shall always be committing a category mistake.

Alongside the multitude of philosophers who have been trying since Plato to replace politics by Reason—while distorting the image of knowledge even though that was what they wanted to straighten out!—there are a few (in our tradition, they can be counted on the fingers of one hand or at most two) who looked the Gorgon in the face without being terrified of her. They are the ones who realized that, with the Circle, something radically *discontinuous* was happening, but the discontinuity was entirely *proper* to the political and must not be confused with any other. The discovery of this hiatus is of course what the Sophists (the real ones, not those who served to make the Philosophers shine) celebrated in their unprecedented capacity to bring out the truth or the falsity of any situation by the simple play of speech. The Philosophers were quite wrong to make fun of what they took to be "indifference to truth," whereas the Sophists explored with their own words the total originality of this contrast, which they brought out and showed the world in the crucible of their frenzied agonistics. What a discovery, indeed, and one that has nothing to do with contempt for truth! If one can make Helen of Troy innocent or guilty without changing anything about the premises, it's really because politics is played without a net, without a court of appeals, without a world to fall back on; it means that between any given opinion and the following one, there is truly a radical disconnect that no artificial continuity can conceal.

What the Sophists discovered is that there is *a truth of curves*, a necessary truth when one has to produce, in the middle of the agora, statements like "We want," "We can," "We obey," whereas we are multitudes, we do not agree about anything, above all we do not want to obey and we do not control either the causes or the consequences of the affairs that are submitted to us. To pass from one situation to another, yes, a *miracle* is required, a transposition, a TRANSLATION in comparison with which transubstantiation is only a minor mystery. Straight talk is of no use. We have to have freedom of movement, we have to be skillful, we have to be flexible, we have to be able to speak freely without any of the other constraints of veridiction. And it is Socrates who has the nerve, confronting Callicles, to propose as an alternative to this formidable requirement the prospect of teaching *just one* handsome young man, away from the agora, sheltered in the corner of some cheap eatery, under a canopy protecting him from the sun, sipping ouzo and talking about what? A theorem in geometry! And this is the requirement that claims to be defining all truth, reason, *against* the other? As if Socratic dialogue could put an end to the pandemonium of the public. As though one could not admire geometry without mixing it up with what it is least well made to measure: the search for agreement in the middle of the agora, still fuming with sound and fury.

And it is the same hiatus that Aristotle identified so well as *rhetoric*, a partial truth that we must certainly not try to replace with epistemology, on pain of making a category mistake with catastrophic consequences. But it was also what Machiavelli (the real one, Nicolas, not the one with the adjective derived from his name) sought to follow under the heading of *fortuna*, a discontinuity so total that it allowed the audacious, the clever, to become Prince where all the legitimate sovereigns—those who follow the rules—have failed. Closer to us, this is what Carl Schmitt detected in the "state of exception" or the "personal moment" (terms that have become, since Schmitt's day, the objects of a dubious complacency). Whatever these expressions may be, they all aim to grasp the break, the step, yes, the TRANSCENDENCE of the political; what we are sketching in here by the overly geometric notion of *curve* that makes it necessary to *distribute* the little transcendences *all along* what is becoming a Circle, against the temptation to go straight. It will have become clear by now that everywhere there are only *little* transcendences.

This definition of the curve also has the advantage of keeping the State of exception from needing an "exceptional man" who would "be decisive" because he would be "above the law. " Schmitt's error lay in his belief that it is only on high, among the powerful and on rare occasions, that the political mode has to look for exceptions. Look at the Circle: it is *exceptional at all points*, above and below, on the right and on the left, since it *never goes straight* and, in addition, it must always *start over*, especially if it is to spread. This *exception* is what *cuts this mode off* from all the other modes that the true political philosophers all have tried to capture. Here is a real Occam's razor that must be kept sharp! It slices, it has to slice, but it also cuts in the sense that it has to contrast with all the other courses of action. It is authentically exceptional, and it is an exception that must be protected against the panicked cries of good old Double Click as well as against the greedy appetites of reactionaries who believe that true leaders and they alone can "allow themselves" exceptions. In politics, each of us, at every moment, is in an exceptional situation, because it's impossible, and because things never go straight, the baton has to be passed further along so there will be a chance to get it back. There's no point piling on mysteries where there is only one, but one that is truly sublime: the ever-renewed Circle of the collective around issues that are different every time.

⊙ AND A PARTICULARLY DEMANDING TYPE OF VERIDICTION, ⊙ According to the principles of this inquiry, every time we manage to isolate a mode of existence, a type of HIATUS (here, the curve, the exception), a TRAJECTORY (here, autonomy, freedom), we also have to be able to define an explicit form of VERIDICTION. We saw earlier how such a demand appeared incongruous in the case of political discourse: either speakers rationalize too much, or they overestimate irrationality. Might it be necessary to give up speaking truths on the pretext of speaking politically? Must one change oneself into a ghoul, as Socrates demands at the end of *Gorgias*, in order to be right, but only after the fact, emerging from Limbo, a shade judging other shades, a phantom judging other phantoms? No, of course not, since, dispersed in institutions, buried in practices, captive in our imaginations and in our judgments, there is a whole *know-how* concerning speaking well and speaking badly, acting well and acting badly in the political realm, which should make it possible to define the felicity and infelicity conditions of the Circle. Let us recall that to make this competence explicit is not to formalize it according to a different enunciative key but, on the contrary, to follow it in its own language.

Now, if we look more closely, we find that no requirement of truth is more terrible, more radical, more unbearable, more educational, more civilizing, as well, than the obligation to *speak truths politically*. Those who do not recognize the force of this requirement really don't deserve to bear the fine name of "citizens. " The demands for verification and falsification are so constraining that one would do anything, understandably, to avoid them—and people indeed do everything to avoid them . . . Of course, the production of objectivity is also demanding [REF]; just discourse in religion carries such a weight that one addresses God only with fear and trembling [REL]. But how could we designate the anguish of expressing opinions politically *without being ready to run through the entire Circle*? Nietzsche proposed to replace moral exigency with the thought of the Eternal Recurrence: "Act in such a way that you can wish for your action to be repeated eternally"; his readers quaked at the harshness of such a demand. What must we think of the yoke of this maxim: "Speak publicly in such a way that you will be ready to run through the entire circle, coming and going, and to obtain nothing without starting over again, and never to start again without seeking to extend the circle"? If there is such a thing as the dignity of politics, the truth of its enunciation, it lies in its having agreed to put itself to the wheel in this way, to have yoked itself to such a grindstone.

We can now measure the immense difference between chains of reference and political Circles without devalorizing the latter in favor of the ⊙ WHICH THE [REF · POL] CROSSING MISUNDERSTANDS. former, but understanding also why mutual misunderstanding is inevitable [REF · POL]. So much work, so much equipment, so many institutions to ensure the service of both! So many hiatuses to leap over! And so many misunderstandings, if we start mixing up their incommensurable requirements. If Callicles sets out to judge Socrates's geometric proof by the yardstick of the Circle, he will misinterpret it as surely as if Socrates claims to be teaching Callicles the art of speaking straightforwardly to an angry crowd. To mark the passage from the ideal of autonomy to the course of the Circle, we have already come across the admirable word AUTOPHUOS, which Socrates flings in the face of the Sophist, in *Gorgias*, as a term of scorn, without noticing that he is accurately defining the virtue of the *logos* that he is purporting to mock: the self-engendering of the Circle.

As for us, we need no longer confuse chains of reference with the Circle; we can speak straight when we need to, and crooked when we need to. Are we not becoming a little more articulated? We can finally respect democracy as much as proof [**REF · POL**]. Since we have become capable of admiring both the APODEIXIS of scientists as well as the EPIDEIXIS of politicians, we need no longer deceive ourselves by embracing Plato's macabre solution, in which he claims to have Reason reign over politics but only a posteriori, from Hades, from the kingdom of the dead! Here is a category mistake that we may be able to bring to an end: access to remote beings, yes; the reign of the living dead, no.

[**POL**] PRACTICES A VERY DISTINCTIVE EXTRACTION OF ALTERITY, ☉   There is indeed a transcendence of the political, but it can be identified only on condition that we refrain from appealing to a world other than its own, to a netherworld, a kingdom of Hades. And there is the difficulty in a nutshell. How can we engender the Circle without immediately giving it any consistency other than that of habit, a sort of inertia that would make it possible to do without the requirement of reprise by putting it off until another day, until the cows come home, until the Great Day arrives? And yet what a relief that would be!

It is clear that every mode extracts from being-as-other a particular form of ALTERATION to which none of the other modes is yet attached. For every mode one has to ask the same question: *without* it, what would be *missing* in the set of values to which we hold? The case of the political mode is actually simple enough: without the Circle, there would be no groupings, no group, no possibility of saying "we," no collecting, quite simply, and thus no collective, either. All the other modes thrust themselves into being and alteration. This one (like law, as we shall see, but in a different way), this one alone, *comes back* to assemble those who otherwise would disperse. This one alone comes full circle. But it comes back through the effect, constantly renewed, of a reprise that has an exhausting aspect, since it cannot, it *must not* rely on some substance, some form of inertia, for that would amount to substituting a different body for the Circle and thus suspending its own movement.

If we must speak about a "BODY POLITIC," we shudder for fear of making a new category mistake. It may constitute a "body," but its corporality resembles no other, precisely because it can never be supported by any other continuity: not that of subsistence [REP], or technology [TEC], or fiction [FIC], or the religious [REL]; and it can never become a living composite mimicking the beings of reproduction [REP · POL]. And yet this isn't for want of efforts to base it on something a little more stable and a little less demanding. From the Roman fable of the Belly and the Members through the "apparatus of the State" to the sociobiology of the new advocates of social Darwinism, we have never stopped *substituting* one type of being for another in order to try to understand this strange work of collection, while masking it.

But these substitutions have always failed, for they have been unable to capture the particular strangeness of the beast sketched by the Circle of representation and obedience. This creature is composed of segments, or vectors, that are transcendent with respect to those that precede and those that follow; there is no continuity between one and the next. Either one leaps, and the political moves along, or else one doesn't leap, and nothing political is said: even if people think they are "taking a stand," even if they think they are "giving orders," even if they are indignant, or suffering, or complaining, even if they are absolutely right to complain about being so badly represented—below—or so badly obeyed—above. If it was terrible not to be able to speak about religious things without converting the listeners one was addressing, the requirement to fill or to empty what one says politically depending on whether one leaps over this chiasmus or not is more terrible still. *Hic Rhodus, hic saltus.* Put your money where your mouth is.

A body that is not one; harmony that never ⊙ WHICH DEFINES A harmonizes; unity that disperses immediately; PHANTOM PUBLIC ⊙ dispersal that must be reassembled at once; different issues, every time, around which people have to assemble because they don't understand one another . . . We have to admit that this Circle, constantly renewed, is really a pretty odd creature. That is why we can sympathize, in spite of everything, with the horror Plato must have felt before the hydra of democracy: he may have gotten the monster wrong, but he was right: it really is a matter of *aliens*. To sketch their emblems, we mustn't hesitate to go back to the invisibles. There is something of the *phantom*, indeed, in this body politic that is not an organism (we shall return to this point when we come to the crossing [POL · ORG]).

The word "phantom" won't surprise us any longer, since we have gradually learned to give each mode its proper weight of being. This phantom is not a mysterious ectoplasm that passes through walls to frighten children; it is the exact definition of the form created by the incessant reshaping of the Circle, provided that the process is not stopped, for then the phantom disappears for good or shrinks down to a glowing point, as when children use a flaming stick to draw shapes on a dark summer night and suddenly suspend its movement. Here is the particular alterity that the political extracts from being-as-other, an alter-ation, an alienation, that no other mode has ever attempted: producing oneness with multiplicity, oneness with all, but doing so *phantomati-cally*, provisionally, by a continual *reprise* and without ever being under-girded by a substance, a durable body, an organism, an organization, an identity. It is for just this reason—Walter Lippmann may be the only person who really got it—that one can respect the ontological dignity of the political mode only by grasping it in the form of a PHANTOM PUBLIC to be invoked and convoked. Neither the public, nor the common, nor the "we" exists; they must be brought into being. If the word PERFORMATION has a meaning, this is it. If there are invisibles that one must take special care not to embody too quickly—for example in the State, that other cold monster—this particular phantom is one of them.

⊙ IN OPPOSITION TO THE FIGURE OF SOCIETY, ⊙   This is why those who have sought to give body to the body politic are so few in number. There is a powerful temptation to *replace* the phantom with something that would be seen in transparency *behind* it, namely, "SOCIETY" (along with NATURE and LANGUAGE, Society is the third unwar-ranted form of padding, as we know, that we must gradually learn to empty of its role as foundation). Here is the trap of the political, playing the same role for this mode of existence that BELIEF plays for religion [REL]. As soon as one relies on something other than the political, this particular mode of enunciation vanishes like Nosferatu with the first ray of sunlight. In this view, politics would be a mere appearance that one could perceive only by discovering behind it the powers of "Society."

The veils of the political would only "express" the harsh realities of the economy or of power relations, while concealing them. Those who express themselves this way are obviously wrong about politics—not to mention that they are also wrong about appearances [HAB], and, as we are about to see, about The Economy as well. To avoid meeting one phantom, they rush into the arms of another, this one really appalling: the one they call "the social." From bad to worse.

Society, clearly, is not a mode of existence. Like Nature, like the Symbolic, it is a perfect example of false transcendence. It is the amalgamation of all the modes and all the networks whose threads the Moderns have given up trying to untangle and which they take as a foundation in order to explain how all the rest holds together—religion, law, technology, even science, and of course politics. Society, as Gabriel Tarde declared, is always what has to be explained rather than what explains, just as substance is in fact the sometimes durable result of all the modes rather than what makes them subsist. Such at least is the starting point of the ACTOR-NETWORK THEORY, which the present inquiry seeks to complete. If there is a Society that supplies explanations, the movement of envelopment of the political mode will remain forever invisible. People will find it preferable to rely on identities that appear to be more durable, more assured, more rooted than the Circle. And thus they will stop maintaining the only mode capable of performing artificial, provisional, immanent identities, the only ones we have at our disposal for producing any COLLECTIVE whatsoever—and, above all, the only ones we have at hand to extend the collective to "all," that "all" of variable dimensions that the project of universalizing political autonomy had oversimplified. Persevering in this error simply leads to losing the habit of speaking politically. We have to choose between two phantoms: either the phantom of Society or the phantom of the political. For politics assembles and reassembles, but it does not make a body, it does not splice together, it does not make agreement. (Still, we shall see in Chapter 14 how to present a more charitable version of the emergence of "Society.")

This phantom is dreadful, of course; it frightens Double Click as much as it fascinates perverse souls and reactionaries of all stripes, but there is no point rubbing it in by making it even more horrible through

⊙ WHICH WOULD MAKE THE POLITICAL EVEN MORE MONSTROUS THAN IT IS NOW.

hybridization with others. This is nevertheless what happens when, through a sort of guilty complacency, the Moderns amalgamate it with the beings of metamorphosis that we met in Chapter 7. And if it is quite true that a large part of the political institution seems to respond rather to the injunction "Eat or be eaten" (if the memoirs of statesmen are to be believed, this would seem to be the most indispensable of maxims), what we have here is, as always, an intersection between two forms of reason that must not be confused with each other [MET · POL]. The term "balance of power" is part of the overly facile metalanguage to which it is time to add its modality so as to avoid confusing [MET] power with [POL] power.

The same parasitic monstrosity is found again in the temptation to seek the assurance of a definitive and unanimous unity in religion, on the pretext that "the time has come"—while this is actually never the case. The reprise of the Word gives even less durable assurance than the requirement of renewing the Circle. It is the [POL · REL] crossing that would now find itself impossible to unfold owing to confusion between the virtual totality of the political with the even more virtual totality of those who are saved by the Word and who form a particular group, the one called a "Church. " Blending the two has created the worst of political theologies, has caused as much confusion as the orthopedics of the political undertaken by Reason [POL · DC] or even—something seemingly even more harmless—the idea of basing all politics on technology [TEC · POL] or law [TEC · LAW]. We have to learn to respect the phantom public without blending it with other invisibles, without making it rely on other substances, without making it even more horrible to look at. It is already hard enough to summon up this phantom; its apparitions have already become somewhat rare.

The reader will understand that no anthropology of the Moderns is possible as long as we do not render to politics the worship whose altars they should never have abandoned. (I am obviously not speaking here about the pathetic episodes of the "Cult of Reason.")

WILL WE EVER BE ABLE TO RELEARN THE LANGUAGE OF SPEAKING WELL WHILE SPEAKING "CROOKED"?

What hope do we have of reviving a life form that may have already gone the way of the religious institution and whose contrast may have entirely disappeared from contemporary experience, either because it has basked in the hope of good governance or sought refuge in contorted cynicism? For the worst case would obviously be to believe that one can approximate the Circle by going sometimes straight ahead—but that is impossible—and sometimes along a crooked path—as if it sufficed to lie and obfuscate to be a true politician! If we could do for political enunciation what we have done for truth and falsity in science and religion, we would allow the political mode to raise its head. We would resuscitate it—*suscitate* it again, call it back to life and action. The kingdom of the living dead would not have become a "glorious body," to be sure, but more modestly, as Lippmann puts it, a phantom public. The world is being populated with a few too many invisibles? Yes, but it *is being populated*, and that's what counts. Before reacting in shock to all these phantoms, the reader should remember what the modern world looked like, with its objects, its subjects, and its representations! Those who reject invisibles give birth to monsters.

If we have redefined the word CATEGORY as what makes it possible to speak well in the agora to those who are concerned with what one is saying about them, it is because nothing is more important for this inquiry than to find the difference between truth and falsity in politics. If there is one area where our inheritance has to be revisited, it is surely that of the hopes placed in politics and its capacity for extension. What will we have to do to situate *appropriately crooked speaking* once again at the center of our civility as the only means to collect the collective, and above all to universalize it? Does the Circle give us a thread like Ariadne's that will let us speak here again of the rational and the irrational but in a well-curved way, that is, in its own language, provided that we don't seek to judge it with the help of a different touchstone? We need this thread, for how could we stand up straight on the agora, with no hope of help from any Science and yet without giving up on reason, about controversial issues that have taken on the dimensions of the planet and in the heat of a crowd that now numbers in the billions?

·Chapter 13·

THE PASSAGE OF LAW
AND
QUASI SUBJECTS

Fortunately, it is not problematic to speak about law "legally" ⊙ since law is its own explanation.

It offers special difficulties, however, ⊙ owing to its strange mix of strength and weakness, ⊙ its scarcely autonomous autonomy, ⊙ and the fact that it has been charged with too many values.

Thus we have to establish a special protocol in order to follow ⊙ the passage of law paved with means ⊙ and to recognize its terribly demanding felicity conditions.

The law connects levels of enunciation ⊙ by virtue of its own particular formalism.

We can now understand what is distinctive about quasi subjects ⊙ while learning to respect their contributions: first, beings of politics [POL], ⊙ then beings of law [LAW], ⊙ and finally beings of religion [REL].

Quasi subjects are all regimes of enunciation sensitive to tonality.

Classifying the modes allows us to articulate well what we have to say ⊙ and to explain, at last, the modernist obsession with the Subject/Object difference.

New dread on the part of our anthropologist: the fourth group, the continent of The Economy.

I T IS NOT A BAD THING, IN THE COURSE OF SUCH AN ARDUOUS INVESTIGATION, THAT THE AN-THROPOLOGIST OF THE MODERNS CAN FINALLY catch her breath as she approaches a more familiar mode, that of law [LAW], which we first encountered in Chapter 1. Still, readers shouldn't feel they've been let off the hook! The difficulties won't really start until the next chapter, when we meet the beings who find themselves almost arbitrarily mixed up in the famous continent of The Economy, the Moderns' pride and joy, the only universal, alas, that they have actually succeeded in extending to all the other collectives. The most daunting challenges will arise when we have finished Part Three and the anthropologist will be obliged to speak *well* of all the values *to those* who are concerned by them.

Unlike the other modes, law does not strike the ethnologist as an insoluble brainteaser, for it promises a fairly satisfying correspondence between a DOMAIN, an INSTITUTION, and a CONTRAST whose specificity, technicity, and centrality seem to have been recognized by everyone. This time, the investigator can take almost literally what her infor-mants tell her about the VALUES to which they hold and for which they are prepared to give their lives. Let us think about the famous "state of law" that they put forward so complacently (actually a composite [POL · LAW]). Unmistakably, law has its own separate place; it is recognized as a domain that can be isolated from the rest; it has its own force, as everyone would agree; and above all—a crucial element for our inquiry—it has its own

mode of veridiction, certainly different from that of Science, but univer-
sally acknowledged as capable of distinguishing truth from falsity *in its
own way*. Even more reassuringly, we don't need to bend ourselves out of
shape to see that, here again, here as always, we need to look to the adverb
to grasp it. There are laws, regulations, texts, issues without number; but
to capture the type of veridiction proper to the law, one has to take things
legally. Whereas we have to go to a great deal of trouble to learn how to
speak "technologically," "religiously," or "politically," it seems that we
agree without difficulty to take law as a PREPOSITION that engages everything that follows in a specific mode that is both limited and assured.
Better than all the others, law thus lends itself to analysis in terms of
modes of existence.

All the more so in that, if law enjoys a rare form
of autonomy (not to be confused with the [POL] form
we have just identified), this is because it never loses
its enunciative key along the way. When lawyers are asked to define what
they do, they string together long sentences in which they unfailingly
use the adjective "legal" to qualify everything they say, without troubling to define it further, without even realizing that they are caught up
in a tautology! "It suffices to recall the classic definition of the legal act,"
Carbonnier writes: "a manifestation of will that is destined to produce
effects *of law*, modifications of *legal* mandates; in other words, which is
destined to introduce a human relationship into the sphere of *law*." Or
Hart: "Such rules *of law* do not imply duties or obligations. Instead, they
provide individuals with the *means* to fulfil their intentions, by endowing
them with the *legal power* to create, through *determined procedures* and *in
certain conditions*, structures *of rights* and duties within the limits of the
coercive apparatus *of the law*."

Such tautologies fail to embarrass these excellent authors. Nor
should they, since they reflect the very originality of the law, that which
gives it its extraterritorial status: it is its own explanation. Either you
are inside it and you understand what it does—without being able to
explain it in another language—or you are outside it and you don't do
anything "legal." Like all the other modes, you'll say? Yes, except that law
is the only one to have offered this sort of resistance to the demand for

⊙ SINCE LAW IS ITS
OWN EXPLANATION.

explicitness imposed by Double Click's hegemony [DC] (and his passion for the EXPLICIT in the sense defined in Chapter 10).

To put it differently, the legal institution does not seem to have suffered as much from the jolts of modernism as the other contrasts. Everyone can experience it: whether we are running into a legal problem, spotting a justice's black robe, reading the fine print of a contract, signing a document before a notary, for example, each of us is well aware that there is something quite particular going on, something frightfully technical, which will establish the difference between truth and falsity in a way that is at once obscure and respectable. Without even being surprised at this, we may say that a statement is true or false in "the legal sense," which means both "in a narrow and restricted sense" and "with a completely original interpretive key that one has to learn to recognize."

Law thus benefits precisely from the form of respectful difference that would have protected all the modes of veridiction we have been reviewing here, if only Double Click had not paralyzed them all, one after another. This is why, starting in Chapter 1, I have relied on law to begin to unfold the modes of existence in the only way that allows us to set them down side by side, the mode called prepositions [PRE], which has served from the outset as a metalanguage for this inquiry—or, better, an *infralanguage*.

IT OFFERS SPECIAL DIFFICULTIES, HOWEVER, ⊙ And yet nothing is ever simple in our study; it would be a mistake to think that it is easy to work out the originality of a key even if that key is recognized by everyone. Precisely because it has been kept at a respectful distance, the legal mode occupies such an autonomous position that it has been entrusted to specialists—lawyers, judges, legal scholars—whose importance, authority, and usefulness are certainly acknowledged, but who have never learned to share the definition of that value with others. If its tautologies have protected it from the judgments of Double Click, they have not made it more comprehensible outside its own sphere. Whereas technology, fiction, reference, religion, and politics have penetrated everywhere (thereby multiplying the risks of cascading category mistakes), law suffers from the advantage of having been kept too respectfully at a distance. Hence this strange experience: while it may be obvious to legal experts that something is "legally true or false," for anyone external to the law—that is, with rare exceptions, for almost everyone else—it is a total surprise to see that law can both take on so much importance and take up so little space.

Hence the continual disappointment felt by those who must suddenly grapple with a legal problem they had not anticipated. "What?" they exclaim. "How could this little obstacle have the

⊙ OWING TO ITS STRANGE
MIX OF STRENGTH
AND WEAKNESS, ⊙

power to stop us?" "Now that you've solved the problem, is that all there was to it?" On the one hand astonishing force, objectivity; on the other remarkable weakness. We feel that force every time we learn that, "because of a simple signature missing on a decree," the appointment of a bank director was blocked; or when we see that a dam construction project essential to the survival of a valley has been suspended owing to "a tiny defect in the declaration of public utility"; or that jobless workers have lost their rights "because they misread the contract that bound them to their employer"; or that one business was unable to acquire another "because of a legal constraint imposed by Brussels." But we feel the weakness every time we despair at seeing that the "legally justified" decision is not necessarily just, opportune, true, useful, effective; every time the court condemns an accused party but the aggrieved party has still not been able to achieve "closure"; every time indemnities have been awarded but doubts still remain about the exact responsibilities of the respective parties. With the law, we always go from surprise to surprise: we are surprised by its power, surprised by its impotence.

This is what is meant, at bottom, when someone declares with a blend of scorn, envy, and respect that law is *superficial* and *formal*. Superficial, necessarily, because it is attached only to a minuscule trace—that of attachments, as we shall see. Formal, in a very particular sense, because it respects the FORMS (a multimodal term like no other) that make it possible, in certain very special and highly technical circumstances, to ensure the continuous passage from one document to another in order to ensure continuity through astounding discontinuities. This is the HARMONIC that it has in common with the passage of reference [REF · LAW]: law, too, slips through forms at a dizzying pace, not so as to ensure the maintenance of a constant through a series of inscriptions—the hurdle race described in Chapter 4—but to ensure the mobilization of "the law as a whole" in a particular case, on the condition that it remain perfectly superficial so as not to be encumbered by everything. A strange capacity that the ethnologist ultimately finds more difficult to trace than that of chains of reference.

⊙ ITS SCARCELY AUTONOMOUS AUTONOMY, ⊙   What makes law so hard to grasp is that as soon as it has been defined as a separate world, carefully delimited by its own tautologies, we notice how *flexible* it is, and with what confounding agility it absorbs all sorts of injunctions from other regions: politics, the economy, trends, fashions, prejudices, media. As a result, just when we think we have discovered it as a particular sphere, with its own regulatory modes, we notice that the legal institution is so porous that its decisions look like so many weathervanes, turning with every breeze. But there is more, something even more astonishing from our ethnologist's perspective: legal experts themselves—lawyers, judges, professors, commentators on legal doctrine—recognize this porosity with a curious mixture of innocence and seeming cynicism: "Well, yes, naturally, law is flexible, and it *depends* on everything else; what did you think?" Hence the symmetrical temptation to treat it not just as superficial and formal but as a mere cover, as a rather clumsy disguise for inequalities. "Tell me what you want to get out of this and I'll find you the legal wrappings that will do the job." In a matter of moments, we have passed from one extreme to the other: we were admiring the objectivity of law, its capacity to make all powers bow before it; now we are indignant that it is so supple, so obsequious, that it has the regrettable capacity to cloak the nakedness of power relations.

⊙ AND THE FACT THAT IT HAS BEEN CHARGED WITH TOO MANY VALUES.   This constant seesawing is all the more awkward in that it is hard to see how to reconcile it with the heap of values that law has been asked to establish. At various historical moments it has been entrusted with the task of bearing morality, religion, science, politics, the State, as if its fine spiderwebs on their own could keep humans from quarrelling, going for the jugular, tearing each other's guts out; as if it were law and law alone that had made us civilized—and even made us human, through the happy discovery of the "Law" without which we would still be subject to the "reign of Nature"! We know perfectly well that each mode claims in its own way to explain all the others, but if we have challenged Double Click's assumption of the role of linesman, it is not so as to make law the general overseer of all the modes. We have to resist all fundamentalisms, including that of law.

It is true that, precisely because of its isolation, law appeared fit to serve as a repository or warehouse for all endangered values. Especially if it managed to corner one of the most polysemic terms there is, the term RULE. Unable to supply this multimodal term with the exponents that would have allowed them to qualify it, the Moderns told themselves that, without law, "there would be no more rules" and "humans would do anything they liked." Once again we see law shift from minimum to maximum and, after having been reduced to a bit of packaging blown around in the winds, it swells up to become our last rampart against barbarism. Ultimately, despite its seeming homogeneity, the legal institution, too, has a proteiform aspect.

Having rejoiced too soon at finding law preformed to fit her analysis, our investigator has to demonstrate great flexibility once again in order to shelter its preposition from all these sequential THUS WE HAVE TO ESTABLISH A SPECIAL PROTOCOL IN ORDER TO FOLLOW ⊙ shifts and to grasp its distinctive features through meticulous ethnographic work. (But how could she doubt this? Doesn't she know that each mode demands its own method? That it is in the very nature of her inquiry to have to change the analysis for each type of veridiction? That there is in fact no metalanguage adapted in advance for grasping a given mode? None, *not even her own* ...)

To grasp law, then, we have to begin by unburdening La Fontaine's "ass carrying relics" and ask it to transport only itself, without trying to serve as carrier for humanity, decency, civilization, truth, morality, the Law of the Father, the whole kit and caboodle. Next, we have to resist being impressed by the mix of total autonomy and total porosity: if we think we can grasp it by the pincers of such an alternative, this means that it must make its way quite differently. Finally, the ethnologist has to acknowledge that law can be strong, objective, solid, decisive, even though its solidity is totally unlike that of the other modes. We know why: if law is a mode of existence, it depends on the passage of particular beings that have their own specifications, their own mode of visibility and invisibility, their own particular ontological tenor or tonality. On this level, at least, there is no reason to be surprised.

⊙ THE PASSAGE OF LAW
PAVED WITH MEANS ⊙

What does the legal experience look like when we follow its particular movement? What we see is the passage from one stage to another of a quite distinctive fluid that is manifested materially in a series of *dossiers* whose content, size, and composition vary according to the stage reached by the affair in question. With one of their extremities, these dossiers hold onto multiform complaints more or less well articulated by plaintiffs who feel attacked for one reason or another in their property, their dignity, their interests, or their lives. But with the other extremity they hold onto texts that, in one way or another, one step at a time, mobilize "the law as a whole." The test for the plaintiff, the legal experts, and the observer, always a long and trying experience, thus consists in confronting, on one side, cases, "facts," feelings, passions, accidents, and crises, and on the other, texts, principles, and regulations that may "take them a long way," sometimes all the way "to the top," to the Constitutional Council, the Supreme Court, the European Court of Justice, and so on.

And in the middle, between the two? This is where a series of transformations, translations, transmutations, transubstantiations unfolds; by degrees, and by paying—sometimes very dearly—for an endless lineup of clerks, lawyers, judges, commentators, professors, and other experts, the passage of law gradually modifies the relation between the *quantity* of facts, emotions, passions, as it were, and the *quantity* of principles and texts on which it will be possible to rule. This proportion of relative quantities is known by the admirable term "legal *qualification*." It would be useless to stick to the facts alone so as to sympathize with the victims or to seek objective truth [REF · LAW]; and it would do no good, either, to consider the principles alone without applying them to anything. To move forward, one has to find out whether some fact corresponds, or not, to a definition that will allow a judgment to be made. The legal professionals are not looking for novelty or for access to remote states of affairs; they are seeking only to stir up this fact in every direction in order to see what principle could actually be used to judge it; they are seeking only to stir up all the principles until they find the one that could perhaps be applied *also* to this fact.

As we saw in Chapter 1, the movement of law follows a path paved all along with MEANS, a seemingly banal term but one that is constantly in use (not to be confused with [TEC] means). Either there is a legal means and it works—the means is sometimes said to be "fruitful"—or else there are no legal means and "there the matter rests." You wanted to stop a factory from polluting? Yes, but here's the problem: you lack the "quality to act," you have no standing; you can make as much fuss as you like, but nothing will happen on the legal level. You wanted to have your French nationality recognized? Yes, but you don't have your parents' naturalization certificates; you can go ahead and alert the media, but nothing will happen on the legal level. You can make what you will of the stalled affair: fiction, religion, science, a scandal, but not law. The linkages of law thus have this distinctive feature: through the intermediary of a particular hiatus they allow means to follow a highly original trajectory in a series of leaps from facts to principles. There is nothing continuous that can link a deliberate misrepresentation and a text, and yet legal means establish this type of continuity, which gives the full force of a principle to a little case of no importance.

And how does it end? As always, through the establishment of an amazing bypass among a set of principles that have just been mobilized in toto to hold the affair together, and through a judgment whose particular quality depends on a lengthy *hesitation* over its attachment to the case. But we must not expect any novelty from this judgment. On the contrary, what counts above all is what is known by the admirable term "legal integrity." Unless, by a reversal of jurisprudence, the affair offers the opportunity—but it will never be more than an opportunity— to modify the principle; and, moreover, even in this case, it will only be a matter of making the body of legal doctrine still more coherent, so that, in the last analysis, nothing will really have budged. Law is homeostatic. It can be flexible, heteronomous after its fashion, certainly, but it has to be able to proceed in such a way that all cases, all deliberate misrepresentations, all crimes, thanks to a minimum of innovation, can be set into relationship through the intermediary of specific cases with the totality of what is valued by those who have drawn up its principles. It is about law that we should say "plus ça change, plus c'est pareil"!

It is understandable that the Romans were rather proud of having isolated this astonishing contrast and decided to entrust to it everything they cared about—and they took pains to shore it up with the sword. What is most astonishing is that we are still so Roman when we judge, even today, in wigs or black robes, in enclosures designed to resemble the ancient basilicas of the Forum Romanum; when judges and lawyers alike still rely on the same word mills and on dossiers, writings, texts, documents whose technology has hardly changed, except for the occasional computer screen. Cicero could take his place in the French Council of State or in the Luxembourg Tribunal without having to do anything except learn French! After taking in all the details of the dossier and all the resources of the doctrine, he would know very well what to do, without relying on any technology but his own words to convince those to whom he was speaking of the quality of the "legal means," which he would expose with talent. He might be surprised by everything in the content of the law, but not by the way of speaking *legally* about these affairs. "*Quo usque tandem abutere, Catalina, patientia nostra?*" "How long, O Cataline, will you abuse our patience?"

⊙ AND TO RECOGNIZE ITS TERRIBLY DEMANDING FELICITY CONDITIONS. Our ethnologist asserts that every mode obliges us to respect FELICITY AND INFELICITY CONDITIONS, a "particularly trenchant" discrimination between truth and falsity. This is what her method requires. Like a mother who loves all her children with the same exclusive and all-embracing love, the investigator has to reconstruct each mode's exclusive manner of demanding truth. This is the only way the pluralism of veridictions can work not to attenuate but on the contrary to sharpen the constraints of the rational—without letting any one of these constraints get ahead of another.

And so, here again, before the law that "works" or doesn't work, before the finesse of what separates good and bad means, before the sharp edge of this knife, the investigator is poised in admiration. One has to have spent long afternoons with judges in the Litigation Section of the Council of State, listened to them chew over—and over and over—the same infinitesimal story of a trash-can or a deportation, from subsection to section, from section to plenary session, from reports to the conclusions of the "public reporter," from conclusions to rulings, sometimes for months on end, in order to feel the subtlety, the delicate balance, of this hesitation. And yet to proceed quickly would be to lie, legally. No, the process has to be repeated, again and again. Without hesitation, there is no law—the affair would simply be categorized, managed, organized. You think they've finished with the report? Here comes the reviewer who takes up the whole affair again, means by means. They're ready to vote? Not yet. The judge's assistant, the president, the commissioner all chime in with their own scruples. So the process is endless? No. As soon as the channels for appeal have been exhausted, as soon as the litigation has gone to the Assembly, then finally "there's a judgment," "the matter is settled," "the case is closed." Has it been well judged? Yes, provided that there has been sufficient *hesitation*.

Of course a researcher, too, will wake up in the middle of the night tormented by anxiety about whether the experiment has been badly set up and will therefore prove nothing. Of course an artist, too, will wake up at night nudged by the sphinx of the work—a trite paragraph, a botched stage entrance, an awkward cut between two sequences. Of course a woman in love, too, will wake up at night hearing an appeal from the lover whom she fears she may not have loved well enough. Of course we shudder at the realization that our indignation has gone for naught, that we have done nothing to extend the political Circle. But not all insomnias are alike. The scruple that obliges the judge, goaded by fear that he "has judged badly," to take up the dossier once again derives from the fact that he has to encounter beings of law—as if there were some objectivity, some legal externality, some proofs, incontrovertible because they have been discussed at length, whereas he knows perfectly well that the slightest breeze, the slightest passion, the slightest prejudice, the slightest influence will drive them away. He must train his body and his mind not only to function at an often disheartening level of technicality—all that law, all those codes—and not only to pay the most meticulous attention to the details of the most insignificant or most sordid affairs but also to become—like the blindfolded Justice that serves as his emblem and whose sword threatens to run right through him—an ultrasensitive scale that nothing must be allowed to disturb and that, *after* having wobbled, *because* it has wobbled, settles into equilibrium. Try this, and then ask

yourself what beings it would be best to associate with in order to be quite
sure that you'll be able to sleep in peace …

THE LAW CONNECTS LEVELS
OF ENUNCIATION ⊙ By isolating the notion of means, the ethnologist of law sees clearly what makes it possible to leap
over the successive HIATUSES between the affair full
of sound and fury, sweat and emotion, and the severe, stable principles
to which it must gradually lead. By following cases, she understands by
what sort of "fractional distillation" one passes from an immense and
complex dossier to the question of a single sentence submitted to a jury
or a tribunal. But she still does not see what makes this miracle possible.
As always in this inquiry, she has to turn toward the particular ALTERA-
TION that the beings of law succeed in extracting from being-as-other and
that gives them their particular tonality of unquestionable objectivity
and their extreme sensitivity.

To detect this alteration, it may be useful for her to rely on the notion
of SHIFTING OUT that we identified in Chapter 9. Every enunciation—every
hiatus, every dispatch—ends up in fact in a SHIFTING OUT that creates an
antecedent and a consequent and, between the two, a chiasmus that
always has to be crossed, whatever the mode, to obtain a particular type
of apparent continuity. This is even the reason why one can never rely
on any SUBSTANCE for the subsistence of a course of action—even habit
[HAB] depends on a particular discontinuity, the one that omits the PREP-
OSITION without forgetting it. But, starting with the technological mode
[TEC], we have seen that the successive levels proliferate while adding
new discontinuities every time: other spaces, other eras, other actors,
other virtual enunciators. And if there is indeed reprise, there is loss,
each time, of the element that preceded the reprise. The world is populated and filled, yes, but it *scatters*.

In fact, all the modes identified up to now have this distinctive
feature: they *pass*, they move forward, they launch into the search for
their means of subsistence. Each one does it differently, to be sure, but
they have in common the fact that they never *go back* to the conditions
under which they started. Even the political Circle [POL], while it always
has to start over, disappears, as we have seen, as soon as it is interrupted,
without leaving any traces but the slight crease of habit. Even the end
time that is to permit the definitive and salvific Presence, so typical of

religion [REL], disappears without recourse if one stops repeating it in completely different ways in order to respect the strange tradition that does not offer the assurance of any inertia. In other words, the other modes *do not archive* their successive shiftings or translations. They leave wakes behind, of course; they begin again, each making use of the preceding ones, but they do not *go back* to *preserve* the traces of their movements. The predecessors disappear once the successors have taken over. This is what they do: they pass; they are PASSES.

Reprising that reprise: the originality of law lies right here. To ensure continuity despite discontinuity, law links to one another the various levels that shifting out keeps on multiplying. As we saw in Chapter 10, fiction multiplies levels of enunciation in yet another way: forward, by dispatching other figures—this is level n + 1—and, as it were, backward, by the creation of an implicit level—this is level n - 1—that would sketch out the author and the presumed receiver of the work. Between the two, a new chiasmus, a new gap that each new dispatch, each new course of action, only deepens. To achieve continuity between an enunciator, what he says or what he does, and those whom he addresses, it would be necessary to *reattach* what the continual movement of dispatches never stops *detaching*. An apparently contradictory requirement. Now the originality of law is that it attempts an exercise as impossible as that of politics or religion, but just as important for the definition of what will be called, for good reason, a "subject of law."

If law [LAW] is so original, if it has even resisted the hegemony of Double Click so successfully that it has always been considered as "true in its genre," ⊙ BY VIRTUE OF ITS OWN PARTICULAR FORMALISM. it is because law alone ensures this *reattachment* of frames of reference. Thanks to law, you can *multiply the levels of enunciation without causing them to disperse.* But this reattachment has a price, one that is always held against it even when its technicity is respected: that of "sticking to the forms." How? The [LAW] form is what could be called an *archive,* or, to stretch the meaning of another legal term, a summons, an *assignation.* It is this aspect of the fabrication of "subjects," this padding with enunciators and enunciatees on which the highly distinctive mode of existence of law is going to insist quite literally. It has in common with habit [HAB · LAW] the fact that by its own discontinuities it *ensures* the continuity in time and space of courses of action that would otherwise always scatter by dint of undergoing the continual shock of shifting out. While in fact there is neither real continuity of courses of action nor stability of subjects, law brings off the miracle of proceeding as though, by particular linkages, *we were held to what we say and what we do*. What you have done, signed, said, promised, given, *engages* you.

This is how law manages to retain traces of *all* courses of action, on condition that it retain as few as possible. Double Click, always starving for information, will be disappointed [**LAW · DC**]. (Poor Double Click is starving for everything: politics, religion, law—won't he end up starving to death?) This is why law is so often said to be formal, even formalist. If it disappoints, it's because it has to remain without content. But it does something better: it reassures. It even manages to circulate everywhere, to mobilize the totality of law for every case, but with a draconian condition attached: it must say almost nothing, except that there have been attachments—of one utterance to another, one enunciator to another, one act to another, one text to another. While it transports no information, it insists on asking whether there is a path from one particular utterance to another, or between a given utterance and a given enunciator.

The originality of law can be identified by a number of features: the very notion of *procedure*; the assignation, the *signature* and its quite distinctive "tremor," since it leaps over the division of levels of enunciation; imputation; qualification, the link between text and case (what does it mean to be a "journalist in the sense of article 123 of the code"?); and even canonical definitions such as responsibility ("so-and-so is indeed the author of this act"), authority ("this person is indeed authorized to sign the acts"), property ("this person indeed has the right to hold that piece of land"). If "legal means" do not resemble any of the other trajectories, it is because of this particular obsession for making visible what the passage of all the other modes fails to disclose. It is as though law—a regime that makes it possible to connect enunciators and utterances by invisible threads—managed to *go back up the slope that the proliferation of enunciatory messages or dispatches kept on coming down*. With law, characters become assigned to their acts and to their goods. They find themselves responsible, guilty, owners, authors, insured, protected. And this authorizes us to say that "without law," utterances would be quite simply *unattributable*.

The diffusion of writing has certainly made these traces easier to follow and to archive, but even among peoples said to be "without writing," the anthropology of law attests to hundreds of astonishing procedures for attaching promises to their authors by solemn oaths and imposing rituals. On this point, writing has only accentuated the habit of already well-established links. Which explains, moreover, why even the most exotic collectives have always been recognized as perfectly capable of producing law.

Law cannot totalize the entire set of existents, of course—no more than religion [REL] or persistence [REP] can. It occupies only small networks; it is necessarily disappointing. But it is not wrong to consider that the entire set of functions that make it possible to link, trace, hold together, reattach, suture, and mend what is constantly distinguished by the very nature of enunciation belongs to the attachment that our tradition has celebrated under the name of law—while blending it into a necessarily more confused institution. We can understand why the Moderns have been able to admire it unreservedly, even sometimes giving in to the temptation to entrust it with the responsibility of keeping us from ever going astray. And yet it does nothing, says nothing, informs about nothing. But what it does it does well: it attaches, it darns, it paves with continuities a world of which it has become the author despite a cascade of shiftings.

In short, law, like politics, is distinguished in several ways: it depends on a way of doing—form, formalism; it implicates the totality—like politics and religion; and above all, it *comes back* to the conditions of enunciation by a sort of hooking effect that is of capital importance for giving consistency to subjects who have become capable of being engaged by what they have said and done. This is why it would be quite useful to continue to nurture our little classification scheme by putting politics [POL], law [LAW], and religion [REL] together in a single group, that of QUASI SUBJECTS.

WE CAN NOW UNDERSTAND WHAT IS DISTINCTIVE ABOUT QUASI SUBJECTS ⊙

At the end of Part Two, we proposed to regroup the beings of technology [TEC], fiction [FIC], and reference [REF] under a single heading, that of QUASI OBJECTS. Each of these three modes, while hinging on the materials it puts forward—technologies, figures, chains of reference—would give rise, through a sort of rebound effect, to virtual positions for subjects to come. This is what semiotics identified so clearly with its theory of ENUNCIATION. This is what allowed us never to begin our analysis with acting, thinking, speaking human beings, humans capable of "creating technologies," "imagining works," or "producing objective knowledge." To put it in the shorthand terms of anthropogenesis: humanoids *became* humans—thinking, speaking humans—*by dint of association* with the beings of technology, fiction, and reference. They became skillful, imaginative, capable of objective knowledge, by dint of grappling with these modes of existence. This is why we have reused the expression "quasi *objects*" to designate both the advent of these beings (they are truly objects) and the still-empty place of the subjects that might come later (they are only *quasi* objects).

Now, the three modes grouped together here are distinguished by the fact that they come to *fill*, as it were, the still-empty form of the implicit enunciator. They are not subjects (we know that the subject has been unmoored; we arrive at the subject without starting from the subject), but these beings are nevertheless *offers of subjectivity*, of critical importance for the definition of our anthropology. They are thus in fact *quasi* subjects. To sum up the originality of this THIRD GROUP in an overhasty sentence, let us say that, while following along the political Circle, humans become capable of opining and of articulating positions in a collective—they become free and autonomous *citizens*; by being attached to the forms of law, they become capable of continuity in time and space—they become *assured*, attributable selves responsible for their acts; by receiving the religious Word, they become capable of salvation and perdition—they are now PERSONS, recognized, loved, and sometimes saved.

This time it is no longer a question, as it was with the SECOND GROUP, of focusing on dispatched or known fabricated objects, but of offering those who fabricate or use them, those who create or who receive the work, those who know or who learn, a new *consistency*. These roles were tacit, implicit, presupposed; they inhabited the limbo of level n - 1, which they could in no way preexist. Now it may be possible to articulate them in their turn, by offering them a role, a function, a figuration. Whereas the second group was *centripetal* with respect to objects, which drew all the attention, the second group is *centrifugal* with respect to objects, which become the *occasion* for assembling, judging, praying.

But who is doing the assembling, judging, praying, and the like? It is not absurd to consider composition via the political circle [**POL**] as one of the modes that will, *on the occasion provided by objects*— ordinary things, affairs, issues—give form and figure to other beings that are attached to them. The Circle, an original mode of existence, makes it possible to bring those other beings into existence. "One," "someone," will give rise to "us," in the plural. "We" are those who gather around things that "we" have in common because they are also things about which "we" do not agree. An astonishing condition of enunciation, since "I" am going to become the one who "makes" those who *represent me* say what I would have said "in their place" and since, at the same time, as the Circle turns, by *obeying* them "I" express what they make me say.

⊙ WHILE LEARNING TO RESPECT THEIR CONTRIBUTIONS: FIRST, BEINGS OF POLITICS [**POL**], ⊙

This is why politics can never be based on a preexisting society, and still less on a "state of nature" in which bands of half-naked humans end up coming together. The exploration of successive alterations goes in the opposite direction from this implausible scenography: we have gradually learned to become "us"—in the plural—by dint of taking on the successive propositions that the political Circle never stops grinding out. *Autophuos* is in fact the *physics* of the self-production of subjects by and for themselves. Without it, we would be quite incapable of *saying* what *we* are or of being what we say.

Doesn't that suffice to engender "persons with a human form"? No, of course not, because the political [**POL**] is not the only mode of existence that participates in the gradual engendering of quasi subjects. Here is where the full power of law [**LAW**] is going to weigh in. Without law, every act of enunciation would disperse possible authors with no chance of ever linking what they say with what they do. The connection is weak, almost infinitesimal, as a Spanish proverb puts it so well: "Take a bull by the horn and a man at his word." If it is so difficult to focus one's attention precisely on the form of autonomy proper to law, if so much care must be

⊙ THEN BEINGS OF LAW [**LAW**], ⊙

taken to unfold its fine cloth without tearing it, this may be because of the strength of weak links. If law holds everything, if it makes it possible to link together all persons and all acts, if it authorizes, by a continuous pathway, the connection of the Constitution to a trivial legal case, this is also because it collects from each situation only a tiny part of its essence. Its tissue resembles an open-weave net. This is what common sense retains from its movement when it qualifies law as cold, formal, fastidious, abstract, empty. Well, yes, it has to be empty! It is suspicious of fullness, of content that would slow it down, make it heavier, *prevent it from setting into relationships*, through its own pathways, what it retains from the world. It can go everywhere and make everything coherent, but only if it lets almost everything slip away.

It does not seek, as knowledge does, to shift the territory onto the map by the vigorous grip of reference [REF · LAW]. It never thrusts itself, as science does, into the risky test of constructing, within powerful calculation centers, reduced models that would resemble the world and make it possible to see the whole world in a single glance. Metonymy is not its strong point. What is a notarized act alongside one's home? How can that fragile sheet of paper be compared to the thickness of walls and the weight of memories? No relation of resemblance, no mimeticism, no reference, no blueprint. And yet, if there is a conflict, an inheritance, a dispute, it is in fact through the dazzling link between this pathetic little piece of paper and the body of texts, through the intermediary of lawyers and judges, that you will be able to prove your claim, authenticate your property—and perhaps keep your home, confounding those who want to put you out in the street. The attachment is minuscule and yet total, the grasp infinitesimal and yet capable of being linked to all the rest.

AND FINALLY BEINGS OF RELIGION [REL]. Clothed by political enunciation [POL], attached to their enunciations by law [LAW], these quasi subjects can still be said to lack a great deal of consistency. Indeed, they still lack the opportunity to become PERSONS. The third regime, the religious mode [REL], is the one that will fill them, at least in our tradition, with a new weight of *presence*. If "I" and "you" are to emerge, a new flow of beings is needed, beings that offer the gift, the present of presence. This time, it is the enunciator himself/herself who is directly addressed: "I am speaking to *you*"; and, of course, *without address, there is no "you."* This mode of existence has certainly been instituted in a thousand different ways throughout history—from the predication of the living God to tele-reality!—but without it one cannot proceed without losing one's sense—this particular sense—of personhood. The institutions that fabricate loves and lovers may have various names, but their form of enunciation, announcement, news, good news, is unquestionably different from all the others. The quality of their veridiction is measured by an infinitely subtle know-how that is instantly lost if one stops replaying in full the appeal to what is present. Instantly, one ceases to address "anyone in particular." A ruthless requirement? Yes, but no more constraining than that of politics [POL] or law [LAW], each in its own genre.

The third group has a common feature, moreover, that justifies calling it that of quasi *subjects*: the fact that the felicity and infelicity conditions for the group always depend on the moment, the situation,

QUASI SUBJECTS ARE ALL REGIMES OF ENUNCIATION SENSITIVE TO TONALITY.

the tonality, almost on the tone of voice—in any event, on form. (For this reason it would not be a bad idea to reserve the term REGIMES OF ENUNCIATION for this THIRD GROUP, for it is definitely a matter of a "manner of speaking.") It was the very fragility of these conditions that led modernism to declare them irrational, or at least irrelevant to truth and falsity alike. And yet what a loss, if we couldn't trace once again the differences between *truly* speaking politically, legally, religiously and *falsely* speaking politically, legally, religiously. And a still greater loss if we were to mix up these forms of truth, if we were to amalgamate them. What an astonishing adventure on the part of the Moderns, to have identified such lovely contrasts and then to have tangled them up so awkwardly, to the point of living as if they had not discovered them, or as if they could do without them! Anthropologists apparently have the gift of tears: they only visit people in the process of disappearing; the World of the Tropics, for them, is always on the wane. Similarly, we never know, confronting the anthropology of the Moderns, whether we should weep with admiration before their discoveries or with pity before the inheritance they have squandered.

Nothing prevents us now from adding the modes of the third group to the chart we began in Part One (see pp. 488–489). There is a danger, of course, in constructing a systematic list, but we have to consider carefully the danger in not categorizing. How, otherwise, could we deploy modes each of which interprets the others in its own way, each of which has to be free to speak in its own language, and each of which requires that each term of the metalanguage be marked with a sign that makes it recognizable? How could we clearly articulate the differences that we risk losing at every moment? The great advantage of every listing by way of a chart is that the unfolding of rows and columns on paper helps us keep track of categories that would otherwise be confused in our minds. Our little chart is nothing but a memory aid, but it suffices to remind us that in empirical philosophy we can now count *beyond* two, and even beyond three . . .

If we refused to set up lists, we would risk tipping once again into dualism and resorting once again to the distinction, as strange as it is commonplace, between those who strive to speak *literally* and all the other discourses designated as *figurative*, discourses that could only lead us astray, we are told, set us on the wrong path, where we would be carried away by the currents of imagination, skill, violence, lies—at worst, the currents of madness, at best, those of poetry. As if, whatever we do, the Bifurcators will always be asking us "Are you speaking literally or figuratively?" We'll rub our hands together in vain: "All the perfumes of Arabia will not sweeten this little hand . . . "

If one wants a dichotomy at any price, it will no longer be the one between speaking literally and speaking figuratively, but between speaking "straight" and speaking *well*. To speak well, as we are learning painfully, is not necessarily to be a gifted speaker, it is to take seriously both the things one is talking about and those to whom one must speak, in such a way as to respect the sense of what they seem to cherish as the apple of their eyes. How? By specifying the type of beings we wish to address on each occasion, by defining their particular type of veridiction and malediction, by identifying the alteration that is peculiar to them. The word and the world start anew, and drift. In this sense, there is thus no such thing as "literal"—except what Double Click says, Double

Click who is precisely neither "going" nor "leading" anywhere. What the canonical distinction between "the figurative" and "the literal" registered so awkwardly, in modernism, was the diffuse feeling that *differences* must exist between, for example, building chains of reference [REF], stating law [LAW], and making forms resonate in materials [FIC]. The Moderns' metalanguage gave them no way to articulate the nuances of these distinctions. On this point, at least, our inquiry cannot be faulted.

If you claimed to be speaking "literally," to what mode would you be alluding? To reference [REF]? But if researchers finally end up going straight, they all know only too well that they proceed by impressive leaps over obstacles. When an engineer is finally effective, it is through the dizzying zigzags of technology [TEC]. If politicians [POL] speak frankly and directly, it is by following twisted paths. And if you settled for "speaking figuratively," to what mode would you be alluding? To that of the beings of fiction [FIC]? But it seems that their demand to be "held," their demand for "style" and "tension," put you under a much greater obligation than one might think in hearing you laud the advantages of metaphor. As if it were enough to "express oneself freely" to produce a work of art! And those who speak about the beings sensitive to the one who enunciates them in order to bring them into presence [REL]—must we say that they speak "in figures and parables"—yes, undoubtedly—or that they speak as literally as can be about what is, what was, and what is coming—which is true as well? Isn't God himself said to "write straight with crooked lines"? And how are we to qualify the formidable drift of lines of force and lineages [REP]? Will you say of life itself that it goes on "literally" or "figuratively"? It would be good to know, for everyone who visits aquariums and zoos and museums of natural history wonders about this. As for the beings of influence and possession [MET], who has ever managed to address them by approaching them head on?

To reduce all these flows of meaning to the single opposition between direct and deviated is like trying to play *The Magic Flute* on a pennywhistle. It would be best to leave the impoverished distinction between "literal" and "figurative" entirely aside. The next time someone asks you if you're speaking straight or crooked, if you still believe in reason, if you are really rational, first insist on specifying the enunciative key in which you wish to respond to your interrogators.

⊙ AND TO EXPLAIN, AT LAST,
THE MODERNIST OBSESSION
WITH THE SUBJECT/
OBJECT DIFFERENCE.
 Another advantage of this rudimentary categorization is that it finally does justice to the distinction that has been so amplified by philosophy between the world of Objects and that of Subjects, of which we have had to be so critical before finding its raison d'être at last. It was not as a result of a simple mistake that the Moderns dug such a chasm. They behaved like an engineer who had tried to use buttresses to shore up a weight that was too much for them. Looking at the chart, we see that the FIRST GROUP corresponds to the modes that completely ignore *quasi objects as well quasi subjects*, whereas the second group brings together the quasi objects and the third the quasi subjects. To focus one's attention on Objects is not the same movement as focusing it on Subjects, this is clear, but there is nothing in this movement that would allow us to trace *the* difference between two distinct compartments of reality. Not to mention that the first group *does not correspond to either of the other two*. In the negotiation to come, it should thus be possible to reserve a place for this opposition, which the informants seem to cherish, yet without agreeing to entrust to it the daunting cleavage that has been obsessing Bifurcators for three centuries—and that obsesses them all the more when they struggle to "get beyond" it or to "critique" it. The slight nuance between the second and third groups can in no way be superimposed on the dichotomy between the *res cogitans* and the *res extensa*. (The Mind/Body problem: here is one puzzle at least that anthropology will no longer have to solve.)

And yet, even though we need no longer ask those ancient buttresses, Subject and Object, to support any weight at all, the anthropologist, having become the guardian of the patrimony, may be able to proceed like a skillful architect who restores an ancient site from top to bottom but manages to save some of its vaults, wall sections, and columns so that observers will at least understand what perilous architectonics once underlay the ruined and now renovated building.

For it is indeed with architectonics that we are now going to have to be concerned.

In any event, the real reason for this rudimentary arrangement lies elsewhere: in the identification of a FOURTH GROUP about which we have carefully refrained from speaking up to now, the most important, the most difficult, too, the most naturalized, the one that is going to link the *quasi objects*

and the *quasi subjects*. If the Moderns had extracted only the contrasts on which we have focused, they would never have emitted the cloud of exoticism into which they have plunged both themselves and the others: neither Orientalism nor OCCIDENTALISM would have confused them to such an extent. By identifying technological innovations [TEC], the splendors of works of art [FIC], the objectivity of the sciences [REF], political autonomy [POL], respect for legal linkages [LAW], the appeal of the living God [REL], they would have glowed in the world like one of the most beautiful, most durable, most fruitful civilizations of all. Proud of themselves, they would have had no burden weighing them down, crushing them like Atlas, like Sisyphus, like Prometheus, all those tragic giants. But they went on to invent something else: the continent of *The Economy*. We have to start all over again.

·Chapter 14·

SPEAKING OF ORGANIZATION IN ITS OWN LANGUAGE

The second Nature resists quite differently from the first, ⊙ which makes it difficult to circumvent The Economy ⊙ unless we identify some gaps between The Economy and ordinary experience.

A first gap, in temperature: cold instead of heat.

A second gap: an empty place instead of a crowded agora.

A third gap: no detectable difference in levels.

All this allows us to posit an amalgamation of three distinct modes: [ATT], [ORG], and [MOR].

The paradoxical situation of organization [ORG] ⊙ is easier to spot if we start from a weakly equipped case ⊙ that allows us to see how scripts turn us "upside down."

To organize is necessarily to dis/reorganize.

Here we have a distinct mode of existence ⊙ with its own explicit felicity and infelicity conditions ⊙ and its own particular alteration of being-as-other: the frame.

So we can do without Providence for writing the scripts, ⊙ provided that we clearly distinguish piling up from aggregating ⊙ and that we avoid the phantom metadispatcher known as Society ⊙ while maintaining the methodological decision that the small serves to measure the large, ⊙ the only way to follow the operations of scaling.

This way we can bring the arrangements for economization into the foreground ⊙ and distinguish between two distinct senses of property ⊙ while including the slight addition of calculation devices.

Two modes not to be conflated under the expression "economic reason."

HAVING REACHED THIS POINT, THE ANTHRO-POLOGIST OF THE MODERNS HESITATES BETWEEN LEGITIMATE SATISFACTION—she has stubbornly managed to maintain the same investigative protocol across such incommensurable modes of exis-tence—and anguish—she has not even begun to understand what tru-ly moves her informants. For she can no longer ignore this huge fact: when she tells them that she is undertaking an inquiry into their "vari-ous modes of existence," they don't think about the exact weight of di-vinities, gods, microbes, fish, or pebbles, but rather about the way they themselves "earn their living," their subsistence. And by this vigorous term they don't seem to be designating the subtleties of metaphysics, the confusions of epistemology, or the targeting mistakes of antifetishism! No, they are designating, rather, the frameworks of their own lives, their passion for consuming, their daily difficulties, the jobs they have or have just lost, and even their paychecks. The investigator cannot bask in the il-lusion that she has completed the inventory of the "values" for which her informants are ready to give their lives. Hearing them talk, it seems to her on the contrary that she has understood nothing so far, that she has spent all her time, as the Gospel suggests, "straining out gnats but swal-lowing a camel" (Matthew 23:24). How could we tackle the tribulations of diplomacy later on, if we had deployed only the modes of existence that the Moderns cherish perhaps *the least*?

If their anthropology is so delicate, this is because it is almost impossible to define the VALUES to which they are attached. To hear them tell it, all questions of substance or subsistence have to be brought back not to the subtle detours of ontology but to a master discourse, that of The Economy. What they mean by living, wanting, being able, deciding, calculating, mobilizing, undertaking, exchanging, owing, consuming, is all entirely situated in this world, on which the inquiry has not yet touched. Might we have spent too much time getting around the RES EXTENSA, when those who call themselves actual "materialists," or, on the contrary, "idealists," invoke not that matter but a different one, infinitely more robust and more widespread, the one that foregrounds the "iron laws" of interest? In practice, it is in The Economy that the informants really learn about facts, laws, necessities, obligations, materialities, forces, powers, and values. The FIRST NATURE was certainly important, but it is this SECOND NATURE that turns out to shape them in a lasting way. It is in and through this second Nature that the informants, whenever their "passionate interests"—to borrow Tarde's expression—are at stake, first encounter the terms "reality" and "truth." So this is where we now have to dig.

But how can we approach The Economy with sufficient dexterity without giving it too much or too little credit? Because The Economy offers the analyst such a powerful metalanguage, its investigation might have been concluded at once, as if everyone, from one end of the planet to the other, were now using the same terms to define the value of all things. Not only would it offer no handhold to anthropology, since it would have become the second nature of an *already* unified and globalized world, but it would have achieved at the outset all the goals taken on by our projected diplomacy, by allowing all peoples to benefit from the same measuring instrument made explicit everywhere in the same idiom. With The Economy, there would always be mutual understanding, because it would suffice to calculate. A quarter-century-long effort to specify the history of the Moderns would have been useless since, from now on, the entire Earth would share the same ways of attributing value in the same terms.

⊙ WHICH MAKES IT DIFFICULT TO CIRCUMVENT THE ECONOMY ⊙

With the ecological crises, the first Nature seems in danger of losing its universality—some go as far as to speak of "multinaturalism"—and yet, despite the scale of the economic crises, the hold of the second Nature has only increased. If our ethnologist wanted to study it seriously, she could do her fieldwork in Shanghai, Buenos Aires, or Dacca as well as in Berlin, Houston, or Manchester. How can she resist the appeal of this genuine universal, perhaps the only one we have in common? A single language, a single world, a single yardstick: "the real bottom line." It is as though Double Click, just when we thought he was going to die of starvation, were finally on the verge of realizing his wildest expansionary dreams.

And yet the investigator feels that she would only be giving in to a manifest exaggeration: she would find herself confronting the same padding against which we struggled so hard at the beginning of this work: confronting the hegemony of Knowledge [REF · DC] that claimed to define (the first) Nature for its own part, and to dictate the universal laws governing the evaluation of everything. The ethnologist's suspicion is moreover reinforced by the counter-narrative, almost as widespread, according to which The Economy is not the basis for the world finally revealed to everyone thanks to the benefits of globalization but a cancer whose metastases have gradually begun to infect the entire Earth, starting from various sources in the old West. In this narrative, the cancer has succeeded in dissolving all the other values in the cold calculation of interest alone. If The Economy is universal, it is because of the deadly ailment, the unpardonable crime known as CAPITALISM, which continues to be a monstrous product of history that has infected nearly all the cells of a body unequipped to resist it. Inventing Capitalism and using it to *possess* the entire Earth: these would be the unforgivable crimes of the Whites.

But isn't this a new exaggeration? Doesn't it attribute too much power to this monster? More seriously still, isn't it a way of agreeing to conspire with Capitalism by taking it too hastily as a cancer with terrifying destructive power? As always, the ethnologist starts to hesitate (that's why her inquiry is taking so long!), because she would like to be able to *get around* The Economy entirely and avoid using its metalanguage, whether it's a matter of speaking of it as a good thing—as the universal dialect of a globalized world—or a bad thing—as the fatality of a world infected with the cancer of Capitalism. How can she manage to respect what her informants say about the troubles they have with subsistence, without believing that The Economy would supply "the unsurpassable horizon" of her investigation? In other words, how can she DISECONOMIZE herself sufficiently to grasp what she is told without adding or subtracting anything? And how can she achieve this new disincarceration through formulas sufficiently compatible with her informants' ways of speaking so that, later on, when she has put on diplomatic garb, she may still be able to reach agreement with them?

If she has to hesitate for a long time, it is because she does not find herself confronting a single contrast more or less well pulled together by a more or less composite INSTITUTION, as she did with the modes we have unfolded up to now, but rather a contrast drawn together by *three modes of existence* that the history of the Moderns has blended for reasons she is going to have to untangle. It is this inter-weaving that explains why she has to resist the temptation of believing either in The Economy or in its critique. As always, it's a matter of resisting the temptation to believe. But here is something quite curious: while it is not completely impossible to circumvent the first Nature, not to believe in The Economy is going to require even slower and more painful efforts at agnosticism. As if the second Nature clings to our bodies much more tightly than the first. How much piety there is, indeed, in this "dismal science," this "abject science"!

⊙ UNLESS WE IDENTIFY SOME GAPS BETWEEN THE ECONOMY AND ORDINARY EXPERIENCE.

Anthropology always has the virtue of being able to reconsider, at additional cost, and as if from the outside, in all its freshness, an experi-ence that proximity, habit, or local prejudices had made unavailable. Let us try to feel the strange *gap* between the qualities with which economic matter is purportedly endowed and the experience that the inquiry is seeking to bring out, in such a way as to reopen the space that will make it possible to accommodate the various modes of existence. (The attentive reader will recognize the method that we have used since Chapter 3 to empty out the space that the first naturalization had filled too quickly.)

A FIRST GAP, IN TEMPERATURE: COLD INSTEAD OF HEAT.

A first gap. You observe goods that are starting to move around all over the planet; poor devils who drown while crossing oceans to come earn their bread; giant enterprises that appear from one day to the next or that disappear into red ink; entire nations that become rich or poor; markets that close or open; monstrous demonstrations that disperse over improvised barricades in clouds of teargas; radical innovations that suddenly make whole sectors of industry obsolete, or that spread like a dust cloud; sudden fashions that draw millions of passionate clients or that, just as suddenly, pile up shopworn stocks that nobody wants any longer . . . and the immense *mobilization* of things and people; they say it is driven only by the simple *transfer of indisputable necessities*.

Everything here is hot, violent, active, rhythmic, contradictory, rapid, discontinuous, pounded out—but these immense boiling cauldrons are described to you as the ice-cold, rational, coherent, and continuous manifestation of the calculation of interests alone. The ethnologist was already astonished at the contrast between the matter of the *res extensa* and the multiform materials that technological folds revealed [TEC · DC]: nothing in the former could prefigure the latter. But the abyss is even greater, it seems, between the *heat* of economic phenomena and the *coldness* with which she is told she must grasp them. Here again, nothing in the icy matter of economic reason allows her to anticipate what she will discover by plunging into these witches' cauldrons the genuine matter of the "passionate interests" that stir up the planet in its most intimate recesses. As if a mistake had been made about the temperature and rhythms of economic passions. As if there were another idealism at the heart of this other materialism.

A SECOND GAP: AN EMPTY PLACE INSTEAD OF A CROWDED AGORA.

A second gap. In The Economy, the question, she is told quite gravely, consists in dividing up rare goods, in parceling out scarce materials, benefits, or goods, or, on the contrary, in making the largest number profit from a marvelous horn of plenty debited from one resource or another. As everyone repeats with imperturbable seriousness, these are the most important questions we can address in common, because they concern the whole world, all humans and all things, henceforth engaged in the same flows of mobilization, in the same history, and in the same common destiny. It is necessary to divide up, distribute, decide. And in recent times, they tell her even more energetically, these questions have become all the more constraining since a scarcity more unexpected and more fundamental than all the others has been discovered: we don't have enough planets! We would need two, three, five, six, to satisfy all the humans, and we only have one, our own, the Earth, Gaia. (There may be other inhabited planets, of course, but the closest ones are dozens of light-years away.) Economics has become the optimal distribution of rare planets.

The ethnologist, surprised, then tries to find out how her informants are going to go about settling such huge questions, how they will do justice to all those for whom the answers are so urgent. She wonders what procedures they will adopt to bring off such feats of decision, division, and distribution, and what instruments, what protocols, what assurances, what verifications, what scruples they will deploy. She is already directing her gaze toward the noisy assemblies where such common matters are going to be violently debated. And there, what is she told? *Nothing* and *no one* decides: "It suffices to calculate." The very place where everything must be decided and discussed, since these are matters of life and death for everyone and everything, appears to be a public square entirely *emptied* of all its protagonists. In it she finds only the incontrovertible result of unchallengeable deductions made elsewhere, away from the agora! It is no longer the difference between extreme heat and extreme cold that stuns her, but the difference between the fullness, the agitation, the commotion she expected and the emptiness, the silence, the absence of all those who are concerned first and foremost. This whole vast engine apparently functions on autopilot. Here where everyone must decide, *no one* seems to have a hand in.

A third gap. When the talk turns to The Economy, her informants assert with respect, one has to approach vast sets of people and things that form organizations of astounding complexity and influence, covering the planet with their reticulations. "Ah!" the investigator exclaims. "Finally something solid, something resistant, something empirical. I'm going to be able to study defined, durable, circumscribed entities called enterprises, apparatuses, arrangements, perhaps even 'nation-states,'

A THIRD GAP: NO DETECTABLE DIFFERENCE IN LEVELS.

'international organizations'—in short, sets that are consistent and, above all, of great size. I've had enough invisibles, enough REPRISES, enough beings of occultation or metamorphosis. Here's something I can get my hands on for a change!"

And she sets out to approach enterprises, organizations qualified under the law as "corporate bodies." She extends her hand and what does she find? Almost nothing solid or durable. A sequence, an accumulation, endless layers of successive disorganizations: people come and go, they transport all sorts of documents, complain, meet, separate, grumble, protest, meet again, organize again, disperse, reconnect, all this in constant disorder; there is no way she could ever define the borders of these entities that keep on expanding or contracting like accordions. The investigator was hoping to get away from stories of invisible phantoms; she finds only new phantoms, just as invisible.

And if she complains to her informants that they have taken her for a ride, they reply with the same unfathomable confidence: "Ah, it's because *behind* all that agitation you haven't yet detected the assured presence of the real sources of organization: Society, the State, the Market, Capitalism, the only great beings that actually hold up all this jumble. That's where you have to go; those are the real substances that ensure our subsistence. That's where we really live." And, of course, when she begins to investigate such assemblages, the gap reappears, but this time multiplied: more corridors, more offices, more flowcharts, other meetings, other documents, other inconsistencies, other arrangements, but still not the slightest transcendence. No great being has taken charge of this ordinary confusion. Nothing stands out. Nothing provides cover. Nothing decides. Nothing reassures. It is immanent everywhere, and everywhere illogical, incoherent, caught up at the last minute, started over on the fly.

ALL THIS ALLOWS US TO POSIT AN AMALGAMATION OF THREE DISTINCT MODES: [ATT], [ORG], AND [MOR].

The poor anthropologist, thrice deceived by the reflection of a specific matter that would be called The Economy! But thrice reassured as to the fact that there are indeed, in the experience, *three distinct threads* that would make it possible to circumvent the question, if only one could manage to follow each strand separately, without immediately submerging them in the transfer of indisputable necessities.

The first gap will make it possible to do justice to the abrupt changes in temperature of passionate interests; we shall call this the mode of ATTACHMENTS (noted [ATT]). The second gap will allow us to fill once again the place we had found empty with a mode we shall call MORALITY (noted [MOR]). Finally, the last gap will allow us to explore the astonishing immanence of ORGANIZATIONS (noted [ORG]).

We are going to hypothesize that, if we learned to respect the contrasts brought out by these three modes, we would be liberated from the second Nature as we were from the first, and we could then explain how the overly composite institution of The Economy had at once revealed them and also, quite simply, *managed them badly*. Nothing more. The anthropology of the Moderns would then have carved out a path—actually, hardly even a trail: a track, a trace by means of which one could escape both from belief in a domain of Economy and from its critique. The chances of success are infinitesimal, but the whole project of an inquiry into the modes of existence depends on this ultimate attempt.

In this chapter, we are going to focus on organization. Of the three modes, this is the one whose experience is at once the easiest to trace and the most paradoxical. Easy, because we are constantly in the process of organizing or being organized; paradoxical, because we always keep on imagining that, elsewhere, higher up, lower down, above or below, the experience would be totally different; that there would have to be a break in planes, in levels, thanks to which other beings, transcendent with respect to the first, would finally come along to organize everything. It is thus a strange experience, known to all, and yet, as always with the Moderns, almost impossible to register appropriately. The ethnologist is thus in her element—the organizer too!

THE PARADOXICAL SITUATION OF ORGANIZATION [ORG] ⊙

As is our habit, we are not going to start with organization as a result, but as a PREPOSITION. This is the only way, as we know, to identify a mode: to ask what it means to act and to speak *organizationally*. As this adverb is too awkward, let us say that we are going to try to follow a particular being that would transport a force capable, in its displacements, of leaving in its wake *something of organization* no matter what the scale; this is the crucial point. We could not have grasped technology [TEC] by starting with the objects it leaves behind; similarly, we shall never manage to grasp what is proper to this new trajectory if we start with what "organizational theory" designates by the term. We would be setting off on the wrong track if we were to start by taking "organized beings" for granted, beings whose dimensions and consistency need precisely to be explained by the passage, the continual slippage, of the *action* of organizing. It is this action that we must thus accompany.

⊙ IS EASIER TO SPOT IF WE
START FROM A WEAKLY
EQUIPPED CASE ⊙

We shall begin by a very limited and very weakly equipped COURSE OF ACTION in order to learn to concentrate on its own mode of extension— a mode that will then enable us to see how one can *change dimensions* without changing the method.

Let us take a meeting between two friends. Paul: "I'll meet you tomorrow afternoon at 5:45 at the Gare de Lyon under the big clock. OK?" Peter: "OK, see you tomorrow. Cheers." What could be more ordinary? For, finally, today, by telephone, they sent each other a little *scenario*, projecting themselves into the near future, and imagining a meeting sketched out in broad strokes by the shared identification of a landmark known to all: the big clock at a major train station in Paris. Paul and Peter each told the other a little story. Here we are, incontestably, in fiction. But if we think about it, it is in a quite singular form, since, between the end of today's phone call and 5:45 p.m. tomorrow, Peter and Paul are going to be held, organized, defined—in part—by this story, which *engages* them [FIC · ORG]. They told the story together yesterday as though they occupied the *roles* that it assigned them. "Peter" and "Paul," now in quotation marks, have become characters in a story of which they are the heroes. They can of course phone each other in the meantime to agree on a different meeting time and place, but if everything goes according to their plan, tomorrow at 5:45 p.m. they will greet each other under the clock at the Gare de Lyon, and the scenario that has guided their steps and controlled their behavior up until the last second will no longer be activated. Delighted to have found each other again, they will tell each other new stories and deal with other business together.

What lesson can we draw from this anecdote? First, that we have to take advantage of the powers of fiction if we are to be able to tell each other stories, make plans, propose scenarios, or draw up programs of action [FIC · ORG]. Here we rediscover the triple shifting out that we encountered in Chapters 9 and 10: it sends figures into another time ("tomorrow, 5:45 p.m."), another space ("the clock tower"), and toward other actors or "actants" ("Peter" and "Paul," finally reunited after a separation). But the defining feature of these narratives is that they have a *hold* on those who tell them. So much so that the narrators find themselves face to face with themselves *in two positions at once*, with a slight gap: *above* themselves, as if they were free at any moment to write the story—Peter and Paul act; then *below* themselves, as if they were not at all free to modify the story— "Peter" and "Paul" have acted. (The possibility of taking the stories up again and rewriting them along the way changes nothing except the cadence imposed on the same course of action.) These narratives are fairly close to what sociolinguists call PERFORMATIVES, since the stories, too, do what they say and engage those who are their authors.

For stories that manage to subject narratives to such torsion, we shall reserve the word SCRIPTS. To designate the dispatching of these paradoxical scripts, which give roles to those who have sent them, and which they must then catch up with in order to obey them, we shall speak of the organizational act, or, better, the *organizing act*.

⊕ THAT ALLOWS US TO SEE HOW SCRIPTS TURN US "UPSIDE DOWN."

Let us note first of all that the trajectories sketched out by such scripts bear no resemblance to the ones traced by chains of reference [REF · ORG]. Even though scripts have a referential capacity, since they designate places and moments to which access is available (the Gare de Lyon, 5:45 p.m., the clock tower), these references are there only to facilitate identification and serve as accessories to direct the attention of the actors designated by the roles. Reference is not their main property, since these reference points are chosen only for their self-evidence and not for their novelty. The goal is not to gain access to remote states of affairs, but rather, since the remote entities are well known and easily recognizable, to use them to simplify the bearings. We rely on the referential indications only to judge the tenor of the script according to a different touchstone: its capacity to define ends, borders, meetings—let us call these *due dates*, in the broadest sense of the term. Their felicity conditions entail knowing whether Paul and Peter, after having lived all day "under" the injunctions of the script of "Paul" and "Peter," actually find each other. At the precise moment of their meeting, the script will have achieved its outcome and disappeared into the void.

We are so used to these practices that we sometimes have trouble seeing the originality of a position that makes us sometimes *authors* of a narrative, sometimes *characters* in that same narrative projected ahead in time—a narrative that disappears as soon as the program for which it was written has been completed.

This positioning of authors/above/characters/below is all the stranger in that neither Peter nor Paul is completely "above" during their telephone conversation nor completely "below" during the day as they wait to meet, both preparing to go to the train station. From one stand-point, nothing prevents them from revising their scripts to take into account the vagaries of their moods or the contingencies of train sched-ules—let us say that they are at once "below" and a little bit "above," in a sort of watchful or vaguely attentive state that prepares them for the "MANUAL RESTART" that we have already encountered with habit [HAB · ORG].

But if Peter and Paul are not wholly "below," neither are they wholly "above" their scripts; their mastery is not complete. It can never be said that they had nothing else to do but meet each other at the Gare de Lyon. At any given moment in time—yesterday, for example, during the phone call—there were dozens of "Paul" and "Peter" characters who resided in *other scripts*, scripts that gave them *other roles* and anticipated *other due dates*. "Paul" no. 2 was expected at the dentist; "Peter" no. 2 was seeing his girlfriend; "Paul" no. 3's boss was waiting for him to turn up at an impromptu meeting, and "Peter" no. 3's mother wanted him to bring a special gift for her granddaughter when he came to Paris... If novelists—at least the classical novelists—ensure what literary scholars call their characters' ISOTOPY (in a detective story, it's the same "Hercule Poirot" with the same mustache and the same shiny bald head from beginning to end), organization for its part does not guarantee any miraculous isotopy.

And this is where things get complicated. When Paul and Peter were deciding to meet, they were practicing a triple shifting out with ease: spatial,

TO ORGANIZE IS NECESSARILY TO DIS/REORGANIZE.

temporal, actantial. But as soon as they hang up their phones, they notice that other scripts written by many other authors have also shifted out, in other spaces, other times, and other actantial roles, the very *same solitary* Peter and Paul who must then—and the weight of this "must" depends on a multitude of other linkages—find themselves at the same hour obliged to "*fulfill* other obligations" in other places and at other times. This is both a radical experience and a common one, as every (dis)organized (dis)organizer knows.

For even if each of these scripts has been written, validated, or in any case approved by Peter or by Paul, *nothing*, absolutely nothing, *ensures* that they are mutually compatible and that they will achieve their outcomes at the same moment while designating the same place and the same role. To organize is not, cannot be, the opposite of disorganizing. To organize is to pick up, along the way and on the fly, scripts with staggered outcomes that are going to *disorganize* others. This disorganization is necessary, since the same beings must constantly attempt to juggle attributions that are, if not always contradictory, then at least distinct. Instead of an isotopy, it is *heterotopy* that wins out. Paul has rushed to the station from the dentist, Peter from his girlfriend; Paul loses his job because he missed the meeting with his boss; Peter has his mother mad at him again. "I can't be everywhere at once." "I only have two hands." "I can't be in two places at the same time." Impossible for any human to unify in a coherent whole the roles that the scripts have assigned him or her.

Since we recognize a mode of existence by a **TRAJECTORY**, the continuity allowed by a confounding passage through a **HIATUS**, a chiasmus, a gap, there is

HERE WE HAVE A DISTINCT MODE OF EXISTENCE ⊙

no doubt about it, the organizing act must be added to our list. Obviously, if Double Click gets involved, claiming that organization—now understood as a result—merely *transports*, unchanged, the force, the roles, and the power defined by another level, transcendent in relation to the humble scripts, this distinct mode disappears at once [**ORG · DC**]. But the trivial example of Peter and Paul helps us see that this type of organization is always, from top to bottom and from beginning to end, a series of breaks that obtain continuity in courses of action through constant *discontinuities*. Organization *can never* work: the scripts always define dispersed beings; they always achieve their outcomes in staggered fashion; one can only try to *take them up again* through other scripts that add to the ambiant dis/reorganization.

Organization never works because of the scripts; and yet, because of the scripts, it works after all, hobbling along through an often exhausting reinjection of acts of (re)organization or, to use a delicious euphemism from economics, through massive expenditures of "transaction costs." From this standpoint, the organizing act is just as constantly interrupted as the movement of the political Circle [POL], or the attachments of law [LAW], or the renewal of religious presence [REL], or the mere survival of a body [REP]. Sameness can never nourish these strange beasts. They require otherness.

The counter-test is easy to find: it suffices to believe in the *inertia* of a transcendent organization for it to begin to shift course! For the *little* transcendence between Peter and Paul and the other "Peters" and "Pauls" does not authorize them in any way to *rely* on a *maxi-transcendence* that would organize them in spite of themselves. Yes, the script will indeed be transcendent with respect to Peter and Paul since, between the moment they reached agreement until the moment they met, it will have "watched over them" and they will have referred to it with more or less anxiety, regularly consulting their watches as the script unfolds, but this doesn't mean that it dominates them. When Peter and Paul refer to it to check the time it has set, they cannot answer the question "Was it set by the script, or by themselves?" In fact, it is in the nature of a script to dominate them after they have dominated it; it is both above and below. With respect to the script, we are always both inside and outside; it positions us both before and after; this is its distinctive mode of reprise.

If the first transcendence defines a narrow break, a hiatus, along a *horizontal* line between the launching of a script and the follow-through, the second carves out a *vertical* chasm between the *level* of all the scripts and the miraculous level of the anonymous agency that writes them. If we were to mistake the transcendences, we would render both the duration and the solidity of the organization incomprehensible. The paradox, indeed, is that, if we are to last, we can never count on what does not pass. For anything to last, we have to count on *what passes*. Isn't that true of all the modes? Yes, but in organization, thanks to the transitory intermediary of scripts, we can see the mechanism at work much more clearly.

Is this mode of existence capable of making explicit its felicity and infelicity conditions, its particular manner of defining truth and falsity in its own language? No question: concerning the quality of the organizing act, we can all go on forever! Just think about the number of times in a single day when we discern what is true or false, good or bad, about the organizational scripts that we submit to or propel. All those complaints, hopes, expectations, disappointments, revisions. It seems that each of us has not just one touchstone but a whole battery of them for judging the good or bad quality of the enterprise where we work, the State that ought to be better organized, our children who ought to have done one thing or another, we ourselves who are so disorganized that time slips through our fingers and we don't understand why . . . If you want proof that in this area we possess a particularly active competence for distinguishing between the true and the false, in the organizational sense of these terms, just position yourself in any office, next to the coffee machine, and listen to the conversations. When it's a matter of *falsifying* the claims of those who purport to be organizing us, we're all new Karl Poppers!

⊙ WITH ITS OWN EXPLICIT FELICITY AND INFELICITY CONDITIONS ⊙

Organization is astonishingly fragile, since at any time we can *miss* the doubling of the scripts "above" which we situate ourselves and "below" which we are situated: "It's not up to me to take care of that, it's your job." "That's way above my pay grade." "I have nothing to do with it." "I'm washing my hands of it; let them cope." And these failures are all the more frequent given that there is no *resemblance* between the script and the beings it organizes; whence new infelicities: "That doesn't fit into the rubrics." "This wasn't in the plans." "There's nothing we can do." "Are they asleep in there, or what?" "This is no longer our problem."

Where organization is concerned, we never stop weighing all courses of action on a highly sensitive scale. There is, first of all, the test of performance: did Paul and Peter meet, *in the final analysis*, at the Gare de Lyon under the clock tower, or not? But this test only validates one isolated script. There is also the test of the consistency of the scripts: is it the case that, at the same time and in the same place, Paul or Peter met with injunctions more or less compatible with all the "Pauls" and all the "Peters" of the other scripts? Organizational consultants, "coaches," managers, and "downsizers" earn small fortunes by tracking down the multitude of "contradictory injunctions" that pull the participants in contradictory directions. And then they all start up again, in a new cycle of dis/reorganizations, commented on, watched over, analyzed, disputed, by a new concert of complaints, clamors, suffering, which other coaches, other managers, other consultants will come in to analyze, decode, mix up, encumber with their flowcharts and their PowerPoints. Organization may be impossible, but we still keep dreaming of organizing well and definitively *at last*—until the next day, when we have to start all over again.

But there is another, much more self-reflexive test: how can we rely on the inertia of an organization to determine the inertia that it lacks? Curiously, in fact, although organization is not an essence, everyone talks about it as if it possessed its own particular objectivity, its own tenor, as if it had its trade, its "core business," its "soul," as it were, and even, sometimes, as they say, its "culture." And there we go: more meetings, assisted by more consultants, more plunges into the archives of the organization to understand "what we are," "what we want," and "what our priorities are." An even more astonishing reprise, since, this time, we come back to the "foundations" of the organization on which we are supposed to "rely," foundations with no foundation whatsoever, but that serve nevertheless as hooks, indices, road signs as we decide how we are going to "continue to be faithful to our vocation." "How to be faithful to the spirit of our founders," or, by a mind-boggling crossing with another biological mystery, we ask ourselves gravely: "What is the DNA of our organization?" [REP · ORG]. A surprising bit of juggling between a mythic past and an imagined future that allows us to toss across the abyss of time a thousand new scenarios attributing roles and functions that are just as scattered, contradictory, caught on the fly as their predecessors, but that will allow the "stable essence" of the organization to last a little longer. An admirable ontology of the organizing act that wouldn't look out of place between Aristotle's *Metaphysics* and Heidegger's *Being and*

Time. Yes, the stable *being* of organization through the unstable *time* of its incessant reprise.

Some will object that it is highly exaggerated to speak of metaphysics or ontology concerning such modest, trivial activities. And yet, if we think about it, no other mode of existence procures the particular

⊙ AND ITS OWN PARTICULAR ALTERATION OF BEING-AS-OTHER: THE FRAME.

type of spatial, temporal, and actantial continuity designated by the term "isotopy." On the contrary, philosophy may well have drawn some of its most important concepts from the organizing act. Every script, in effect, defines a FRAME, a framing, in the wholly realistic sense of the term: Peter and Paul, once they have been *placed under* the script that they have drafted in common, indeed find themselves *within* something that frames them. A capital innovation, for no other mode thus ensures *borders* to the entities that it leaves in its wake. Of course, these borders vary with each script, and it is as impossible to delimit them definitively in space as it is to ensure that they will achieve their outcomes at the same moment in time. Nevertheless, they do have the function of establishing limits, functions, definitions.

Whence the HARMONICS with many other modes. With law, of course, since law offers continuity to the concatenation of levels, something that could not be obtained in any other way [LAW · ORG]. With habit, blessed habit, which veils the prepositions slightly and thereby ensures somewhat stable courses of action [HAB · ORG].With religion, too, since each of these two modes bears, in its own way, on the *end*, in the sense of completion—though it is never completed—for religion and in the sense of *limit* for organization [REL · ORG]. That Churches have used and abused this link does not keep us from recognizing that The Economy has been conceptualized in the notion of a divine Script for a millennium—and it still is, as we shall see.

If the plurimodal term ESSENCE has any content, it is virtually certain that one of its features depends on the organizing act. It is as though, with organization, we were discovering the *beings of framing* that only come into view, curiously, if we abandon the idea that above the scripts there exists a frame within which we could place them. To the first paradox—scripts obtain duration through what does not last—we must add a second: it is because the frames come from inside the scripts that they manage to frame them. The frames are what achieve the effects of continuity, stability, essence, inertia, *conatus*, whose importance has never ceased to inspire all hope of stability. Even if behind them there is no substance that would be situated "below," no other world on which to "rest" the arrangements, they nevertheless offer the possibility, by dint of renewal, return, rectification, complaint, and obedience, of making something last, something that finally has borders, frontiers, mandates, limits, walls, ends: in short, something that actually begins to look like what the philosophy of being-as-being was looking for in vain, something that is going to supply the composite term INSTITUTION with one of its most important features.

It doesn't hold together on its own, and yet there is something small and something large in it, something that encompasses and something that is encompassed, something structured and something structuring, something framed and something framing. It is by dint of small discontinuities—the hiatus of what cognitive scientists call "the execution gap"—that we end up obtaining assured continuities, without any assurance whatsoever. Each of us knows perfectly well that we cannot place our trust at any given moment "in" organization, and yet we know that we can also rely on it. There are, finally, essences—provided that they are maintained continuously enough. If we agree not to separate ourselves for a second from the continuous flow of scripts, being-as-other ends up also, through an excess of attention, vigilance, and precautions, providing blessed essence. It is thus the generosity of being-as-other that offers being-as-being, its old adversary (or its old accomplice?), the provisional shelter that the latter sought hopelessly in the definitive and in the substratum.

So WE CAN DO WITHOUT PROVIDENCE FOR WRITING THE SCRIPTS, ⊙ Why is it so hard to concentrate on a mode of existence that is so ordinary and so widespread? Because we always are at risk of yielding unwittingly to the temptation of entrusting the establishment of consistency among all these scripts to a *second level*. It is at this point that the everyday experience we have just evoked threatens to turn upside down. We start from a gap inherent to the nature of the organizing act, and then decide to take it either as the experience proper to this mode of existence or as a *mistake* that must be corrected by turning toward an

Organization—this time with a capital O—that we expect, thanks to the miraculous writing of a Script—this too with a capital S—to *make compatible and coherent* something that can be neither. Here we encounter the usual slippage that leads from mini- to maxi-TRANSCENDENCE. We begin to believe in *Providence*, or in any case in an anonymous and mute Author capable of projecting in advance all the scripts attributing all the roles to all the "Peters" and "Pauls" in the world to verify that they all reach their outcomes at the same moment and in the same place, or that they are at least arranged in cadence in such a way that they will remain mutually compatible.

Here is a branching not to be missed: the first path leads us toward an IMMANENCE that is quite ordinary but in practice difficult to register accurately; the second leads toward the hope of a transcendence that has had decisive importance in European history. In the first case, there is only *a single level* of analysis; in the second, there are *two levels*, and a radical break between the two. It is because they have so consistently taken the second path, despite the refutations offered by experience, that the Moderns can be said to *believe in Providence*. Not the one that the Church Fathers thought up in the form of the "Economy of God's Plan," a divine *dispensation*, but in a form that, even though secularized, is nevertheless infinitely too "believing" to consent to remain on a single level. It is about organization that Nietzsche ought to have wondered "why we are still pious."

This is no time to believe or to tremble: there is indeed *aggregation*, but it is understood quite differently according to whether we stay on one level or on two. It can be said without exaggeration that this category mistake is what has perverted the fine word "rational."

⊙ PROVIDED THAT WE CLEARLY DISTINGUISH PILING UP FROM AGGREGATING ⊙

For the Peters and the Pauls, what intervenes to "limit their margins for maneuver," as it were, is the *piling up* of scripts assigning inconsistent roles that pull apart all the "Pauls" and all the "Peters." And we can easily understand why, since, by dint of piling up dentists on top of friends, mothers on top of mothers-in-law, bosses on top of girlfriends, grandchildren on top of projects, all these scripts end up merging into an undifferentiated mass that *resembles* a phenomenon of a different order, one that would be transcendent with respect to the scripts. But it is only a matter of resemblance. If Peter and Paul both sigh and curse "Paris life," "consumer society," and perhaps life in general, this in no way implies that what they are designating is made of *some other matter* than their meeting at the train station. If they had the time, they could dissect at least in principle—but other scenarios rob them of the time they are rushing to save!—everything that is falling on their shoulders like a destiny. Or rather, this Destiny could be divided up into thousands of discrete little destinies each of which would integrate a different proportion of Paul's and Peter's decisions. We may as well admit it: "Peter" no. 4, yesterday, "Peter" no. 5's mother the day before, "Paul" no. 6's girlfriend ten months ago, "Paul" no. 7's boss in the office, and so on, all dispersed in time and space, all this is beginning to weigh pretty heavily. All the more so given that of all these "Pauls" and "Peters," none is the right Paul and the right Peter, both of whom are limited to the bodies left by their lineages [REP · ORG]. A single body of one's own to bear the weight of a thousand characters: here is a truly rare phenomenon.

The other version of aggregation amounts to believing that the proliferation of different scenarios is only the manifestation, the expression, the materialization, of a *higher level* that would be made of an entirely *different fabric* from the first. This level would be the source both of the contradictory impulses received by the participants and of the hopes for order, logic, or coherence that would allow them to get away from the pathetic confusion of the scripts. It would no longer be a matter of a simple aggregation, then, but of a *transmutation* through which the humble scripts would themselves become envoys, delegates, representations, manifestations of a higher level from which they would have been enlisted. The millions of small dispatchers—Peter and Paul on the telephone, the boss's secretary with his calendar, the angry mother, the offended lover—would have given way to what we shall call, to attenuate the weight of the word Providence somewhat, a DISPATCHER or a METADIS-PATCHER. Such anonymous dispatchers are drawn on as on a bank account with unlimited funds every time someone calls on Society, the Market, the State, or Capitalism to "bring some order" to the confusion of organizations. It is as though, thanks to aggregation, there has been an increase in rationality in relation to the simple pileup.

The problem is precisely that we must avoid trying to bring order too quickly to these muddled scripts. Here is where the inquiry has to resist the most deeply anchored methods of the social sciences.

⊙ AND THAT WE AVOID THE
PHANTOM METADISPATCHER
KNOWN AS SOCIETY ⊙

If we were to "replace" the scripts "in a framework," "retrace the context" "in" which the scripts are found, we would lose sight of the movement peculiar to the organizing act and thus lose all hope of understanding the very operation of framing, of continuity, and, as we shall see, of change of scale. The category mistake here would be doubled, since it is by believing it is correcting a first mistake—remaining with a particular case instead of "replacing it in its framework"—that sociology leads us into a second mistake—taking the frame as the thing that will explain the case!

This is the point at which the peculiar phantom of SOCIETY appears, that sui generis being that would arise suddenly (Tarde says "*ex abrupto*") amid the scripts to *take over* from their contradictory injunctions, in order to dispatch the roles, norms, outcomes, and tests of veridiction on the basis of a higher, anonymous, rational Script—one capable, in any case, of providing an explanation. Society is the appearance of a "collective being" that must not be confused with what we have been calling from the outset a "COLLECTIVE," since it has none of the means, none of the *collectors*, that would allow it to assemble. Moreover, its consistency is so uncertain that some sociologists insist, on the contrary, that one must "stick to individuals." As if there were INDIVIDUALS! As if individuals had not been dispersed long since in mutually incompatible scripts; as if they were not all indefinitely *divisible*, despite their etymology, into hundreds of "Pauls" and "Peters" whose spatial, temporal, and actantial continuity is not assured by any isotopy. And here the great machinery of social theory starts to function, with its impossible effort to locate the respective roles of "individuals" and "Society" in the completion of a course of action.

"No question," says our ethnologist: "with the Moderns, we never run out of surprises. Here they are, taking 'Society' and 'the individual' to be solid aggregates, whereas neither corresponds to their experience, the first because Society is an overly transcendent aggregate, the second because it is an overly unified aggregate." But as she thinks more about it, she notices that these two ectoplasms in fact correspond to a category mistake—forgivable enough, in the end—concerning the nature of the organizing act. As always, when one finds oneself before a trying dichotomy, it is because one has missed the precise gesture of another pass. As soon as this gesture is restored, the contradiction vanishes.

In fact, as a function of the script, depending on the moment, we find ourselves either "below" or "above" a given scenario; yes, by dint of occupying incompatible positions and maintaining schedules that don't intersect, we end up feeling that a destiny beyond our control weighs on our shoulders. The singular dichotomy of sociology arises here. Sociology starts with two phantoms, Society and the Individual, which exist in part, but only as *momentary segments* in the trajectory of scripts. Thus all we have to do with this organizing act is what physicists have done with the wave-particle duality. The individual looks a little more like the "above" sequence; Society looks a little more like the "below" sequence. But the resemblance is not really striking, and the alternation between the two is rendered imperceptible. Instead of striving to find the proportion of Individual and of Society in each course of action, it is better to follow the organizing act that leaves these distorted, transitory figures behind in its wake. It always comes down to following ghosts.

⊙ WHILE MAINTAINING THE METHODOLOGICAL DECISION THAT THE SMALL SERVES TO MEASURE THE LARGE ⊙

There is aggregation; there is no break in level. There is mini-transcendence; there is no maxi-transcendence. There is piling up; there is no transmutation. There is one level; there are not two. Now that we have recognized the course followed by scripts and perceived the danger of abandoning this course for a different level, we are going to use this thread to grasp the intermediaries that allow such a change of scale to occur. This will bring us to the heart of our task: understanding why Economics has managed to define aggregation and the metadispatcher in its own way.

The reader is probably astonished that we claim to understand the organizing act with the help of such a trivial example as a meeting between two friends. Yet the whole question now comes down to knowing whether, by focusing our attention on the course of scripts, we can do entirely without a metadispatcher and treat the *large* as a *fragile, instrumented extension of the small*. The choice of measuring instrument is the crucial question: the trivial case is what will give us the scale for all the others, rather than the organizations—enterprises,

States, markets—that would allow us to situate, arrange, order, or position the small cases. Everything hinges on this inversion of scale. The adverb "organizationally" leads to the verb "organize," which leads to the noun "organization"—and, in particular, to the ones called "market organizations."

It is only because we no longer presuppose the existence of a second level, "greater" than the scripts, that we can identify the very important phenomenon of relative SCALING. If we could begin with such a totally insignificant example, it is because, when Peter set up a meeting with Paul at the Gare de Lyon, they both put themselves *under* a script as surely as if Bill Gates were to redesign Microsoft's managerial flowchart, or the Paris Stock Market adopted a new automated auctioneer, or a new official accounting principle allowed French legislators to vote on the budget by program rather than by ministry. This is no time to be afraid, even if the author of *War and Peace* lost his nerve when he thought he could explain the chaos of the Battle of Borodino after the fact with a vast mechanistic metaphor that ruined all the effects of confusion caught on the fly that he had just described so brilliantly. By relying on the illusion of a Mega-Script written in advance by a Clockmaker God, Tolstoy made contingency a necessity; the War turned out to be curiously *pacified* by the ineluctable unfolding of the great mechanism of divine Destiny. As for us, we have to remain within the chaos of the scripts, resisting any temptation to pacify them too quickly.

⊙ THE ONLY WAY TO FOLLOW THE OPERATIONS OF SCALING.

What is in question with this new mode of existence is not learning how the small is included in the large but *how one manages to modify the relative size* of all arrangements. If we can start from such a small-scale example to define the giant organizations whose proliferation struck everyone during the last century as we were discovering the reach of these new Leviathans—warring nation-states, multinational corporations, tentacular networks, worldwide markets—it is because of the capacity of scripts to *connect* with other scripts. It is precisely here that we must not confuse organization as a mode of existence and "organizations" as a particular domain of reality that is supposed to cohabit side by side with others, for example, "individuals," "norms," and the whole apparatus of the sciences termed, a bit too hastily, "social." There are not, in the world, beings large or small by birth: growth and shrinkage depend solely on the circulation of scripts. To put it still another way, scale is not an invariant.

As we saw when we tried to restore the political mode [POL] to its rightful place, we had to resist the temptation to take difference in size as explanatory in itself. "Behind" politics there was nothing, and certainly not already-constituted "groups." It was the change in aggregation that had to be explained by the movement, itself truly circular, of the political *logos*. The regrouping came from the Circle; without it, there was simply no group at all. Similarly, it is impossible to explain organizations by something that is deemed to be there already, and deemed "greater" than the scripts that engender the provisional maintenance of organizations. If there is something "enlarging," it is that a new *being* is circulating, as original in its genre as the political circle [POL], which makes it possible to make anything it grasps change size [POL · ORG].

In other words, the script would "transport" difference in size. This is what is meant when we talk about a PROJECT while focusing on one of its distinctive features: its capacity to generate a before and an after, by lining up means and logistics behind it. Wherever projects begin to circulate—most often traced by tacit, language-based scripts, whether inscribed in practice or in writing—differences in scaling arise. One feels oneself below and above, framed and framing, inside and outside; one feels caught up in a rhythm that determines a before and an after. When this movement is properly understood, it is no longer hard to see that every modification in the quality or the instrumentation of these scripts is going to *modify the scale* of the phenomena that it collects.

THIS WAY WE CAN BRING THE ARRANGEMENTS FOR ECONOMIZATION INTO THE FOREGROUND ⊙

This time, we are ready to go to work. Only if the ethnologist insists on staying on a single level can the *apparatuses* and the *arrangements* through which scaling is obtained emerge. Her obstinacy pays off: the key phenomenon of ECONOMIZATION now comes to light. Economics as a domain does not precede the disciplines capable of economizing, since it is formatted by them. In Michel Callon's powerful expression: "No economics, no economies." When one is dealing with economic matter, one has to be prepared to pile performatives on top of performatives, like the tortoises in the fable, "all the way down"!

The method is basically the same as for the first Nature: to rematerialize knowledge [REF], we had to disentangle a totally idealized matter, the famous RES RATIOCINANS that preoccupied us in Chapters 3 and 4. Then and only then, as we recall, could the chains of reference required for access to remote beings appear in all their thickness, their ingenuity, their cost, their fragility—in all their beauty, too. Similarly, only once we limit ourselves to a single level—there is nothing in the aggregate except piled-up scripts—can we succeed in shifting to the foreground the *materiality* of the arrangements through which scripts pile up and fuse together. Conversely, as soon as we make the mere supposition that there is an economic matter to be studied *by* economics-as-a-discipline—and not produced, secreted, formatted by it—all these arrangements disappear. Once again, the materiality of materials counters the idealism of materialism.

Let us take the example of stockbrokers on the New York Stock Exchange around the middle of the nineteenth century, proceeding slowly and with solemnity, behind closed doors, to exchange information on the contracts that are coming due thanks to the confidence that each of these gentlemen has in the word of the others. The public outside scarcely hears, from week to week, the stock quotes on which they might rely. Here we are in a situation as "small" and almost as ill-equipped as the made-up anecdote about Peter and Paul. Now introduce into this organization, in 1867, an instrument called a stock ticker, the result of a rather clever technical composite between a telegraph, a printer, and an accounts book. What will happen? Everything about the coordination of these stockbrokers will change: from now on, they will be subjected to the continuous rhythms of printed and registered quotes that become visible in real time throughout the world on a perforated roll of paper tape (later projected onto screens). A new phenomenon emerges, one hitherto unknown, so closely has the extraction of values been mixed until then with verbal exchanges between people of the same world. The fluctuation of prices grasped *for themselves* and *continuously* takes on its own consistency, external to all speech; it is objectified, densified, accelerated. One of the objects of the science of economics is born of this equipping. Everything has indeed changed size, but the change would be incomprehensible without the intervention of the *ticker*—extended today to our portable phones. There is no change of size in the sense of the appearance of a higher level, and yet there have been major changes of scaling. Two entirely different definitions of the relation between small and large: one through insertion and embedding, the other through connection and collection.

To follow economization is to add up the impressive sum of devices like the stock ticker. To tell the truth, we are very familiar with the paths through which economics transits: account books, balance sheets, pay stubs, statistical tools, trading rooms, Reuter screens, flowcharts, agendas, project management software, automated sales of shares, in short, what we can group together under the expression ALLOCATION KEYS, or under the invented term VALUE METER—since it *measures evaluations and values* (we shall soon see how). It is through their intermediary that we can pass from a script that organizes the meeting between Peter and Paul to one that displaces the Red Army, or pays the million and a half employees of the French National Education system, or shifts billions of stock quotes in a few microseconds. By changing arrangements? Yes. By changing dimensions? Yes. By changing level? No.

⊙ AND DISTINGUISH BETWEEN TWO DISTINCT SENSES OF PROPERTY ⊙ The whole difficulty lies in not endowing these value meters and dispatch keys with virtues they lack. In fact, they all have the curious function of distributing both *what* counts and *those* that count. To capture this oddness that is so typical of the instruments of economization, we still need to get across, in the notion of PROPERTY, the echo of the movement of scripts that is required for the coordination of courses of action and for the distribution of the parties involved. In the statements "this is mine" and "that is yours," we grasp something of the organizing act, but not necessarily as Rousseau described it: "The first man who, having fenced in a piece of land, said, '*This is mine*,' and found people naïve enough to believe him, was the true founder of civil society." Jean-Jacques must have led a very solitary life; he must never have played in an orchestra, passed a ball, missed a stage entrance, served dishes at a social event, or hunted from a blind. Otherwise, he would have understood that "this is mine!" and "that is yours!" precede the right to property as well as "civil society" by hundreds of thousands of years. These are in fact the little phrases that make it possible, as soon as the course of action becomes a bit complex, to launch the scripts of those who are cooperating in a common task. "Be ready, the next call is yours," says the director to the actress in her dressing room; "It's *your turn* to play," the impatient partner exclaims; "This one's mine," cries the hunter as he lines up the wild boar in his gun sights. Well before it served to mark off plots of land and building sites, the expression "it's mine" served to trace limits to instances of cooperation that could not have been "brought to an end" otherwise. It is hard to see how "civil society" would have gained by "pulling up the stakes and filling in the moats" of such enclosures. If there were not people "naïve enough" to wait for the signal and believe in these distributions, there would have been no humanity at all. It is thus not in this sense of property that we should seek the origins of inequality. If Rousseau is ultimately right, it is at the next stage.

Let us suppose in fact that certain particularly well-equipped scripts make it possible to create limits visible enough so that those who obey them can say, with some reason to be believed: "We don't have to worry about that; we're quits," and this really curious expression in the mouths of future owners: "It isn't *up to us* to worry about that." We would find Rousseau's famous exclamation, but somewhat modified: the usurper is not the "first man who, having fenced in a piece of land, said, 'This is mine,'" but the *second*, who thought of saying: "It's *not up to me* to take care of you"! And if he managed to enclose something, it is not only with a picket fence and ditches, but with an even more credible act: that of an Accounts Book that places everything it does not take into account *outside* the enclosure and everything it does take into account and that properly belongs to him *inside*, the inside counting less than the outside. An admirable expression, *quittance*. Economists have provided the outline of this enclosure, moreover, with a couple of truly perfect expressions: *externalize, internalize*. What must we take into account? What do we no longer need to take into account? "What? Am I my brother's keeper?"

Let us now go one step further by supposing that the quantity of scripts increases, the number of elements to be taken into account grows vertiginously, the rhythm of due dates becomes more and

⊙ WHILE INCLUDING THE SLIGHT ADDITION OF CALCULATION DEVICES.

more intense. This time, we shall have to make the questions raised by the entangling of scripts and projects *measurable, accountable, quantifiable,* and thus *calculable*: how are we to allocate, distribute, share, coordinate? The scripts are still there, but *equipped with devices* that will necessarily produce quantitative data, as the stock ticker example shows. Without equipped scripts, such interweaving would be impossible: we are too numerous, there are too many quasi objects and quasi subjects to put in series. There are too many contradictory injunctions. The speed of transactions is too great. We would get lost.

When the disciplines of economization arose, they manifested themselves by the overabundance of these quite particular types of "quali-quanta" that connect the two senses of the French expression *prendre des mesures*: "taking measurements" and "taking measures"— hence the term "value meter." These sorts of quantities look more like *proportional shares* in property ownership charges—which are unmistakably quantities—than like the number of red dwarfs in a galaxy or the numbers on a dosimeter. Economization produces MEASURING MEASURES, which must not be confused with MEASURED MEASURES. It is in the very nature of the figures produced, and because they calculate, that value meters fuse values with facts—and, fortunately, since it is their main function: they divide up who owes how much to whom for how long a time.

Economization spreads by way of this equipment, just as reference, as we have come to see, spreads through instrumentation and the quite particular ideography of INSCRIPTIONS [REF · ORG]. The scripts' trajectories cadence, order, and collect, but in no way result in access to remote states of affair. The stock ticker does not measure prices in the sense of reference: it gives them rhythm and pace, it visualizes them, arranges them, accelerates them, represents them, formats them in a way that brings to light both a new phenomenon—continuously fluctuating prices—and new observers and beneficiaries of these prices, new exchange "agents," new entities "agenced" or "agitated" by these new data. And with each apparatus we see the emergence of both new (quasi) objects and new (quasi) subjects. This aptitude to produce new capacities at both extremes is something we also saw with chains of reference, which are capable of leaving in their wake a knowing subject endowed with objectivity as well as a known object likewise endowed with objectivity. Similarly, the trajectories of the scripts, by equipping themselves, produce at their two extremes a type of agent—the exchange agent has indeed *been exchanged* for the ticker—and a type of arrangement—the prices have changed form and nature. These are the two sides of the same coin.

If the Book of the first Nature was written in mathematical characters, so was the Book of the second Nature, as it happens, but this is not a reason to confuse the Great Book with the Bible, or to believe that a God is its author. It is by not respecting this TWO MODES NOT TO BE CONFLATED UNDER THE EXPRESSION "ECONOMIC REASON." distinction that the economists, by a new MALIGN INVERSION, threaten to *rarify* what ought to be superabundant; to dry up the horn of plenty. Fortunately, the anthropologist of the Moderns, seasoned by the harsh trial of her lengthy inquiry, no longer finds it difficult to resist confusing two modes on the pretext that they resemble one another: the same calculations, the same numbers, can serve both reference and organization, even though they haven't been dispatched by the same prepositions [REF · ORG]. Even if all the instruments that cover the chains of reference have a performative dimension, their function consists in making constants pass from FORM to form, not to divide up property rights or orchestrate the concatenation of due dates and quittances The term *calculation* and even *calculating devices* should no longer lead us astray. That the equipment of economization allows calculations does not mean that it *ceases to be performative* for all that. If the data produced by value meters are calculable, it is for reasons that *have to do with their nature as organizational scripts* and not at all, as we shall see, because they refer back to some quantifiable *matter* toward which they would procure privileged access. Well before being a particular domain that would form the basis for the world, as it were, and would explain all behaviors and all evaluations, the disciplines of economics now appear to us, from the standpoint of the organizing act, as the concatenation of equipment invented over time to allow us to follow the multiplicity of scripts and if possible—but we know that it is impossible!—to put an end to their inconsistencies with a new script that is to dis/reorganize the earlier ones.

The branching point is delicate, for it is exactly at this point that the innocent and indispensable work of *economization* is going to tilt into The Economy, in the sense of a metaphysics of inclusion and exclusion determined in the name of Reason—itself detached from its referential chains [REF · DC]! It is at this very point that the term "rational," of which economists have been making ill-considered use for three centuries, becomes poisoned. Not only does it not allow us to follow, as we are doing in this inquiry, the reason proper to each of the modes, but in addition it confuses two modes that should be kept carefully apart. Where is Occam's razor when we really need its keen blade?

We see that economics is very ill equipped to follow the thread of experience. It had taken the INDIVIDUAL as its basis and had endowed it with innate rationality! Now, when we follow the thread of the scripts, this anchoring point appears inappropriate to any foundation: the individual is second and secondary, entirely dependent on the multiform roles left in the wake of the courses of contradictory injunctions that can never assemble these roles in a round, solid, full subject. But we also see what is improbable in the expression "rational calculation." The agents that already receive help from numerous beings on the outside must, in addition, beg from the value meters to which they are connected the alms of fragments of approximate reasoning. Without an apparatus for calculating, no capacity for calculation. The expression "limited rationality" is rather weak for describing such a thoroughgoing reversal of perspective! We can understand that it is impossible to sketch out the immense scenography of a universe that has finally become rational thanks to the advent of Economic Calculation. What do we find in its place? Quivering little beings wandering around groping in the dark while waiting to receive something from the passage of scripts: sometimes fragments of projects, sometimes allocations of preferences, sometimes suggested roles, sometimes quittances.

We can sympathize with the surprise felt by the anthropologist of the proximate: "What?" she wonders. "Are there really hundreds of millions of people who are struggling to believe that they live in this world—and it's the same world as mine, after all!—as if they were individual agents calculating rationally? And they have managed to extend this implausible cosmology to billions of other beings? What ethnographic discovery can top this one?" But of course she knows that this is not the fruit of a tragic illusion of the Moderns about themselves: she is beginning to see to what apparatuses, institutions, networks, schools, and arrangements one must be connected, what apps one must learn to download, in order to believe in something so contrary to all experience, so manifestly utopian.

Provided that she can grasp that other mode that The Economy fuses with organization to obtain the idealized matter it uses inappropriately to shut down its calculations of interests and passions rather too quickly. If this inquiry has succeeded in sorting out the first Nature, it has done so by unfolding the amalgam of reference and reproduction [REF · REP]. What is this new crossing, then, that she will have to learn to respect in order to de-idealize The Economy and rematerialize it at last by restoring its taste for immanence? Perhaps we shall manage to untie this new Gordian knot on which all the rest depends without having to slice it open.

·Chapter 15·

MOBILIZING THE BEINGS OF PASSIONATE INTEREST

Whereas the whole is always inferior to its parts, ⊙ there are several reasons for making mistakes about the experience of organization: ⊙ confusing it with the Political Circle [POL · ORG]; ⊙ confusing organization with organism [REP · ORG]; ⊙ ballasting scripts technologically [TEC · ORG]; ⊙ confusing unequal distribution of scripts with scaling; ⊙ all this leads to an inverted experience of the social.

By returning to the experience of what sets scripts in motion ⊙ we can measure what has to be passed through in order for beings to subsist ⊙ while discovering the beings of passionate interest [ATT].

But several obstacles to the depiction of this new experience have to be removed: first, the notion of embedding; ⊙ then the notion of calculating preferences; ⊙ then the obstacle of a Subject/Object relation; ⊙ fourth, the obstacle of exchange; ⊙ and fifth and last, the cult of merchandise.

Then a particular mode of alteration of being appears ⊙ with an original pass: interest and valorization ⊙ and specific felicity conditions.

This kneading of existents ⊙ leads to the enigma of the crossing with organization [ATT · ORG], ⊙ which will allow us to disamalgamate the matter of the second Nature.

A S LONG AS WE TURN AWAY FROM THE EXPE-
RIENCE OF SCRIPTS, IT IS IMPOSSIBLE TO DIS-
TINGUISH THE WORK OF ECONOMIZATION
from the metaphysics of the Economy. The notion of
INDIVIDUAL captures neither the experience of being *under* the scattered
scripts nor that of being situated *above* them. As for the notion of SOCIETY,
it registers neither the experience of finding oneself under a pile of contradictory scripts nor that of rewriting them partially in places that are
well circumscribed in each case and always equally "small." The larger
is smaller than the small, which is not small but distributed . . . Starting
from the Individual or from Society is a hopeless undertaking, then; it
would lead, literally, nowhere—and in any event there is no path! A mistake has been made in the scaling of associations. Yes, and a sizeable one:
a mistake in *size*.

The anthropologist knows that she must no longer be surprised at
the distance between the experience of the organizing act and the official version of it provided by the science of organizations. And yet she
cannot help being constantly blocked by what looks to her like an inversion of the relative size of the phenomena to be observed: it is as though
an organization were an individual entity with clearly defined borders
and a complete, well-rounded identity that had been inserted "inside" a
set much larger in size. Yet the small and the enclosed are not what go into
the large and the open; rather, the dispersed, the tentacular, the multiple
are what sometimes turn out to be caught on the fly, always in confined

spaces, by the lasso of scripts that are then going to delegate other roles to other characters; this may keep the scripts a little less scattered, but only for a time. The relation is not one of small to large, but rather one of an undulation that imposes its syncopated rhythm on beings, sometimes imparting a sense that those beings are following orders from on high— as if they were "below"—and sometimes inciting a "MANUAL RESTART" of the organization—as if they were now "above," looking over the whole scenario and able to rewrite it.

To rediscover the sense of proportions, behaviors, and vehicles, and to keep from considering scale as the element that remains invariant, we must now carry out our struggle in reverse. Whereas up to now we have had to clear out space so that invisible beings could circulate in a too-quickly-saturated world, we now need to empty the world of the ectoplasms that we think we detect every time we view as a break in level what is only a pileup or a connection among scripts varying only in equipment, quantity, due date, or cadence. Here again, here as always, by failing to recognize the partial originality of a distinct mode of existence, the Moderns have set about "padding" by inventing Society, which is as incapable of extending its reach everywhere as is NATURE or LANGUAGE. In practice, the social—the trajectory of associations—leads to an entirely different experience from that of sociology. This should no longer surprise us, since PSYCHOLOGY, as we saw in Chapter 7, led in fact to a place entirely different from the unfathomable depths of the human spirit.

⊙ THERE ARE SEVERAL REASONS FOR MAKING MISTAKES ABOUT THE EXPERIENCE OF ORGANIZATION: ⊙

At the same time, the analyst is well aware of all the good reasons that have kept the Moderns from identifying the organizing act in its full originality. In the first place, there is the temptation to mix it with politics [POL · ORG]. By a fine interpolation, in speaking of Society, they proceed as though the injunctions of scripts resembled the Circle described in Chapter 12. It is quite true that the "we" sketched out by the course of political enunciation always appears "larger" than the "I" expressed in the "we," since this "we" "represents" its principals more or less faithfully depending on the quality of the Circle and the rapidity of its renewal. If one confuses these two rhythms—the reprise of the

⊙ CONFUSING IT WITH THE POLITICAL CIRCLE [POL · ORG]; ⊙

415

Circle of representation and the reprise of scripts, some of which reach their term while others are relaunched—one can indeed feel that one is "in" a group whose greater weight lies heavily on the shoulders of each of its "members." With the unanticipated consequence that social science runs the risk of missing both politics and organization, mixed up as they are in the same apparent circularity.

And yet in both cases the relation of the part to the whole is not that of a shoebox in a pile of shoeboxes. At any moment, depending on the quality of the political or the organizational reprise, this "whole" may again become *smaller* than its parts. How many omnipotent dictators have become, from one day to the next, scared little fugitives? How many powerful CEOs, early one bleak morning, have had the impulse to strap on golden parachutes and leap from their skyscrapers? What politician, what director, what representative, what delegate can fail to see these sudden shrinkages in size? Relative scaling is always the consequence of the circulation of these two modes. Neither the political body nor the organization can count on any sort of inertia to persevere in being. The whole is smaller than its parts, and if it captures certain aspects of the parts it is only for as long as it keeps moving, connects, gets a grip on itself and starts over from scratch.

The sociology of "the social" (as opposed to the sociology of ASSOCIA-TIONS) may have been right to see the social as one of the major phenomena of the human sciences, but it was a mistake to define it tautologically, and an even greater mistake to seek to extend it to all the modes. Scripts do not present themselves as tautologies (we make Society, which makes us; we are held from the outside by norms to which we nevertheless aspire) unless we forget the slight temporal *gap* thanks to which we never find ourselves "above" or "below" a given scenario at exactly the same time or with exactly the same capacities. Unfortunately, the notion of tautology completely misses this sinuosity, which is so particular to scripts. And even if it managed to follow that mode of extension, it would still not be able to serve as a yardstick for politics, religion, law, or psyches—not to mention the first or second Nature. Once again, we observe the tendency of each mode to propose a hegemonic metalanguage for speaking about all the others; a quite innocent tendency, but one from which this inquiry aims to protect us [ORG · REP]; the political body is a phantom, yes, but it is not an ectoplasm like the phantom of Society. Greek and Latin must not be confused here: AUTOPHUOS is a mode of existence; Society sui generis is not.

So we have to give up seeking to add something that would be already present, already large, enveloping, framing, since that would amount to concealing the particular movement of scripts. Fortunately, this new temptation has already been well recognized: it is the one that tilts organization toward *organicism*. The Large Creature that would be "behind" an organization proceeds only from a confusion between the beings of reproduction and those produced by PROJECTS [REP · ORG]. The confusion goes in both directions, moreover, since "nobody knows what a body can do." What is true of politics is also true of organization: it is already so monstrous that it should not be further disfigured by other monstrosities.

⊙ CONFUSING ORGANIZATION WITH ORGANISM [REP · ORG]; ⊙

A casual remark that a living body is "organized," or that a multinational company reacts as an organism would, or that such companies are going to disappear "like the dinosaurs," presupposes that the problem of maintaining themselves in existence is already *resolved*, as though this could be done *without* the passage of the slightest script. In other words, one can never consider an organization (state, corporation, network, market) as a body, for want of knowing just how a body [REP] manages to hold itself together! Conversely, it is because there is no other world, above or behind an organization [ORG], whose particular way of creating something "above," "below," "before," and "after" we are going to be able to detect—and the same problem arises whether we are dealing with IBM, futures markets, or DNA. In other words, the mini-TRANSCENDENCE proper to this mode of existence (and also to the political mode [POL], moreover), can be detected only if no other maxi-transcendence is introduced parasitically. We no longer live in the time when Senator Menenius could bluff the Roman populace into ending their strike with his naïve fable about the "organic" solidarity that its active, hungry Members ought to maintain with the lazy, satisfied Belly [POL · ORG]. Aggregation as pileup, yes; aggregation as Large Being, no. The rematerialization of economic disciplines comes at this price.

What makes such a conclusion implausible in spite of everything, the investigator feels, is—a third reason—the gradual introduction of materials that are no longer oral or written injunctions to modify relative size in a lasting way. How can someone maintain that organizations are not "larger" than individuals when she is the superintendent of a hundred-story tower whose highest level, which she has never dared to enter, holds the office of the CEO? Who can fail to feel tiny while approaching the Pentagon? Facing the Egyptian pyramids, how could the most narcissistic general, before starting a battle, not relate his own small size to the "forty centuries that loom over him"? Placing oneself "under" a script has a different weight if the script is relayed by injunctions of stone, concrete, or steel, each one of which puts you "in" a precise place without any possibility of turning back. "Getting put in one's place" has a different meaning depending on whether that place is a scenario consisting of a few words exchanged on the phone, as with Peter and Paul, or a place weighed down by a sarcophagus of assorted materials. Relative SCALING, when nothing can reverse it any longer, does seem to have become a difference in size between "the big things" and "the little things."

And yet the anthropologist feels entitled to maintain her position, because she is well aware that speaking about technology doesn't suffice as a way of distancing ourselves from the social. This is even, as we know, the origin of the actor-network theory in sociology: what is called the "technological dimension" of a situation is always a script that nevertheless retains its nature as instruction, injunction, inscription, distribution, even if it was drafted a very long time ago, by organizers who have long since disappeared, for generic beings whose positions are punctuated by different due dates [TEC · ORG]. The fact that there are no more servants in that seven-story Haussmannian building in Paris doesn't mean that Balmain's script (the architect's name is inscribed on the façade) won't keep operating. Balmain decided in 1904 that the elevator for the owners would only go up as far as the sixth floor. Today, all the students who have to climb the back stairs find themselves "under" a script so dispersed in time that to be "above" it one would have to go back to the Balmain agency in 1904, or else spend a fortune on technology to expand the reach of the elevator that serves the privileged residents. Here there was indeed a decision to *ballast* the difference between masters and servants by all the weight, effectively irreversible, of a difference in stairs. The meeting between Peter and Paul at the Gare de Lyon had the lightness of a telephone call, and it faded into nothingness after their successful encounter. There haven't been servants for a long time, but Balmain's scenario still weighs on the top-floor tenants.

And yet, from the viewpoint of the beings of organization, all the scripts remain comparable—except for the spatial, temporal, and actantial SHIFTING OUT—if we take the word "relative" in the ordinary sense of that which depends on *relations* dispersed in space and time, associations traced by networks of the [NET] type. The fact that we can distinguish lightweight utterances from heavily ballasted ones, utterances with short-term outcomes from those with long-term due dates, reversible utterances from those that are difficult to reverse, does not keep us from recognizing that they all have performative effects. Similarly, as we saw in the preceding chapter, the quantitative and calculated character of VALUE METERS does not take away their nature as ALLOCATION KEYS; nor does the fact that the arrangements known as socio-technological pass through gradients of diverse materials keep them from fulfilling their function as distributors. Moreover, this is what accounts for the obsessive efforts made by every institution to make the distribution of roles, places, and functions as definitive as possible. Pouring the injunctions into a more solid material: this is what gives dispatchers the duration that is always lacking, by definition, in scripts, because scripts always have to be renewed when they reach their due date.

Still, the fact that this relation is always relative to a given distribution of forces is something we notice every time an enterprise has to occupy spaces set up for a different one: supposedly irreversible decisions have to be reversed, most often by dint of reorganization, but also by pickaxe or dynamite. Starting in the 1960s, for example, it took Paris metro engineers some fifteen years of colossal construction work to achieve an interconnection between the subway system and the railroads—an interconnection that a decision by the Socialist majority at City Hall under the Third Republic had made definitively unfeasible, by making sure that its easily reversible vote would be taken over by

tunnels so narrow that the private railroad companies could never penetrate the network of the new metro. Definitively, as the elected officials
believed? Let us say provisionally definitively. By following the socio-
technological arrangements, our investigator gradually learns to stop
introducing artificial distinctions between the gradients of materials
offered by a slogan, a flowchart, an accounting technique, or a concrete
wall [ORG · RES]. "It's not carved in stone?" Yes, it is, and in brass and cast
iron—but, carved or not, these are all still scripts.

⊙ CONFUSING UNEQUAL
DISTRIBUTION OF SCRIPTS
WITH SCALING; ⊙

The fourth reason for being mistaken about
the experience of organization is that the statement
"the whole is always smaller than its parts" is transformed into "the whole is greater than its parts,"
and this transformation plays an important role in the identification of
inequalities. By inverting the inversion, do we not risk losing a particularly convenient—and morally indispensable—system of coordinates?
Just as the relations between high and low, large and small, are brought
forth irresistibly by the walls, the architecture, of many organizations—
elegant central stairways, gigantic halls, majestic pediments, private
entrances—so the feeling of "smallness" depends crucially on the
number of opportunities one has to revise a script. Those who never have
access to the narrowly circumscribed spaces where scripts are rewritten
are "small"; those who, without being physically larger or morally superior, frequently enter and leave offices—for it is almost always a question of offices—where roles are redistributed and rewritten are "large":
"bosses," "masters." The fact that what is at stake is another series of
instructions and inscriptions that dispatch the roles between those
who "execute" and those who "decide," the fact that this series too is *flat*,
and that its scaling is entirely relative, doesn't ward off the irresistible
impression of absolute and irreversible difference. What common sense
rightly calls *domination*. And it is on this often overwhelming difference
that so many denunciations of power and inequalities insist endlessly:
yes, the distribution of scriptwriting is indeed unequal; yes, some beings
are "below" much more often than others. Everything is flat, but in the
last analysis there is indeed an "above" and a "below."

And yet if the investigator can't keep from stubbornly following
the scripts, it is because she often uncovers, in her insistent focus on

domination, a risk of once again freezing the differences that she is seeking to *relativize*. If she insists so strenuously, it is because she wants to gain access to the precise mechanisms through which a relative difference ends up cloaking itself in absolutes. It thus seems essential to her to hold fast to the suppleness—that is, the contingency—of inequalities, never to give them more strength, durability, solidity than what they have acquired through those provisionally irreversible decisions, never to refuse to transform the masters into inhabitants of a transcendent level that would dominate the poor wretches "below" for real.

Registering variations in the balance of power is possible only as long as one strives always to start from a situation, virtual if necessary, in which the balance is equal at the outset. Even if the ethnologist occasionally risks being accused of heartlessness, of "underestimating the weight of inequalities," or even of "ignoring power relations," the only way to help overturn inequalities, in her view, is to refuse to yield to any illusion about their relative size. And thus never to slip surreptitiously toward the "bad" transcendence, the one with two levels. The scales are already uneven enough not to be charged in advance with a whole weight of injustices. Here is her mantra: when domination is at issue, whatever you do, *don't add to it*.

For the whole to appear superior to its parts, then, we see that a certain number of conditions are required: the rhythm of the organizing act must be invisible or suspended; there has to be confusion between organization and the political requirement of composing a group; organizations have to be endowed with a biology that is farfetched both for them and for living organisms; the movement of technological detours must be lost from sight; efforts must have been made to freeze relations of domination; finally, as we saw in the previous chapter, belief in the supreme wisdom of a METADISPATCHER—a religious or secular Providence—must not be open to discussion. One might think that such an accumulation of conditions would make the experience of a two-level social reality unlikely. Yet just the opposite is true: the experience of the social on one level, the social finally flattened, finally continuous, finally immanent, appears impossible and even scandalous, although it is the most common experience—but also the hardest to decipher. How to

⊙ ALL THIS LEADS TO AN INVERTED EXPERIENCE OF THE SOCIAL.

account for such an inversion? If this "whole superior to its parts" isn't a product of experience, where does it come from?

BY RETURNING TO THE
EXPERIENCE OF WHAT SETS
THE SCRIPTS IN MOTION ⊙

As she asks herself this question, the investigator notices that she has not yet followed faithfully enough the ins and outs of the experience with which she began. She has understood perfectly well how Peter and Paul manage to coordinate their scripts, and she has grasped what was original about that coordination [ORG]. But she has not yet gone back far enough to figure out just *why* Peter and Paul wanted to meet in the Gare de Lyon. She has captured the rhythm of their organization, but not the *energy* that got them going. She sees that they are constantly being bombarded with characters that delegate to each Peter and each Paul myriads of "Peter doing this" and "Paul doing that," but she has nonetheless chosen as her starting point a Peter and a Paul without quotation marks, as if they existed *in and of themselves*, as if they were always points of emission and reception of scripts. As if the initiative came from them.

Now, as she extends her inquiry, the analyst quickly perceives that the Peter and the Paul who met at the train station are still more *distributed*, less the indivisible individuals of social theory, than she thought; she sees that other things agitate them and give them impulses that make them *agents*. She notices, in fact, that if they have been impelled to organize this meeting, it is because they are "attached" by a common "desire," the desire to create a software program that, according to them, is already "attracting" the "attention" of many persons "lured" by the potential "earnings" and by the "beauty" of this innovation, whereas the program "worries" certain other persons who have been "alarmed" by the "risks" and the "consequences" of the invention. This is what *interests* our protagonists. It is thus toward this focal point of their attention that the investigator must direct her gaze, since it is what got them moving. With respect to this focal point, the organization of the meeting counts in the end only as an accessory temporarily in the service of something else. Just as a script uses reference as a simple instrument for establishing landmarks [REF · ORG], so a script serves now as a tool for another type of setting-into-motion.

She begins to see that there is, *beneath* and *beyond* Peter and Paul and, so to speak, slightly *ahead* of them—but not above or below them—another linkage that she must now learn to follow. What is given in experience is not merely Peter and Paul "below" and "above" scripts, it is also Peter and Paul *attached* to *things* that interest them *avidly* and that agitate them while exceeding them. So she now needs to concentrate on what agitates the two characters. If she were to follow only the concatenation of scripts, she would not become capable of following the flow that forms its real *content*, for the actors. How can she describe such connections, such overflows?

It suffices to raise the question to see another type of network emerge little by little. The same thing might be said about this one as about the network that links the beings of reproduction: that it is at once the most important of all, the most obviously percep-

⊙ WE CAN MEASURE WHAT HAS TO BE PASSED THROUGH IN ORDER FOR BEINGS TO SUBSIST ⊙

tible in experience, and the most difficult to characterize without getting it wrong from the start. Its *patuity*, to borrow Souriau's term, its particular form of presence, does not make it any more visible, for it appears in the form of an ever-lengthening list of beings through which it is necessary to pass in order to subsist. But this time it is not a matter of the risk taken by lines of force or lineages in order to insist, maintain themselves, pass, reproduce, endure; what appears is the surprising series of detours necessary to *attach quasi subjects and quasi objects* and *make them hold together in a lasting way*. We no longer have Peter and Paul as bodies launched into the risk of becoming, but everything through which the beings that have Peter and Paul *as segments* in their chains of associations must pass. No longer chains of associated humans—as we met them in the THIRD GROUP—and no longer chains of nonhumans lined up one after another—as in the second group—but rather the capacity to concatenate humans and nonhumans in seemingly endless chains. To risk a chemical metaphor, if we note humans as H and nonhumans as NH, it is as if we were now following long chains of polymers: NH-N-NH-NH-H-H-NH-N-NH-H-H-H-H-NH, in which we could sometimes recognize segments that look more like "social relations" (H-H), others that look more like aggregates of "objects" (NH-NH), but where attention would be focused on the transitions (H-NH or NH-H).

They certainly form networks [NET] that can be identified by the registration of trials all along a COURSE OF ACTION. If I am "well off," as they say, the series of gestures, actions, and products through which I have to pass to buy my brioche at the fine Kayser bakery in the Maubert neighborhood can be traced. But if am hungry and poor, I am going to have to beg for a few coins in order to get myself a scrap of bread; this course of action also defines a series—pleas, coins, scrap—that defines quite precisely *what I had to pass through* to subsist today. The difference between the two courses of action does not lie in the series of heterogeneous steps I have to take, but in the *sign* that will affect the passage from one to the other—negative in the case of the scrap (it will be a matter of *ills* and poverty), positive in the case of the brioche (a matter of *goods* and affluence). By proceeding this way, we draw a sort of grid on which the starting point matters little. If we wanted to define the baker from whom one person buys a brioche (without giving it a second thought) and another begs (very painfully) for a scrap of bread, we would still find ourselves facing a series, differently composed, to be sure, but similarly heterogeneous, that would define the entire set of *goods* the baker has to procure (flour, sugar, kneading machine, assistant) and the *ills* he has to endure (getting up early, putting up with the cashier's chatter, paying his taxes, voting for the bakers' union representative, and so on).

This grid can thus be read starting from any point and in any direction. Let us note that we find the baker on the path of the poor man and the rich man alike; we find the poor man on the rich man's path and vice versa, and both on the baker's path; the rich man may find himself crossing the poor man's path (where he has an opportunity—which he may ignore—to alleviate the poor man's troubles). Let us note, too, that both the scrap and the brioche are found on the paths of all three: a series of this sort describes what is possessed (or not) just as much as those who possess (or not). We can describe the same situations without worrying any longer about whether we desire what is desirable or whether it is what we desire that becomes desirable. The series is entirely *reversible*. As this example illustrates, in any case, any given element is defined by the set of *aliens*, of others, through which it has to pass.

⊙ WHILE DISCOVERING THE BEINGS OF PASSIONATE INTEREST [ATT].

How can this type of network be qualified? The verb "to be" cannot capture the grid. In his last book, *Psychologie économique* (1902), Gabriel Tarde—yes, Tarde again—set us on the right path: everything changes if we agree to choose the verb "to have." From the verb "to be," Tarde says, we cannot draw anything interesting that would involve interests, except identity with the self, the "easy way out" of substance; but from the verb "to have," we could get a whole alternative philosophy, for the good reason that *avidity* (unlike *identity*) defines in reversible fashion the being that *possesses* and the being that is *possessed*. There is no better definition of any existent whatsoever beyond this list of *the other beings* through which it must, it can, it seeks to pass. But for the sign (but the sign is everything!), poor and rich are in the same boat, for what we do not possess attaches us as much as what we can easily acquire for ourselves. In this sense, we are *altered, alienated*. It is as though, here again, a philosophy of identity and essence—of being-as-being—had played a trick on us by concealing the avidity, the pleasure, the passion, the concupiscence, the hook, of *having* and *had*. This philosophy would have forced us never to confess our attachment to the things capable of giving us *properties* that we didn't know we had. This is what makes us move, if we are poor, from one end of the planet to other, and it is why, if we are rich, we displace the most coveted goods while shoving the "bads" off onto the others.

If we remain faithful to the principles of our method, we can recognize particular beings here, even if they seem very difficult to institute for themselves: beings *of passionate interest*, or *interested passions*, as you prefer, that oblige us to register the presence of a new preposition, the next-to-last in this inquiry. We find ourselves squarely in front of a new mode of existence that, since we cannot count on a recognized term, we shall call *mobilization*, or, better yet, ATTACHMENT [ATT].

What makes the extraction of this contrast so difficult is that it resonates as much with the beings of metamorphosis [MET · ATT] as with those of reproduction [MET · REP]. The same exteriority, the same surprise before abrupt transformations, the same uncertainty about the person targeted, the same brutal alternation before the enthusiasm of being carried away by energizing forces and the depression of being subjected to forces that exceed us in all directions. So many passions in this coveting, this envy, these obsessions. So much danger of becoming possessed for real—the thirst for gold has driven more than one conqueror lost in the depths of Amazonia into madness; similarly, the abundance of choice in Zola's *Ladies' Paradise* perverted well-to-do women shoppers, who lost their heads in the turmoil resulting from the vast quantity of goods on display in a new department store. And when we speak of the misdeeds of capitalism, don't we have to turn toward sorcery in order to understand the black magic that leaves us helpless in the face of its abuses? And yet these are not the same beings, for the beings of metamorphosis are as completely unaware of quasi objects as they are of quasi subjects, and do not seek to hold them together in long chains. They pass. They transfer. They mutate. And yet there is no doubt about it: they lie in wait, they watch over the linkages of goods and bads, ready to turn them, at any moment, into entirely different passions.

BUT SEVERAL OBSTACLES TO THE DEPICTION OF THIS NEW EXPERIENCE HAVE TO BE REMOVED: FIRST, THE NOTION OF EMBEDDING; ⊙

How can we characterize the beings of attachment? The simplest solution, to follow what sets Peter and Paul in motion, would be to say that they have "economic interests" to which a certain "social dimension" must be added. But here the ethnologist would be breaking down open doors, by discovering flows of production and exchange and reinventing socioeconomics as if The Economy were lying in Society's bed, or "embedded in social relations."

This first obstacle is easy to set aside. First because, as we are beginning to understand, The Economy does not correspond to any attributable experience, but above all because there is no such thing as "Society" that could serve as a duvet, a canopy, a hotel room where The Economy could be tucked into bed, nothing that could even serve as a frame for embedding it. From the standpoint of this mode, what is called "Society" is only the effect of scripts piled up helter-skelter, scripts whose precise nature and whose type of stacking have been lost from sight—and to which has surreptitiously been added, through confusion with politics, a giant Dispatcher, a Metadispatcher, a Providence that would attribute places, roles, and functions without any way for us to know in what offices it exercised its wisdom or through what vehicles it transmitted its injunctions, its formats, its standards. Since we are trying to peel away the onion-like layers of which The Economy is composed, we are not likely to make the mistake of picturing a sui generis "Society" that would

exceed The Economy or serve as its outer covering. To speak of socioeconomics would only amount to piling up category mistakes one on top
of another. If there are not two levels but just one, if we have gone to so
much trouble to restore the experience of flatness to the social, it is not
so as to add the transcendence of "Society" to that of "The Economy." If
we're going to free ourselves, we may as well profit from the occasion
and free ourselves from both. The social world is flat. Freedom comes at
this price, and perhaps liberalism as well—not to be confused with its
antonym, obtained by adding the little prefix "neo" ...

Let's not rush, either, to smooth over the link-   ⊙ THEN THE NOTION
ages by declaring that Peter and Paul have "made   OF CALCULATING
calculations" and have managed to express "their   PREFERENCES; ⊙
own preferences," along with those of their prospec-
tive clients and their competitors Yes, of course, they have drawn up a
"business plan": they have it in their briefcases, and they may discuss it,
if they have time, at the *Train bleu* bar at the station, going over the "bullet
points" of the PowerPoint presentation they have designed to convince
their "business angels" (the surroundings may be French, but the English
language is compulsory in this context). Still, these calculations do not
support any particular course of action. They are rather what Franck
Cochoy calls "QUALCULATIONS": they refine judgments concerning the
intersection of scripts, but they are ultimately incapable of *untangling* the
passions needed to launch the same scripts. What we are trying to grasp
is what has set them into motion, what *moves*—in both senses—Peter and
Paul and all those whom their software attracts or worries. When calculations appear, they are there to reinforce, emphasize, amplify, simplify,
authorize, format, and perform the distribution of agencies, not at all
to *substitute* for the experience of being set in motion, moved, attached,
excited by things that are different on each occasion. To give calculations
their rightful place, we must let passions—calculations of passion—take
their rightful place as well.

Is the experience one in which we are moved   ⊙ THEN THE OBSTACLE OF A
by things or one in which we are moved by our   SUBJECT/OBJECT RELATION; ⊙
desires projected onto things? Here is another
obstacle that would oblige us, in the Peter-and-Paul experience, to separate out what derives from the software they have invented, and what

derives from the inventors themselves, from their sponsors and/or their competitors. This time our ethnologist feels right at home. She has tangled with the question of Subject and Object enough so that she no longer needs to shut herself off in the impossible choice that has caused more than one of Buridan's donkeys to starve to death: am I interested in this item because of its "objective qualities," or is it because I am interested that I find all sorts of good qualities in it, or, finally, is it because others have "influenced" me in order to "make me believe" that I am interested in it? Having gone to so much trouble during this inquiry to introduce quasi objects and quasi subjects, we can now harvest the fruits of all these "quasis."

If there is a question we no longer have to raise, it is whether interest stems from the individual, the object, or the influence of the milieu. We shall simply say that Peter and Paul, along with their friends and enemies, find themselves linked, attached, bound, interested, in long chains of quasi objects and quasi subjects whose surprising unfolding gives their experience all its piquancy. Interest arises impromptu. And it attaches *people* and *things*, more or less passionately. The "careers of objects," the "social life of things," the terms hardly matter; what counts is capturing the real Copernican revolution through which we finally entrust ourselves to the vast movement of these chains so as to receive new capacities and properties from them, on the side of the quasi subjects, and new function and uses, on the side of the quasi objects. If we have been able to bring off this reversal for divinities, demons, angels, and gods—they are given to us, they come to us—the acknowledgment that these long alignments of interesting things are what make us act shouldn't really cost us very much.

After all, this is what the etymology of interest indicates: a mediator par excellence, it arises *between* two entities that do not know, before it arises, that they could be attached to each other. Here again, a "break in continuity" has to be introduced to prolong the course of action. The addition is reversible. What is an object? The set of *quasi* subjects that are attached to it. What is a subject? The set of *quasi* objects that are attached to it. To follow an experience, it would be useless to try to retrace what comes from the Subject or from the Object; rather, we must try to find out by what new break, what new discontinuity, what new TRANSLATION interest makes the (quasi) subject of a (quasi)object—and vice versa—*grow*.

"A translation? No, not a translation but an exchange, a simple exchange of equivalents: a mere DISPLACEMENT!" And here is the fourth obstacle that

⊙ FOURTH, THE OBSTACLE OF EXCHANGE; ⊙

has to be removed. The exchange of *equivalents* would have an immense advantage: nothing would happen any longer except flows of causes and effects, without any transformation or metamorphosis. This is certainly Double Click's little weakness—maintaining that there are only transports without transformation—but it is surely not the experience that the investigator is seeking to describe [ATT · DC]. Fortunately, her ear is sensitive enough to recognize the grunts of her Evil Genius behind the thickest of masks. We are supposed to believe that she is breaking down open doors in discovering the "world of exchange," the privileged territory of economic matter, while only adding to it some tremolos on emotion, avidity, and the passions aroused by these exchanges. Whereas what interests her, on the contrary, is the break introduced at every point by the emergence of value: things budge, bubble up, so many things happen that one *can no longer do without them*, without these things.

In the word "exchange" there is, above all, "change," yet what we feel in it above all is, on the contrary, the abyss, constantly surmounted, of the *nonequivalence* from which *valuation* arises. Before Peter and Paul found themselves attached to their "attractive," "attaching" software, there was no common measure between it and them. Now, they are defined by it and vice versa. You were walking by that store without thinking about it; how does it happen that you can no longer get along without this perfume when ten minutes earlier you didn't know it existed? You bought that plastic-wrapped chicken without attaching any importance to it; what unexpected discovery has left you really disgusted with it now? That store owner has been struggling unsuccessfully for months to attract nonchalant customers whom nothing excites, and now a slight change in product placement is filling the shop with clients eager to clean out her stock. Yes, of course, later on, when a whole set of calibrated instruments has been set up, these nonequivalents will be turned into a simple exchange of equivalents, but before they can all be worth "the same thing" and before "nothing more happens," they first have to *be worth* something. So many changes, before there are exchanges. So many valorizations, before there are equivalences. These changes and valuations are charged with alterations, surprises, excitations, uncertainties, as they come to *animate* customers. How many souls it takes to stir all this up!

⊕ AND FIFTH AND LAST, THE
CULT OF MERCHANDISE.

"No, here you're really exaggerating: the anthropologist who had promised to resist exoticism is now getting carried away and starting to mistake us, economic agents subject to the rule of the market and in the grip of merchandise, for real savages. Modern market exchange can't be confused with the old bonds of social dependencies and entanglements. Reread your classics: don't you know that all those old bonds have been 'dissolved in the icy calculation of interest'? All that is over and done with—for better or for worse." And yet if exoticism raises such an obstacle along the path of the experience of attachments, it is in the other direction: why do we believe that we are so different from the others on the pretext that we have changed scales? For the ethnologist has done her homework. She has carefully studied the marvelous discussions in economic anthropology about the "careers of objects" in the Andes, gift-giving in the South Pacific, the obligation among the Iroquois never to detach oneself entirely by a transfer of strict equivalence so as never to be quits. In short, she has perfectly grasped the impossibility of driving The Economy out of its bed of social, moral, aesthetic, legal, and political relations. The literature is quite lovely, but it always comes from far away, from long ago, from other times, and it is often marked by a wrenching nostalgia. As if all these studies had forgotten to be symmetrical and had believed themselves obliged to map out the antipodes only in contrast with modern market organizations. So our investigator looks around her, and she investigates, she asks questions, and she is astonished. She has to admit that she no longer knows who is at the antipodes of whom.

She is told that "over there" people don't sell goods "like little bars of soap." But she knows something about the history and chemistry of soap; she sees the incredible complexity of such inventions; and, as she has read Richard Powers's fine novel *Gain*, she also knows what precious ploys marketing departments have to use to maintain their market shares. Who among the critics of merchandising would know how to manufacture, package, market, and sell the smallest little soap bar? She hears melancholy talk about "consumer society," but she visits supermarkets where the shelves are lined with information about "sustainable development" and "responsible agriculture"; packages are covered with legal notices, technical information, coupons, bonuses, telephone numbers for customer support and complaints. Quits? Is anyone ever "quit" of such contracts? Thanks to a friend, she was able to follow the organization of a Telethon and could observe that the sales and accounting mechanisms of a public charity are as well equipped, as complex, and as professional as those of enterprises selling good for profit. So where is the fetishism of merchandise? One day in Bologna, she was hurrying to buy some salami when she found herself among some "slow food" militants who forced her to slow down by laying out for her the long series of ties that attached them to Lombardy peasants, to the gastronomic culture of Emilia-Romagna, to the worldwide solidarity of poor farmers. Politics in a sausage! If she turns toward the shamed financiers, what does she discover? "Icy calculations"? No, little groups of speculators inebriated by the flashes on their Reuters screens, doped up by testosterone, as baroque in their rantings, as inventive in the chains of mediations they put in place, as Pacific Islander participants in a kula exchange. Cold, their calculations? For those who are subjected to them remotely, yes, without question; but in the witch's cauldron we know as Wall Street or the City they boil and burn.

The ethnologist of markets is astonished: do her informants practice a *merchandise cult*? Would they prefer to mask under the idea of pitiless mercantilization—whether they are delighted or indignant at the prospect—the astonishing complexity of the attachments among people, goods, and bads? They look as though they are not interested in this at all, in theory, whereas they plunge into it again and again in practice; they never manage to pull themselves away. Always the same trenchant question at every stage of our inquiry: how do the Moderns manage to such an extent to ignore officially what they do in practice? How do they manage not to recognize, upstream, the number of entities through which it is necessary to pass to produce any goods whatsoever, and, downstream, the number of unanticipated consequences thanks to which, like it or not, they find themselves attached to each of these goods, hindered by each of these bads? On each occasion, at every link in the chain, they succeed in covering up stunning discontinuities, complex heterogeneities that no continuity makes it possible to hide.

If there were only equivalences that are transported in an exchange without anything else "happening" or "passing," as those who talk about *commodification* believe, what explains the millions of hours passionately invested in the process of *adjusting* goods and people? Why are there tens of thousands of marketers and merchandisers and designers and packagers and publicists and accountants and lawyers and analysts and financial backers? What do we really know about the mutations of exchange that would allow us to claim that they have separated us forever, by a radical break, from the timeless entanglements whose testimony is preserved as precious by economic anthropology and archaeology? Here again, here as always, the investigator must learn to avoid exoticism, and especially Occidentalism, that of a finally rational Economy that would have covered over all the rich imbroglios of the ancient world with the cruel materialism of merchandise.

THEN A PARTICULAR MODE OF ALTERATION OF BEING APPEARS ⊙ Must the ontology of modes of existence really lower itself to include the appetite for organic salami or the packaging of soap bars? Well, yes, of course, because these are the most common experiences. Fortunately, this mode is not hard to grasp: we simply need to reverse the usual relationships and take everything economic anthropology views as *"exceptions* to the modern market economy"—exceptions that would now be found only in ancient cultures, from that standpoint—as the general case *that describes us very precisely—except for scale.* What alteration is in question here? The most basic one, the one that defines for a given existent not only *the other existents* through which it has to *pass in order to subsist* but those that it *can no longer do without* and that have to *remain linked* to one another in a chain in order to subsist.

It is this linkage that will allow us to recognize the setting into motion, the mobilization of things and people alike, on ever-broader scales, and that has struck all observers when they study the history of what is called "the life of exchange," going back sometimes as far as the Neolithic era: the immense rustle of souks, fairs, markets, ports; what Fernand Braudel described so well as a "world-economy" and from which the disciplines of economization are going to draw only a few features. This mode of existence can be recognized by the fact that "things move," "things shift," "things heat up," "things connect," "things line up." It is as though all existents, altered, thirsty, impassioned, alienated, energized, were setting out along a very strange type of network, since it strings together completely heterogeneous beings, apples and oranges, ducklings and swans, interposing among them subjects that receive *properties*, each time new and unanticipated, from these interpositions.

The fine words "commerce" and even "consumption" would define fairly well the follow-up to the "careers of objects" made familiar by the anthropology of ancient cultures. It suffices to run through the grid outlined above, which registers the list of beings through which a quasi *subject* would need to pass in order to subsist, along with that of the quasi *objects* through which given humans must pass, seek to pass, can pass. Let us take a quasi object: it is (it has) all those that are attached to it. Now take a quasi subject: it is (it has) all those that have attached it. The point of departure doesn't count; the only thing that matters is the mingling of attachments that define either the objects—through the likes and dislikes, the obligations and detachments, the passions and coldness they have aroused—or the subjects—through the likes and dislikes, the obligations and detachments, the passions and coldness they have aroused. This chart marks no difference between (superfluous) desires and (necessary) needs, no more than between so-called material goods and those that would be "nonmaterial." When Tarde wants good examples of "forces of production," he turns to books, luxury items, fashion, and conversation! The only thing that counts here is the number of *aliens* that have to be multiplied along any given path, and whose presence is felt only in tests of innovation or privation.

We now understand the importance of defining attachments as "passionate interests": interest, as we have seen, is everything that lies *between*, everything through which an entity must pass to go somewhere; as for passion, it defines the degree of *intensity* of the attachment. Add, for each slot in the grid, a *plus* sign for goods and a *minus* sign for bads, and you'll begin to unfold the immense matrix of obligations, the background, the layer that defines us more exactly than all the other modes—and that still never really has the right to emerge as such in modernism, except on the margins and in the other cultures, and, of course, in the most ordinary daily practices. There is nothing imaginary about this matrix, since we can map out its labyrinth by following the subtle distribution of the various trades in a souk, the intersecting participations of companies in Silicon Valley, the lines and columns of an interindustrial chart in a Financial Ministry office at Bercy, as well as in the intermingling of parts of all American automobiles in a junkyard in Queens. This matrix is our world.

There are passes; there are alterations; there is novelty in alteration. There is thus a separate mode of existence that can be defined canonically by three features: a genre of being—mobilization, interest, valuation (there are no conventionally accepted terms)—but also a type of transcendence and a particular regime of veridiction.

⊙ WITH AN ORIGINAL
PASS: INTEREST AND
VALORIZATION ⊙

So what is the leap, the HIATUS, the gap that this mode of attachments [ATT] covers in a dizzying bound? It is the incommensurability between goods and the means for appropriating them, the heterogeneity of attachments and the sudden irruption of value. We experience this constantly, and it always surprises us. We didn't know that we were so interested by these goods; we didn't know we were so helpless in the face of these desires; we didn't know that this thing was worth so much, or so little. Of course we experience this constantly, but our experience is constantly suspended, blocked, biased, broken by the inversion of cold and hot, of appearance—equivalence—and reality—nonequivalence.

This is why marketing experts know so much more than those who want to economize about the heat of displays and about the difficulty of luring the always-too-nonchalant consumer. But "merchandisers" know even more than marketing professors. And those in charge of supermarket sales shelves know even more than the "merchandisers." This admirable inverted pecking order means that the secrets of passionate interests are in fact no secret at all, provided that we keep going further down toward the practical experience of those we look down on (as we saw with political practice [POL]). Those who know they are really attached, in terms of goods and bads alike, are the consumers, the producers, the craftsmen, the tradesmen, the manufacturers, the fans, the unemployed, the tasters, the innovators, all those who press, mill, steep, toss, and grind the vast matrix of interests and passions. Double Click can certainly continue to model the bonds of supply and demand, as if this were one of Newton's laws, but one thing is sure: if he were to become a tradesman or an industrialist, insensitive as he is to minuscule variations in attachment and taste, he would go bankrupt right off the bat!

These beings that traverse us so tirelessly have to be called *valence, energization, investment.* Let's say that value is the always risky lengthening, for quasi subjects, of the number of columns they have to cross, or, for quasi objects, the number of lines. To *valorize* is to register the appearance of differences, either by interposing a new line—we discover on the occasion of a new "product" that we have become capable of new interests or new passions, new attachments—or by interposing a new column— we discover on the occasion of a new "demand" that other combinations of materials and service can be brought into existence. If there is one thing that cannot be reduced to a transfer of equivalences, it is in fact the irruption of value—a new line, a new column, a new *alien*, a new alteration—or *devalorization*, the sudden disappearance of a line or column.

The ethnologist has a pretty good sense that she mustn't rush headlong into a definition of value. But she tells herself that perhaps, by restoring the beings of passionate interest to their rightful place, she could get along without a definition. For in the end, what does it mean to attach oneself to "labor value" if not to traverse the long series of entities necessary to the genesis of goods? What is "market value" if not another course to follow, this time in the series of beings that can be substituted for those goods? Now, neither these necessities—those of production— nor these substitutions—can get along without the irruption of new, totally unpredictable beings that always *precede* equivalences. The icy calculations of The Economy melt everywhere before the wildfire of passionate attachments. It is impossible to do the anthropology of the Moderns without inverting the inversion, without putting things right, attachments below and calculating device above.

Someone will object that there are no felicity ⊙ AND SPECIFIC FELICITY
and infelicity conditions proper to this regime, that CONDITIONS.
it is situated completely apart from truth and falsity,
that attachments are not rational, that detachments are even less so. As

if we didn't know with stunning precision the difference between *enjoyment* and *indifference*! As if the immense murmur of *discernments* between goods and bads weren't arising from the entire planet! How astonishing, these Moderns, who talk only of rationality, objective knowledge, and truth, as if in the search for truth—yes truth—the difference between wealth and poverty did not count! How much hypocrisy there is in the love-hate feelings for the goods of this world! How well we understand the Gold Coast natives who retorted to those who had come to convert them: "Is Gold the name of your God?"

"What can I do to have the means?" Can you count the hours we all spend answering questions like this, the energy we expend to explore them? I know what I want but I don't have the means; I have the means, but I can't find what I want; I have the means, but I don't know what I want; I have the means and I know what I want, but others want it too. And we would leave all this knowledge undifferentiated, on the pretext that we don't find in it the proof of a "rational calculation"? To enjoy, to have, to possess, to profit—all these verbs displace us from one end of the Earth to the other and make all the existents of the planet mobile, one after another, and this would not be a mode of existence worthy of respect? The process of weighing our most vital interests, of evaluating everything that ensures our subsistence, would not be susceptible to truth and falsity? This evaluation may resemble no other, but it would be too implausible for it not to distinguish between falsity and truth, between evil and good, in its own way.

We have said this about every mode, it is true, but what touchstone can compete in subtlety with the distinctions that each of us can establish between goods and bads, the successes and failures of a new chain of attachments? What is the discernment of a confessor or a psychoanalyst worth next to the "shopping know-how" of a featherbrained twenty-year-old who is capable of comparing two underwear fabrics and distinguishing the difference that will make some notions store in China rich or bankrupt? How are we to qualify the know-how of an entrepreneur capable of giving unanticipated twists to interests? Shouldn't we admire the genius of trucking engineer Malcolm McLean, who was able to modify the habits of shippers by the unanticipated introduction of a standardized container that goes directly—quite indirectly—from producer to consumer without any time lost for reloading? A new example of continuity obtained through the intermediary of a new discontinuity! And a new example of a change in size thanks to the box that "made the world smaller and the world economy bigger," to borrow the title, scarcely exaggerated, of a book that tells this story. And consider Steve Jobs's response, admirable for the form of its wisdom, as inventive as it was trenchant, to the question "Do you do marketing studies for your products?" "It's not the consumers' job to know what they want." With each articulation of courses of action, the surprise of value emerges.

We are beginning to see that what prevents us from capturing the felicity and infelicity conditions for this mode is the surreptitious slippage toward the idea of a calculation according to different scales, different assays. As if one could calculate the evaluations *coldly*, as it were, and above all *continuously*. Here we see once again the investigator's initial surprise: nothing in this vast kneading is defined by coldness; on the contrary, *everything here is hot*. Nothing here is defined by calculation, or at least not by the part of calculation that would presuppose equivalences and transports without transformation [ATT · DC]. "To value" would be an intransitive verb like "to rain," without object or subject, a form of unpredictable event. Yes, *to value* always precedes *to be equivalent*, as *change* precedes exchange, as *translation* precedes mere displacement. Even later, when habit has covered over these countless discontinuities by the smooth transportation of equivalences, it will still not really be possible to calculate. Distribute, yes; calculate, no— or in any case, *not yet*.

The ethnologist is searching for words, but we sense well enough what she is trying to do: she does not want the irruption of value to simplify THIS KNEADING OF EXISTENTS ⊙
the massive phenomenon of the Moderns. The idea of a finally rational Economy must not be allowed to conceal the formidable kneading of the world along with the scale of the unanticipated transformations that have ended up mobilizing the entire Earth, from the polar ice cap to the most intimate properties of matter. What a poor anthropologist she would be if she didn't seek out those humble words capable of grasping the scale of such a kneading of things and people! What passions have to be introduced into the equation to capture these creations of nonequivalent,

surprising, incommensurable values, these destructions of values on ever more gigantic scales! Don't the Moderns invent new needs, new desires, new objects, new markets every day, destroying old needs, old desires, old objects, old markets as they go? Don't they keep on *complicating* the vast matrix through which we become more implicated, more surprised, more interested, poorer or richer, every day? And we should proceed as though there were one source of value—land, labor, money—that could be *displaced* without any transformation and that, thanks to their similarities, would "explain" their wild inventions? While all the agents are trying to "pass" by way of long concatenations of goods and people, with each link in the chain coming as a surprise, nothing would be happening but the circulation of indisputable necessities?

⊙ LEADS TO THE ENIGMA OF THE CROSSING WITH ORGANIZATION [ATT · ORG], ⊙

But this mobility, this avidity, these passions, these interests, aren't they precisely what concerns The Economy? Not exactly; this is the whole problem. Here is where we have to slow down. Take special care not to rush. Look closely: we're almost there. To understand scaling without changing levels, we have to understand how the beings of attachment and those of organization are linked without plunging them into anything else—especially not into some MATTER. It is by exploring this [ATT · ORG] crossing that we may be able to follow the distributions without appealing to any metadispatcher.

Let's take the example offered by William Cronon in one of the best books ever written about changes in scale. He shows us how networks of attachments and scripts, once they are strung end to end, give an Indian village on the shore of Lake Michigan the dimension of a "Metropolis of Nature," his term for the world-economy we know as Chicago. And what he succeeds in braiding together better than anyone are precisely the successive innovations that bear on new valuations as well as on new accounting equipment. The farmer who took his bags of wheat by cart and by boat and followed them with a tender gaze all the way to the flour mill soon learns a quite different definition of value when he gets off the train and sees, first with panic, his slashed-open bags mixing their golden contents with the contents of all the other bags in the buckets of the grain elevator, which has just been invented. At first, he views with great suspicion the little paper receipts he gets in lieu of the miller's banknotes.

How could he guess that with slips of paper a formidable futures market will make Chicago the center of the world for all scripts having to do with grain? Chicago expands, and the grain market becomes "abstract" owing to the extreme concreteness of railroad lines, grain elevators, and paper bonds. A vast socio-technological network modifies the relative size of a multitude of agents and the kind of beings that oblige them, interest them, and make them act. Supported by attachments, the scripts speed up and impose their cadence. Floods of quantifications surge forth. What is going to happen?

Have we entered The Economy? Not yet. The paper bonds remain scripts, and the futures market keeps on starting over, punctuating due dates and quittances, distributing property rights—and speculating wildly. For us to enter The Economy—and ⊕ WHICH WILL ALLOW US TO DISAMALGAMATE THE MATTER OF THE SECOND NATURE. thus to *exit* from the arrangements of economization—everything depends on the notions of calculation, arrangement, and discipline that are going to be put to work, and then of their comparison with the linkages of forms [REF] (as defined in Chapters 3 and 4). To put it somewhat too provocatively, the inquiry has to ask whether economics is a *science* on the same basis as those called "natural"—the sciences "of Nature"— or whether it is the *discipline* of a very particular form of calculation that would *not have the goal* of establishing chains of reference [REF].

The attentive reader will surely have noticed the parallel we are seeking to draw between the beings of reproduction [REP] and those of attachment [ATT]. They are already alike in that both modes are defined by the series of beings through which it is necessary to "pass" for subsistence. They differ, of course, since the beings of reproduction precede both quasi objects and quasi subjects, whereas the beings of passionate interest trace the links, always surprising ones, between objects and subjects, goods and people, whose proliferation the previous modes have never stopped fostering [REP · ATT]. But where the resemblance is strongest, it is because the two modes serve as *raw material* for the invention of Nature—for the beings of reproduction—and for the invention of what we have called the second Nature—for the beings of attachment.

In Part One of this book, we saw how "Nature" had amalgamated two modes, that of reproduction and that of reference [REP · REF], in order to stir up The Economy without criticizing it. The matter of the *res extensa* appeared to us at that point as a fusion, an interpolation, between the requirements of knowledge and the requirements of reproduction. We now face the question of whether we are going to be able to use the crossing of the two modes [ATT · ORG] to change the direction of The Economy without criticizing it. Or, more precisely, whether we are going to be able to identify the subtle detour through which the economic disciplines are going to end up engendering the idealism of (the second) matter. For the second time, matter would be born through the amalgamation of at least two modes, that of attachment and that of organization [ATT · ORG] on the basis of an idea diverted from knowledge [REF · ORG]. This time, the amalgamation would not generate the RES EXTENSA known to the so-called natural sciences, but a new matter simply "known" by a new Science, that of economics. Then the harsh necessities of a second materialism would follow the harsh necessities of the first. For the second time in this work, it is the amalgamation of "matter" that opens up or closes off access to any interpretation of the Moderns and any definition of reason. Historically, EPISTEMOLOGY and Political Economy arose together, and have never stopped relying on one another; together, they must learn to withdraw from the stage.

Fortunately, just as we have learned to distinguish the leap or gap of reproduction from that of reference [REP · REF], nothing prevents us any longer from bringing out, without getting it mixed up with the others, the leap or gap of organization, whose truth and falsity we all define for ourselves. And so we can also see what must still be added to the situation to *smooth over* these two types of discontinuities in order to come up with an entirely different definition of what *gets it right* in the calculations. Unquestionably, several felicity conditions combine to decide on the rightness of a calculated evaluation. A lot of lines have to be written to make sure that no one is writing anything below the *bottom line*. We shall have to attach ourselves to a third and final mode.

The operation requires skilled and agile fingers, but there is no other way to *diseconomize*, while identifying as precisely as possible what the economic disciplines are going to add to the beings they embrace. What would have been the use of restoring the immense imbroglio of passionate interests, giving them back their heat, their violence, their entanglements; what would have been the use of capturing the strange rhythm of scripts and their capacity to modify the relative scaling of everything they distribute, if it were only to settle for a simple critique of Economics? We can do better, by sketching out a place worthy of the indispensable disciplines of economization. Even more than the first empiricism, the second needs to learn to restore economic experience.

·Chapter 16·

INTENSIFYING
THE EXPERIENCE
OF SCRUPLES

Detecting the [ATT · ORG] crossing ⊙ ought to lead to praise for accounting devices.

However, economics claims to calculate values via value-free facts, ⊙ which transforms the experience of being quits ⊙ into a decree of Providence capable of calculating the optimum ⊙ and of emptying the scene where goods and bads are distributed.

While the question of morality has already been raised for each mode, ⊙ there is nevertheless a new source of morality in the uncertainty over ends and means.

A responsible being is one who responds to an appeal ⊙ that cannot be universal without experience of the universe.

We can thus draw up the specifications for moral beings [MOR] ⊙ and define their particular mode of veridiction: the taking up of scruples ⊙ and their particular alteration: the quest for the optimal.

The Economy is transformed into a metaphysics ⊙ when it amalgamates two types of calculations in the [REF · MOR] crossing; ⊙ this makes it mistake a discipline for a science ⊙ that would describe only economic matter.

So The Economy puts an end to all moral experience.

The fourth group, which links quasi objects and quasi subjects, ⊙ is the one that the interminable war between the two hands, visible and invisible, misunderstands.

Can the Moderns become agnostic in matters of The Economy ⊙ and provide a new foundation for the discipline of economics?

Our investigator is beginning to breathe more easily, because she sees that she is approaching the end of her troubles. At the same time, though, she is brought up short by the difficulty of what she claims to be doing: describing acts of calculation so that calculation cannot replace *description*—which is nothing other than the unfolding of SCRIPTS. This is the only way she has found to divert the SECOND NATURE so that, in place of The Economy, the delicate networks of the disciplines responsible for economization can appear with nothing added, nothing removed. She is beginning to believe in her program, to the point of letting herself be seduced by an aphorism: "Up to now the economizers have only *performed* the world; now the task is to *describe* it!"

If she was led to inquire into the beings of organization, it was because she couldn't see where the second level was coming from; she couldn't locate the METADISPATCHER, the "whole greater than its parts" that her informants nonetheless designated as the general framework within which all actors resided. Now, by refraining from formulating a supplementary hypothesis about a transcendent level, by following only the movement of scripts [ORG], she is beginning to see where the effects of scaling and framing are coming from.

If she went to the trouble, next, of bringing the beings of interest [ATT] to the foreground, it is because she did not see, either, how one could reconcile the experience of passions and attachments with the supposed coldness of economic calculations. At this point, she has begun to understand how the addition of calculation devices will allow the "life of exchanges" to be identified, though calculation itself will never be substituted for the surprising evaluations that connect long chains of (quasi) objects and (quasi) subjects, goods, bads, and people. She does feel that she has accounted for COURSES OF ACTION: these are the movements through which *passionate interests and scripts are connected*, thanks to two types of linked discontinuities.

And yet she feels that she is not yet faithful enough to the experience of these flows. If she were right, in fact, the apprenticeship of economization would not be a great mystery; it would never have given rise to what has to be called the metaphysics of The Economy. Courses of interests, like those of scripts [ATT · ORG], would simply have been equipped with devices, abacuses, benchmarks, instruments, arrangements, models, in short, VALUE METERS, to help the actors get their bearings in an ever-increasing number of linkages and thus obtain the ALLOCATION KEYS acceptable to the various parties. It would suffice, at bottom, to sing the praises of *accounting* and of its Great Book. This Book is still present in people's minds, no one could forget it: it is the Accounts Book, the Balance Sheet, the book we are all obliged, just as we are, rich or poor, bankers or beggars, heads of state or peons, to leaf through in order to find out what we *are worth*, what we *are capable of*, and what we *owe*. This instrumentation would never have been turned into a great history of Reason. The first Nature would never have been followed by the second.

⊙ OUGHT TO LEAD TO PRAISE FOR ACCOUNTING DEVICES.

Let us recall that, in an admirable metaphor, Galileo asserted offhandedly that "the book of Nature was written in mathematical signs." A very characteristic metaphor for the Moderns, since it calmly acknowledges the confusion whose importance it denies at the same time: if this is a book, where are the INSCRIPTIONS, the stylus, the author, and the printing press? If indifferent, anonymous Nature writes without writing on any medium of support, why talk about a Book? Thanks to that tenuous metaphor, borrowed from the Book of God, by claiming to see in things themselves only the displacement without distortion of power relations and causal chains—an eminently worthy intention, by the way—we could never tell if we were dealing with the dizzying leap made by existents in order to persevere in being or rather with that other, equally dizzying leap of technologies of inscription that make it possible to reach remote states of affairs [REP · REF]. In the same hybrid notion of MATTER, the secret of existents that allows them to persist in being and the secret of scientists that allows the world to inscribe and publish itself were interpolated. The movement of falling bodies became a mere avatar of the physical law that gives us a hold on them. If the distillation of that amalgamation has required so much effort on our part, it is because the Books of physics, chemistry, or biology are little known by the wider public, and because the Galilean metaphor has worn so thin that we still mistake knowledge for persistence itself [MET · REF].

It ought to be easier to reveal the idealism of the second Matter than that of the first, since the books, the ones that make it possible to *economize*, are still perfectly visible: the red or green ink is still fresh; we bank on spreadsheets whose columns and lines are still blank. We may have had to take endless precautions to learn to disentangle the trajectory of Mont Aiguille from the trigonometry of the geometricians, but we can no longer confuse an account book with the behavior of *what is counted*, at least not for more than a few seconds. We don't need to be experts to counter the metaphor as it weighs on that other "Matter," the one supposedly comprised by the "laws of economics"; it's enough to be a lowly clerk, a scribbler, a low-level accountant, a "scrivener." No one imagines that the management agent in charge of calculating expenses owed by each apartment owner has to be left alone in his office to calculate, like a true scientist; in any case, as everyone knows, if you leave him alone, it's at your own risk.

Bringing the arrangements for calculation—or QUALCULATION—into the foreground ought to be all the easier, given that everyone can sense the installation and extension of new value meters whose standardization has produced the same effects of quali-quantification as the preceding ones, whose use had been invisible. Researchers who measure themselves against their colleagues by comparing citation scores on Google Scholar, like university presidents obsessed by their place in the rankings, are not intoxicated by figures to the point of confusing the spread of these devices with a measure of excellence, in the referential [REF] sense of measure. They are quite aware that what is at issue are standards that make it possible, here again, here as always, to prepare distributions and redistributions among shareholders. The bankers who have to recover from the setbacks caused by the insertion of Black-Scholes-Merton equations in the automated system for calculating price options in the world of finance are well aware that these equations are more like the management agent's calculation of percentages than like the laws of thermodynamics—and that the computerized algorithms shouldn't have been left to do their calculating on their own! Diseconomizing thus ought to be an easier operation than "disepistemologizing" our heads.

Yet this is not at all the way The Economy with a capital E presents itself. It insists on something else, something more decisive and more radical; it not only offers the fragile aid of mixed-up formatting, but it claims to register, for all those who have anything to do with it, indisputable facts in relation to which the question of values seems *not* to arise. Thanks to this Economy, questions of valorization [ATT] and the due date of scripts [ORG] are going to become *facts*, and even—an astonishing phenomenon that will distinguish them still more sharply from chains of reference [REF]—"facts that no longer need to be discussed." We can say that there is nothing here that ought to astonish our investigator: isn't any science defined by an empirical grasp that is "indifferent to values"? But it is precisely this insistence that prods the ethnologist of the Moderns toward incredulity. How does it happen that we find here again, in an even more assertive form, the same canonical distinction between facts and values to which we didn't succumb when it was a question of identifying the grasp proper to objective knowledge? If chains of reference do indeed permit the slow fabrication of FACTS, they never begin by making them indisputable. Quite to the contrary, they authorize researchers to bring in things themselves as witnesses in the *arguments* they are having about those very things. Now here, by a singular reversal, *the question of value tips into the question of facts*, simple, raw, stubborn, material, obtuse facts. It is as though the Great Book had still more power than that of (the first) Nature. It seems to have even more *credit* than the laws of physics. And this is quite normal, since it is the book of Credit, in fact! What trust, yes, what credit, are we to grant it? What value must be placed on a "value-free" science of values? If the

HOWEVER, ECONOMICS CLAIMS TO CALCULATE VALUES VIA VALUE-FREE FACTS, ⊙

beings of reference were "indifferent to values," it was not because they had freed themselves from values, or that they were *indifferent* to them, but only because they were *different*: neither immoral nor amoral, their mode of veridiction was so distinctive—passage through forms to maintain constants—that it simply did not intersect in any way that of known things—things that followed their own paths quite differently.

Confronted with the injunctions of The Economy, the analyst finds herself facing a somehow polemical, exaggerated, almost threatening version of the distinction between facts and values. As if the double "metretics" presented in the very setup of value meters—in the double sense of taking measurements and taking measures—had disappeared to the benefit of a measure of the referential type [REF]. The transformation is total, since it is no longer a question of facts that are *no longer* to be discussed, but of facts that would be off limits to discussion from birth, as it were, because they *must not* be discussed. A strange deontology according to which "one shouldn't" call facts into question.

This distorted use of the schema of equipped and rectified knowledge is all the stranger given that the gradual lengthening of chains of reference, as we have seen, has the goal and often the effect of *reducing the distance*, little by little, between the knowing subject and the known object. By what strange upheaval have we begun to argue on the basis of knowledge in order to *increase the distance* between those who are the first to be concerned by the scripts in which their roles, their obligations, their requirements are spelled out in detail? Either economics has discovered a source of absolute certainty, *superior* to that of all chains of reference, or else we are dealing with a supplementary enigma, a new category mistake.

⊙ WHICH TRANSFORMS THE EXPERIENCE OF BEING QUITS ⊙ It is as though, in the invocation of the second nature, there were a supplement of certainty that is too equivocal, too insistent, to be plausible. How can this supplement be defined? By returning once again to the results of economic anthropology and reinstituting, by contrast, the often terrifying experience of *being quits*. Ever since Mauss's essay *The Gift*, all Europeans who plunge into this literature can only recoil in panic before the imbroglios that they find described there. "But then," they sigh, "those poor wretches will never get out of it, they'll always be bound, attached, indebted, hooked, enmeshed, entangled." Those who express such astonishment benefit unthinkingly from the immense advantage that the austere disciplines of economics have procured for them: prolonged hardening exercises have accustomed them to being "quits" with respect to those with whom they enter into transactions.

The contrast extracted by European history seems to break in fact with all the anthropology of the "other cultures": we seem to think that we've found the way to get ourselves out of such imbroglios by adding to them their exact opposite: "And now we're quits; I owe you nothing; we have exchanged equivalents; goodbye!" In this view, we have maximized procedures that keep us from always *owing*, always *depending*, always *giving back*. Formidable inventions for the exchange of equivalents among those close to us who become strangers with respect to whom we have learned to be quit of any other tie when the due date comes and quittances have been handed over. These are very recent inventions, but at least since Locke, thanks to numerous Robinson Crusoe–style adventures, they find themselves placed at the very beginning of the story, at the beginning of History, as if we had begun with the Market and will end up with it. As if we had invented a type of *private* property from all those imbroglios in which the "other cultures" seem, on the contrary, to get stuck. There is really something excessive, something forced, in this obsession with endings, with finance, with "closing the books." Especially coming from those who, believing themselves quit of all the old attachments, suddenly see on the horizon the unexpected creditor asking them to reimburse very quickly the bonds drawn over two centuries—bonds drawn against the treasure of the Earth.

We can see our investigator's dilemma: if she is content to bring out the little supplement of the disciplines of economization, she will have concealed the real poisons of The Economy; she will have manifested a sort of quietism. She will have participated in the avoidance of Modernism: treating those close to us as strangers to whom we owe nothing because we have discovered a source of absolute certainty, indifferent to values. Now it is on this very point that we are going to be able to take full advantage of our lengthy effort to do without any metadispatcher whatsoever in our descriptions.

⊙ INTO A DECREE OF PROVIDENCE CAPABLE OF CALCULATING THE OPTIMUM ⊙

If the discipline of economics passes for a Science, it is because it adds something that no natural science can give: the certainty of an OPTIMUM *finally calculated by a higher agency that aggregates and unifies all the scripts.* This is the danger of the second level whose implausibility we have noted: by changing levels, shifting to a level on which all scripts are aggregated, the Moderns claim to be gaining in rationality. Yet it is here that they head straight into unreason.

While the first Nature could be unified and ordered, its testimony, no matter how edifying, could never be convincing, since in one and the same breath it was declared to be *indifferent* to humans and *amoral.* This made it hard, obviously, to have Nature dictating moral laws. The second Nature has an entirely different character: when it gets things right, the facts it uncovers are not only simply indisputable, they are marked with the seal of Providence. A paradox: all values are defined by value-free facts! It is this astounding crossing that the anthropologist has to confront if she wants to understand the sole originality of the Moderns: they believe in Providence in the singular mode of material economic facts that bear a formal *resemblance* to those of reference—but it is only a resemblance, this is the key. So we have to take one more step to understand what Karl Polanyi defined as a "secular religion": the religion of the calculated optimum.

⊙ AND OF EMPTYING THE SCENE WHERE GOODS AND BADS ARE DISTRIBUTED.

What had most astonished our investigator, at the beginning of Chapter 14, was that instead of a noisy, agitated public square where all the parties involved engaged in lively debate about what concerned them most directly, she had found only an empty place. To deny attachments [ATT] is one thing; to deny that there have always been organizations whose scripts remain visible without regard to calculations [ORG] is quite another; but to claim that the optimum can escape *scruples* is most astonishing of all. The idea that one can deny suffering humanity this drop of water to quench its thirst, the *collective hesitation* about what is better and what is worse: in the final analysis, this is the most astonishing of all the Moderns' traits that we have studied up to now. The expression is often used metaphorically, but it is as though there had never been anything, in fact, but a "moral economy," as if we had never really left behind the divine Dispensation celebrated by the

Byzantines under the name of *oekonomia*—but without ever daring to make it, as they had, a veritable religion.

When a rationalist's fist pounds on the table in the amphitheater to declare to the relativists that "the laws of physics are there, Gentlemen, whether you like it or not!" it's touching, because we can see what he is trying to say despite the rationalists' calamitous epistemology: he would like the world to which we are at last gaining access through the paths of scientific objectivity not to be made by human hands, not to depend on "my own will," and yet he would like us to be able to get *closer and closer* to it in order to grasp it. How could one disagree? But when a manager pounds his fist at the end of a PowerPoint presentation and the last bullet point dictates the last bottom line as if to say that "there is nothing left to discuss," that there is no other possible solution, that those most directly concerned must *distance themselves* from direct inspection of all the scripts, the situation is infinitely more enigmatic. Whereas we were expecting to find the agora full of animated, concerned agents, the place is empty. *How could the stage of evaluations and distributions have been emptied?* This is what we have to understand.

What confirms the investigator's suspicions is the extent to which the appeal to MATERIALISM differs in the two cases. The laws of physics, however encumbered with epistemology they may be, have always been a fount of unanticipated possibilities; why is it, then, that the "laws of the second Nature" are so often presented as rules of renunciation and impotence? When an engineer, a scientist, an artist, a craftsman plunges into his materials, everything suddenly becomes possible to him because these materials are composite, they give ideas, open endless possibilities, reveal unsuspected capabilities [TEC · REF]. Economic matter, on the contrary, has the distinctive feature that, when one appeals to it, one finds oneself bound by transfers of indisputable necessities. One can no longer do anything. One's hands are tied: "There is no other possible policy." Here is proof that the materialism that has been distilled from economic matter must produce poisons that the other matter did not hide.

If we really had to set out to "physicalize" economics, then it would look more like physics, real physics, and it would take on the cobbled-together, ingenious, equipped, multiform aspect thanks to which one would succeed in setting up delicate experiments most of which would be fortunate enough to be able to fail. Physics has never been accused of using indisputable facts to harden hearts. Dostoevsky showed his astonishment at this when, in *Crime and Punishment*, he had someone respond to a protagonist who wanted more compassion in social relations: " . . . compassion? But Mr. Lebeziatnikov, who keeps up with modern ideas, explained the other day that compassion is forbidden nowadays by science itself, and that that's what is done now in England, where there is political economy." It's hardly surprising that we can no longer agree about economics and that no calculations come out right anymore! To rediscover fairness, then, we first have to confront the "moral question" head on. Since we have to talk about a science of values, we may as well talk about it openly.

WHILE THE QUESTION OF MORALITY HAS ALREADY BEEN RAISED FOR EACH MODE, ⊙ "So, here you are, starting to moralize all of a sudden! Could this be because you're coming to the end of your project and you want a supplement of soul, a treat, a sweet note, like dessert after a copious meal?" Before ironizing about our inquiry, the reader will perhaps acknowledge that I have been "moralizing" from the outset, in the sense that I have brought out the felicity and infelicity conditions for *each mode*. Every instauration implies a "value judgment," the most discriminating judgment possible. Consequently, the "moral question" is not being brought into this inquiry *after* all the questions "of fact" have been dealt with. It has been addressed from the start. There is not a single mode that is not capable of distinguishing truth from falsity, good from evil *in its own way*. Even though I have of course deliberately exaggerated in treating each mode as if it possessed the finest touchstone of all, and the most refined sense for telling truth and falsity apart.

Beginning with reproduction [REP], which maintains the greatest difference of all between success in reproducing itself or definitive disappearance! The beings of reproduction are neither moral nor immoral, nor even amoral, since without the other beings that are their successors, descendants, or outcomes, they would vanish. And yet this is a rather nice way to moralize, and particularly decisive. Moreover, the mode of existence of chains of reference—that of "facts," precisely—is perfectly capable of deciding in its own way between good and bad [REF]: how many values there are in the discernment, always to be begun anew, between a verified utterance and a falsified utterance! In law [LAW], detecting the difference between judging well and judging badly is also, it seems, a way of "moralizing," one that keeps quite a few judges awake at night. As for those who feel themselves so powerfully designated by the requirement of salvation that they sense what separates presence from absence, resurrection from death, or, as in the ancient image, Heaven from Hell, there is no doubt but that they also know a thing or two about good and evil [REL]. Even engineers, neglected by traditional morality, see a huge difference between good and bad setups, between what is effective and what is not, between a good gadget and a bad one. Do they not also "moralize" in their way? And works of fiction—don't they create obligations for you, "lay trips" on you [FIC]? Don't they grab you? Aren't we all capable of discriminating, in the most insignificant film or novel, between what is well constructed and what is not? How could we deny that, in the renewal and abandonment of the Political Circle, there is one of the most important sources of what is called morality [POL]? It is hard to overlook the difference between political courage and political cowardice. And isn't there, even in habit [HAB], a big difference between forgetful automatisms and attention to seeking the good ever more skillfully? Who will ever be able to describe meticulously enough what makes the difference between black and white magic, that tiny difference, so hard to pinpoint, between the therapist who cures and the one who bewitches and condemns [MET]?

In short, all the modes participate in what could be called the institution of morality—if there had ever been such a thing. We have to go down long lines of Bifurcators before reaching Kant, who expects humans deprived of world to "add" values to beings "deprived of ought-to-be." Before him, and in the rest of the world, there hadn't been a single existent that had failed to exclaim: "It must," "It mustn't," measuring the difference between being and nonbeing by this hesitation. Everything in the world *evaluates*, from von Uexküll's tick to Pope Benedict XVI— and even Magritte's pipe. Instead of opposing "is" to "ought to be," count instead how many beings an existent needs to pass through and how many alterations it must learn to adapt to in order to continue to exist. On this point Nietzsche is right, the word "value" has no antonym—and especially not the word "fact."

Perhaps there is a way to avoid the foolishness of accepting the truism of moral philosophy, which believes itself capable of *opposing* "is" and "ought-to-be"! As this entire inquiry attests, deprived of other beings, any existent whatsoever would cease at once to exist. Its very existence, its substance, is defined by the supreme duty to explore through what other beings it must pass to subsist, to earn its subsistence. This is what I have called its ARTICULATION. All those who oppose being to having-to-be are thus addressing strangled, decapitated, eviscerated beings deprived of any means of meeting their own needs, souls in pain surviving in the bleak limbos of moral philosophy. And as we have seen, this articulation is expressed even in stones, cats, mats, and pipes, all those poor "objects" that seem to have no role but to serve as foils for the lofty morality of human subjects. But the strangest moment of all is when moral philosophers claim to be opposing "is" and "ought-to-be" in economic matter! Whereas the whole problem lies precisely in making beings proliferate so that they may have some slight chance of proceeding in such a way that the calculations of evaluation and distribution end up coming out right.

⊕ THERE IS NEVERTHELESS A
NEW SOURCE OF MORALITY
IN THE UNCERTAINTY
OVER ENDS AND MEANS.

And yet the fact that ordinary language points to situations as composite as this with the term "morality" does not oblige us to conclude that we have no way to detect *morality-bearers*. We have recognized psychophors [MET]; why not *ethophors*? In addition to all the differences between good and bad proper to each mode, there could exist a supplementary sense of good and bad that would explain the nuance to which we all seem to hold under the rubric of "moral experience." Now, if we have properly understood the treasure of alterities that being-as-other keeps in its bosom, an enigma is posed to every existent: "If I exist only *through the other*, which of us then is the *end* and which the *means*? I, who have to *pass by way of* it, am I its means or is it mine? Am I the end or is it my end?" This is a problem of the FOURTH GROUP that we shall no longer be able to escape when we start to follow the course of action that attaches long chains of means and ends; and the longer the chain, the more tormenting the question. Especially if we restore to it increasingly, as the modern parenthesis shuts down, the multiplicity of nonhumans that the ecological crises thrust together in all sectors of The Economy.

That tree, this fish, those woods, this place, that insect, this gene, that rare earth—are they my ends or must I again become an end for them? A gradual return to the ancient cosmologies and their anxieties, as we suddenly notice that they were not all that ill founded.

In the philosophy of being-as-being, this question did not come up: conserving one's own identity meant attaching oneself to a substance that lay beneath all attributes, aiming at permanence while forgetting the transitory. The end could be only the substance itself, that which suffices unto itself, that which is causa sui. All morality was thus ordered on the basis of a tautology, whether that was being itself, or, in the modern period, "self-established" moral law, cut off from any world. One could argue endlessly about the solidity of this base, but one couldn't imagine any other project but that of establishing morality on as indisputable a foundation as possible. One had to pass, as people said, from "mere immanence" to transcendence. And since no one has ever discovered a second level in this realm either, moral philosophy has become a vast complaints bureau for addressing grievances concerning the immorality of the world, the "loss of bearings," the necessity of an "indisputable principle," the obligation to have an "external point of view" in order to be able to "judge nevertheless," in order to "escape relativism" and "mere contingency." Here is a project apt to induce doubt about morality itself.

If the bearings we have used up to this point have been well understood, morality can only be a *property* of the world itself. To seek to found it on the human or on substance or on a tautological law appears senseless when every mode of existence manages excellently to express one of the differences between good and bad. And yet, in addition to the moralities scattered throughout the other modes, there is indeed another handhold. This handhold, as we now understand it, is the *reprise of scruples about the optimal distribution of ends and means.* If every existent remakes the world in its own way and according to its own viewpoint, its supreme value is of course that of existing on its own, as Whitehead says, but it can in no case shed the anxiety of having left in the shadows, like so many mere means, the multitude of those, *the others,* that have allowed it to exist and about which it is never very sure that they are not its *finality.* And it is obviously not a question of human beings alone. Only the Kantians leave to that poor subject the crushing burden of becoming moral *in place* of the whole world—and what is more, without world!

If there is a regime that must take up again all that holds the quasi objects and the quasi subjects together, it is indeed the mode of *beings that are bearers of scripts and morality* (noted [**MOR**]). Once again, the Subject/ Object opposition would make it impossible to detect a mode of existence that nevertheless becomes audible as soon as we begin to identify the tribulations of the quasi subjects and the quasi objects, as soon as the key question arises, the question that economics—this is where its own greatness lies—has been bold enough to raise: "How do you know that what you have here is an *optimal* combination?"

A RESPONSIBLE BEING
IS ONE WHO RESPONDS
TO AN APPEAL ⊙

By losing the world that they thought they had left to these doubly ill-conceived amalgams of matter, the Moderns deprived themselves, as it were, of "any sense of morality." They lost track of the thread of experience, or rather they broke it off violently, claiming that to gain access to a moral sense one had to "distance oneself from the details of particular cases" and abandon them to ethics or deontology, so as to undertake a search for "moral principles" that would allow "justification." The quest turned out to be all the less fruitful in that the Moderns didn't even take the trouble to benefit from the precious distinctions introduced by each of the other modes. Whence the importance, for our inquiry, of extracting this contrast anew, giving it once again its own ontology.

In any situation, provided that one *comes near enough* to follow it *as closely as possible*, one must be able to detect the traces left behind by the particular pass of morality-bearing beings. Just as a geologist can hear the clicks of radioactivity, but only if he is equipped with a Geiger counter, we can register the presence of morality in the world provided that we concentrate on that particular *emission*. And just as no one, once the instrument has been calibrated, would think of asking the geologist if radioactivity is "all in his head," "in his heart," or "in the rocks," no one will doubt any longer that the world *emits morality* toward anyone who possesses an instrument sensitive enough to register it.

Just as it was infinitely simpler to consider that if we are frightened, it is because beings exist that actually frighten us [**MET**] (instead of believing that we are "made anxious" by "nothing"), just as it was more objective—yes, objective—to recognize the presence of beings that address us to save us [REL], it is still infinitely more elegant, or in any case more empirical, to understand that when we proclaim proudly "I *answer for that! It's my responsibility!*" it is because beings have come to us and *called* us. Without that, it is hard to see what the expression *feeling responsible* could mean. Only moralists claim to feel something without some odor having wafted by to tickle their subtle moral sense. In the most ordinary experiences, responding always has to mean answering an external *appeal*. Otherwise those who feel RESPONSIBLE would all be deranged souls to whom voices speak in profound silence.

Feelings without respondents: it is not with these that the second empiricism populates the world. Travelers who fly over the Greenland glaciers have the upsetting experience of seeing that what passed for a magnificent decor on previous trips has changed *meaning* and is now becoming something on which the survival of airplane travel depends in part, and something on which we depend in part for our own survival. How could we register that experience by saying that it is merely the projection of "values" onto something that is "inert" and "without intrinsic value"? Those who don't understand that glaciers, too, have acquired a "moral dimension" are depriving themselves of any chance to accede to morality; to seek to be moral *without moral beings* is like seeking to reproduce without having offspring [REP] or hoping to believe in God without letting the angels of salvation reach us [REL]. It is indeed materialism, ours, that of the second empiricism, that requires that there be *in things themselves* something that extends the moral sense—or rather, let us say something that permits the acquisition of a more refined moral sense.

This is what makes the idea of a universal morality *without a universe* so strange! One might as well build a port far from any river or sea. To the Moderns' credit, they have always considered that they did not have a monopoly on morality [MOR] any more than on law [LAW]. They have always recognized that other peoples also had "their" law, even if it was strange and unwritten, and that these others were never so savage as to be "completely without a moral sense," even if their

⊙ THAT CANNOT BE UNIVERSAL WITHOUT EXPERIENCE OF THE UNIVERSE.

criteria for discernment appeared incomprehensible and they some-times had "abominable customs."

The Moderns' claim to stand apart from the other cultures comes from elsewhere: they have sought a *universal* morality. They can be accused of ethnocentrism, of hegemony, of self-righteousness; they can be mocked for their smugness, and their continuing failure can be pointed out. The fact remains nevertheless that they have committed themselves to this endeavor, and they have drawn their moral philosophy from it. But for morality to be universal, *it has to have access to a universe.* In simultaneously claiming to be seeking the universal and cutting them-selves off from the world—reduced to facts devoid of having-to-be—the Moderns hadn't the slightest chance of succeeding. They could only end up with moral pluralism, which they call, with mixed cynicism and despair, "moral relativism."

This changes if we give them back a conduit to a multiverse capable of being deployed according to the particular tonality of morality, among others. There would be no more moral pluralism, but a *plurality of existents* whose assemblage would have to be optimized by reconsidering, for each particular investigation, the compatibility of ends and means. We would then understand the Moderns' wild ambition, their great unhappiness, their constant anxiety, and also their genuine virtue: they have tried to raise the question of the optimum with the universe, without ever being able to suspend the expansion of their scruples. Making fun of them, losing their inheritance, abandoning the task of optimization, would be a betrayal. They must not be left to steep in their impossible meta-physics; what is perhaps the most precious of all their contrasts must be gathered up in other institutions. They may have invented the optimum, and immediately lost it owing to their ill-placed confidence in the help of the providential Economy, but there is no obstacle to refounding the strange and paradoxical value known as optimism. By restoring ontolog-ical dignity to the beings of morality, it may be possible to do some diplo-matic work and to understand quite differently the contrast the Moderns have sought unsuccessfully to extract.

WE CAN THUS DRAW UP THE SPECIFICATIONS FOR MORAL BEINGS [MOR] ⊙   If they really exist, these beings, they must have SPECIFICATIONS that distinguish them from the requirements and obligations of all the other modes.

These specs, as we know, include at least four requirements: What hiatus allows us to detect the mode? What type of beings does it have? How does it differentiate truth from falsity? What particular aspect of being-as-other does it draw on to differentiate itself from the other modes?

To define beings it suffices, as always, to follow the thread of experience—here, the thread of what happens to us when we feel tormented by a moral scruple. Nothing changes, and yet everything changes, for everything has been *taken up again*, but by an original type of reprise: "Have I done the right thing or the wrong thing?" The moral being reconsiders all existents in the light of a new questioning. Although every mode is self-referential with respect to the previous ones, this one accumulates all the self-referentialities, as it were. Beings that bear SCRUPLES ask a question, after the fact, that no other mode has yet posed in this way: "Were we right? Perhaps we have to start all over again. *Let's start all over again.*" All the other modes hurl themselves forward, they utter themselves. Except for law [LAW], which seeks to build an archive of successive disengagements (indeed, this explains the numerous resonances between law and morality [LAW · MOR]), and of course, organization, the agency of limits and ends (and this explains the numerous resonances between organization and morality in the form of RULE, a multimodal term par excellence [ORG · MOR]).

But moral beings—and thus things themselves—extract a unique contrast: "And what if we had taken ends for means, or vice versa; what if we were mistaken about the distribution of beings?" When morality comes on stage, rationality takes up its pilgrim's staff once again. "I was *right*, and yet maybe I was *wrong*." "I know, I know, but still . . ." A scruple that is the exact opposite of what the moralists often take to be the expression of a "moral position," a position that is often judged by its intransigence, by the absence of any reprise, and thus by the *lack of scruples*. Speaking "morally" engages one in an entirely different way from speaking *about* moral problems: once again, the adverb leads to a different proposition from the one associated with the corresponding DOMAIN.

How can we distinguish, in this pass, an original form of veridiction and malediction? If we consider the requirements and obligations of the eleven preceding modes, there seem to be enough ⊙ AND DEFINE THEIR PARTICULAR MODE OF VERIDICTION: THE TAKING UP OF SCRUPLES ⊙

differences already to nourish moral scruples. And yet no, we experience this intimately on a daily basis. It does not suffice to be simply troubled, vaguely uneasy: we have to *commit* to a new movement of exploration in order to verify the *overall quality* of all the links. Here we have a unique requirement. Common sense comes into it without difficulty: it is not enough to be concerned "without doing anything," as in the refrain "I think about it, and then I forget." Such crocodile tears that involve "no commitment" have always been viewed as proof of profound immorality. What torments us is the commitment to a new adventure of *verification* of the thing about which the first scruple was just a fuzzy starting point. This is what defines the truth of the moral sense, or rather its gradual validation. The particular, almost technical pass of morality lies in giving itself the means to go still further in the groping that makes it possible to validate or falsify what the initial uneasiness had only sensed: "Hell, they say, is paved with good intentions"; how can we know whether we have been mistaken or not? By *starting over*. Starting over is the only Purgatory to which we have access and from which we can exit only by groping, feeling our way.

This is what makes it possible to spot the infelicity conditions without difficulty (as is the case with the political mode [POL]): they all come from the *suspension* of the REPRISE, from the abandonment of cases, indifference toward any technical arrangement of proofs. We stop worrying; we suspend our scruples. We use the rich infrastructure of organization to proclaim: "We're quits forever." Worse still, we start fleeing into the search for principles, we seek an external, allegedly transcendent viewpoint from which we shall no longer have any means for verifying whether we are right or wrong. Some even find in the religious mode's requirement of salvation and in its end times a pretext for ending all exploration, even for denying the very necessity of any compromise [REL · MOR]. "What's the point in being moral, since I'm saved?" In taking this position, one is betraying religion as much as morality. The malediction of *negligence*. The word CASUISTICS is perhaps too light to rehabilitate the strong sense that has to be given to the attention, the vigilance, the precautions with which one must deal with each individual case.

To put it differently, everything in morality is OBJECTIVE, empirical, experimental, negotiable, everything presupposes the sublime exercise of *concession* or *compromise*, and even, yes, of *being compromised*—which allows one to compromise oneself, to make promises to more than one other. When you set out as a moral being, everything is *yours*, everything concerns you, preoccupies you, worries you. If, as we saw earlier, the sense of ownership is expressed as "Now it's mine," when "It's not mine to worry about" resonates, we can be almost certain that we have spotted some new *impropriety*. The originality of this mode of being lies right here: it knows no limits. This is what opposes it so strongly to the preceding mode, to organization, to the beings of framing [ORG · MOR]. As soon as it sees a limit, it has scruples about not trying to surpass it. By limiting itself, by believing itself to be quits, is it making a terrible mistake?

Here is where we meet the third feature of moral beings: their type of transcendence, the leap into existence that they have extracted from being-as-other, their very particular form of ALTERATION: ⊕ AND THEIR PARTICULAR ALTERATION: THE QUEST FOR THE OPTIMAL. *everything must be combined insofar as possible even though everything is incommensurable.* It is necessary to reach the optimum even though there is no way to optimize that optimum by any calculation at all, since, by definition, the beings whose relations of ends and means must be *measured* do not and must not have *any common measure*, because each one of them can *also* be counted *as an end.* (Here, at least, Kant succeeded in registering moral quality, even though he reserved it to humans alone.) Goods and evils cannot be weighed against one another, and yet *they must be weighed*.

The genius of the Latin language gives the same etymology for *calculation* and *scruple*. The little pebbles that are used for counting can find themselves lodged in shoes and pressed into flesh! Take out one of these pebbles, one of the terms of this contradiction, and you lose all moral sense: make all links measurable and you tip into utilitarianism by putting an end to scruples; give up the idea of making them commensurable so as to optimize them, and you lock yourself into a local version of good and evil; you set limits to the distribution of what counts and what does not count; you start to moralize; you will have confused the possession of "strong moral convictions" with the exercise of moral scruples. You'll find yourself definitively driven out of the "Kingdom of Ends."

As with each mode, we rediscover the opposition between good and bad TRANSCENDENCE: carefully preserving the distance between the commensurable and the incommensurable makes for transcendence; extracting oneself from situations to seek the "external" viewpoint that alone makes it possible to "judge" situations that otherwise would remain merely "factual" is the classic example of bad transcendence that tips you outside the experience, or, better, outside moral experimentation, and leads to MORALISM.

THE ECONOMY IS TRANSFORMED INTO A METAPHYSICS ⊙   Here is where we may succeed at last in defining the three modes that The Economy, the knot of all knots, had tangled together. The optimum *must be calculated even though it is incalculable.* There are two ways of approaching this paralogism: either by proceeding as though one could calculate it by imitating the mode of reference without ever getting there [REF · MOR]; or by assembling, in a form that must always be renewed, *those beings* that are directly concerned and that are tormented by the scruple that they may have been once again mistaken in the distribution of means and ends. In one case, the stage is emptied: "Move along! There's nothing to see here; we're quits"; in the other, the stage fills up. In one case, there are indisputable facts/values; in the other, around arrangements for calculation, debate *begins*, again and again. In one case The Economy and Nature alike speak mysteriously and dictate their decrees; in the other, poor humans learn to live without any Providence. Once again, a contrast not to be missed.

⊙ WHEN IT AMALGAMATES TWO TYPES OF CALCULATIONS IN THE [REF · MOR] CROSSING; ⊙   How can one pass unawares from one of these versions to another? This shouldn't surprise the reader: through a blunder, this time a perfectly intentional one, on the part of our old nemesis Double Click, who, unlike Tom Thumb, is going to mix up his pebbles [MOR · DC]. Careful, this time we're on the verge: all it takes is a small, a very slight, an infinitesimal mistake about the *nature of the* CALCULATION. Through a certain ambiguity about the very instruments that scientists use; through a slight clinamen that introduces a hesitation between the movement of the *inscriptions* that are necessary to reference and the movement of the *accounts* that are necessary to attachments [REF · ATT], of the *scripts* necessary to projects [REF · ORG], the *balance sheets* necessary to the renewal of the optimum [REF · MOR]. As economists are number people, the solution they have provided for the problems of passionate interests [ATT], organization [ORG], and the optimum [MOR] has appeared indecipherable, because it has deliberately been mixed with the trajectory of the same calculations present in the work of reference. This is the cryptogram that we are going to be able to decipher at last.

In all cases, we find ourselves confronted by inscriptions, instruments, signals, books, equations, models, but their meaning, their direction, is completely modified by the preposition. It is not at all because we count, measure, enumerate, and evaluate, it is not at all because we make models that we shall end up with objective knowledge in the referential sense of the term. Calculation devices are used in all cases, but what we have here is a simple *homonymy*. With numbers and models and even theorems one can gain access to remote states of affairs, to be sure, but one can also learn to *divide*, to distribute, to share, to determine proportions, to make measurements and "take measures." It's not the same thing at all. The whole key to learning the economic disciplines consists in mobilizing the habits of numbering to make them serve the calibration, the formatting of the accounts, and scripts through which interests, roles, and functions are *distributed*. But if you let Double Click intervene, he'll prove to you through the result of an indisputable calculation that I owe you nothing, that it isn't "mine." The Evil Genius obsessed with epistemology becomes the Demon, the Divider, of economics. Reason is invented so that the thread of reasons will be lost. Here is the origin of the inequality that Rousseau thought he had found in the indispensable enclosures; it is here that the discipline of economics risks bearing the sign of Cain on its brow: "Am I my brother's keeper?"

It is useless to complain that the practices of economization are not "objective" enough. Economics does not deflect, for interested purposes, knowledge that *otherwise* would have headed toward objectivity on its own [REF]. Will our anthropologist turned diplomat be astute enough to say this in a way that is not critical? In The Economy, what is at stake is not at all, has never been, objective knowledge, but attachment, organization, distribution, and morality. Economics does not aim at reference any more than Law does. If the difficult training in economics had the ambition of truly referring, it should have deployed the multiplicity, the heat, the incommensurability of attachments [ATT], the exhausting rhythm of scripts [ORG], the incessant renewal of the optimum [MOR]. It wouldn't have waited two centuries before beginning to do its anthropology; it would have *started* there, and it would have become a great science of passions and interests combined. Economic science would have been coextensive with anthropology or the history of exchange; economics departments would have produced countless volumes on the strangeness of our interests, the embeddedness of our values, and the difficulty of summing them up, along with ever subtler experiments to bring the greatest number to share in the reprise of the optimum.

⊙ THIS MAKES IT MISTAKE A
DISCIPLINE FOR A SCIENCE ⊙

Now, the economic science visibly pursues other goals. The category mistake would be to believe that, like physics, chemistry, or biology, it has taken objective knowledge of "economic matter" as its object. We are very familiar with this mistake; we have run into it virtually every step of the way throughout our inquiry: it is the epistemological mistake, the one that consists in believing that knowledge of the referential type—and that knowledge alone—must define the whole of our existence in order to serve as ultimate judge for all the other modes of veridiction [REF · PRE]. But if the economic science does not aim at objective knowledge, then it is irrational, untruthful, or at the very least superfluous! This is the reproach addressed to it by those who find that it "lacks soul"; it *ought* to know and not just format; it *ought* to be a science and not just a *discipline*.

And yet there's no point in complaining; economization has always had other functions besides knowledge. There are many other goals in existence besides access to remote beings. The rational is woven from more than one thread. If The Economy does not manage to institute these three modes of existence while differentiating them, it shares this difficulty with all modern institutions. Let's not forget that in terms of daily life it is the second nature, Sloterdijk's "Crystal Palace," rather than the first that we really inhabit. There shouldn't be anything astonishing, then, about the fact that to survive in this nature we need very different resources. After all, what is really in question, in the etymology of these arrangements whose status is henceforth uncertain, is habitat, *oikos*.

If the "cameral sciences" are not exactly sciences, it is because they belong to a wholly different register from physics, chemistry, or pedagogy. They are more like spiritual exercises, like the challenging discipline of yoga, like self-control and the control of others. Confusing the discipline of economy with reference makes no more sense than asking religion to transport you to the realm of remote beings [REL · ORG], or expecting to come to terms with a loss thanks to a legal judgment [MET · DRO]. The arrangements of calculation have never had the goal of knowing objectively—moreover, they never would have covered over the abyss of dissimilarity, those three hiatuses that we have just recognized. In a sense, they have always *done better*, or in any case done something else: they have made it possible to express *preferences*, to establish *quittances*, to trace *ends*, to *settle* accounts, and perhaps even, if we only knew how to divert them, to help calculate the optimum anew. The calculations of the economic disciplines have never had the goal of "knowing," if we understand by this the trace of the chains of reference [REF]. They have something better to do: they have to set *limits* to what would otherwise be *limitless* and *endless*; they have to offer instruments to those who must distribute means and ends. Let us say that they *format*, they put into form, they give form, they *perform* relations *starting from the raw material* of attachments, scripts, and scruples. Here lies the whole importance, and even, if you will, the entire greatness, of these life forms.

The voices raised in complaint against the coldness, the indifference, the insensitivity, the abstraction, the formalism of the discipline of economics are virtually meaningless if they do not seek to refer but to heat up, to frame, to cool off, to debate. What has happened, then? How have we managed to miss the importance of equipping scripts with calculation devices in order to coordinate division, sharing, distribution, allocation, in other words to delimit—to bring to an end, achieve finality, that is, to *finance*—the proliferation of innovations, attachments, valorizations, commitments, imbroglios of goods and people—in other words, to warm them up. How have we been able to confuse the nature of what calculates with the nature of what is calculated?

⊙ THAT WOULD DESCRIBE ONLY ECONOMIC MATTER.

Might the disciplines of economization thus be entirely innocent of the errors of The Economy? Alas, no, for we have to recognize that the little *mistake in calculations about the nature of calculation* is going to lead to a fatal destiny. If this epistemological passion has painfully afflicted the first Nature, it has transformed the second even more. It became death-dealing when, through a new amalgamation, the second Nature borrowed the [REP · REF] category mistake from the first, imagining a "matter" that had definitively lost the polysemy we have gone to so much trouble to bring out. Instead of seeing economization as something that allows the unfolding of courses of action—allocation, distribution, division, sharing, purification—it defines itself as that which *forbids* all organizations to envisage themselves as pileups of scripts that are all subject to discussion, all to be rewritten. Carried away by its dream of resembling knowledge, it goes even further: it believes it can *escape* all organization, any slipping in of scripts, any renewed debate over the optimum; it believes it can calculate the intermingling of quasi objects and quasi subjects *automatically* and thereby discover indisputable laws of transformation that have the same type of causality as the "laws of Nature." Instead of attachments, organizations, and optima ([ATT]; [ORG]; [MOR]), there would be something that "holds together on its own," something that escapes all interference and intervention. There would be a metadispatcher no longer dependent on any script. The long history of materialism would end, in a stupefying way, with a great providential narrative! Providence has been extended to the Kingdom of Ends. A God of the bottom line; that took some doing . . . After the first Nature, the second nature; after the first supernatural, a *second supernatural!* No question: the Moderns will never stop surprising us.

SO THE ECONOMY PUTS AN END TO ALL MORAL EXPERIENCE.

If we are right in our qualification of moral beings, we understand the catastrophe that the claim to make the optimum calculable by making the expression of value "a mere expression of fact" can represent for Economics. The metadispatcher will never again be just one script among others [ORG · MOR]. Denying the prodigious warmth of attachments [ATT] is of no great moment, ultimately; the Earth of goods and bads will continue to turn on its axis. No merchant, consumer, innovator, or entrepreneur will attach the slightest importance to this denial; they will continue to do business, stock their shelves as before, pushing the stunning inventiveness of their enterprises further and further. Omitting the importance of organizational arrangements [ORG] is already more serious, since it will no longer be possible to know, when people speak of economics, whether they are referring to the disciplines, the long process of learning the formatting and framing, or rather to that which is formatted and which, by definition, overflows in all directions. But claiming to short-circuit moral expression, to suspend the exploration of scruples, to interrupt involvement in the distribution of ends and means, on the pretext of "closing the books," as they put it—here we have something like a form of intellectual leprosy. This is the stigma that can be seen on the faces of the Moderns and that eats away at them from within. Organization was respectable. Calculations were respectable. Models were respectable. Economics-as-a-discipline was respectable. But the interpolation of organization and morality is not [ORG · MOR]. This is one crime too many. This is the crime that makes them monsters whom one can never look straight in the face. One cannot be driven by the search for the optimum, manifest a healthy optimism, and simultaneously claim to have discovered it by a calculation that would short-circuit its reprise by distancing those whom it concerns most directly. One cannot both deny Providence and reintroduce the supernatural of The Economy. If one cannot serve *both* God and Mammon, it is because one must not serve *any* Gods while believing that they are transcendent.

As we have just seen by going through the three modes of existence belonging to the FOURTH GROUP, the one that mingles QUASI OBJECTS and QUASI SUBJECTS in three different ways, it is as though nothing in The Economy as it has been instituted really allowed us to do justice to the experience of the Moderns. There is no economic sphere any more than there is Society, Language, Nature, or Psychology. It is no accident that the disciplines of sociology and psychology are the ones that are having the most difficulty extracting themselves from Modernism—a situation described by the mild euphemism "crisis of the human sciences." If, on

THE FOURTH GROUP, WHICH LINKS QUASI OBJECTS AND QUASI SUBJECTS, the contrary, there is one thing that must be anthropologized and redistributed, it is this continent of The Economy, so as to extract from it only what really counts, that is to say, literally, *what counts*, the arrangements that make organizations traceable and tractable. That is what justifies the quality, the effectiveness, the importance of economic calculation. Not to be supplemented by the slightest whiff of metaphysics.

And yet, by a sort of mistake in civilization, an ill-formed institution, The Economy, has been entrusted with the task of collecting three contrasts, all three bearing on the mutual entanglements of humans and nonhumans, but with no possibility of durable instauration for any of them. Once again, to express one contrast—"careful, we have to calculate, without that we'll be overwhelmed by the number of attachments, the incoherence of the scripts, the incalculable calculation of the optimum"—it has been deemed appropriate to suppress others. For in fact it isn't easy to do justice to these three formidable discoveries of Modernism: the burgeoning mutual entanglements of goods and people on a planetary scale [ATT]; the mutual embeddedness of scripts through inscription arrangements that allow changes of scale and the moving of the whole Earth [ORG]; and, finally, the extension of scruples to the entire universe [MOR]. Whereas the first mode multiplied entanglements, the second does almost the opposite (one learns to be quits by learning the right of proprietorship!), while the third forbids us ever to limit the expansion of doubt about the right way to achieve the optimum. Understandably, it is not easy to find *habitats* capable of sheltering all these discoveries at once—and to do justice at last to the etymology that economics has to share from now on with ECOLOGY. This is not a reason for the West to stick with the most badly-put-together of institutions. All the more so in that the third contrast, that of morality, is the one that suffers the most.

In a sense, this can't surprise us, on the part of those who have already confused the passage of the world with the grasp of data necessary to gain access to remote states of affairs [REP · REF]. And yet it must not cease to astonish us, for, much more than with epistemology, it is a matter of the most down-to-earth, the most rooted, the most vital of contradictions. If there is one thing to which one must not give in on this Earth, it is the idea of a Providence that would come in, without any other action on our part, to make the best distribution of that to which we hold the most, and most firmly. The entire modern experience has stood up against living under the control of an indisputable metadispatcher. No one can make us believe the contrary; *we know that it is false.* The question is how the Whites, who thought they could teach the rest of the world the "pure, hard rationality of economics," are still so imbued with that "secular religion." Why do they continue to believe in another world above and below this one, a world that would not be the result of an organization but the unfolding of a series of decrees to which we can only assent? In other words, why have the Whites never extricated themselves from the old idea of an Economy formulated by the Greek Fathers, which designated the economy of Salvation, that is, the distribution of the salvific work of God in the world, even while believing themselves to be materialists and atheists?

It must be said that they haven't had any luck; they've never tasted *economic freedom*. Not once in their short history have they managed to get away from the simple question "Which tyrant do you prefer? The one with the **INVISIBLE HAND OF THE MARKETS**, or the one with the visible hand of the **STATE**?" They have

⊙ IS THE ONE THAT THE INTERMINABLE WAR BETWEEN THE TWO HANDS, VISIBLE AND INVISIBLE, MISUNDERSTANDS.

never had to choose except between plague and cholera (even after eradicating actual plagues and actual cholera . . .). They have never imagined that there might be no hand at all! They have never believed it possible that they could escape from all tyranny, from all maxi-transcendence. Has any war of religion done more than this one to cover first Europe and then the entire planet with blood? Who will have the courage to publish the Black Books of all these symmetrical crimes? The two camps behave as if there existed somewhere, above or below us, in any case elsewhere, a metadispatcher so powerful, so omniscient, so far beyond all organization, all intervention, all interference, all humble local revisions of scripts, that one could follow it blindly and trust it automatically. The only difference is that on one side the metadispatcher is placed *before* all courses of action—this is the case for the State—and on the other side it is placed *after*—on the horizon of Markets. Yet the very idea of a metadispatcher entirely cancels out the deployment of the organizational mode of existence [**ORG**]—not to mention political circulation [**POL**]. If there is already, elsewhere, above, below, before, or after an organism that has *already* been *composed*, then there is no more to be done.

Those who read today, with the benefit of hindsight, the various manifestos that have justified the systematic assassination and despoliation of tens of millions of poor humans, and who consider the fragility of the charges for which the victims were immolated to those Molochs, those Mammons, ought to be weeping tears of blood, hoping without hope that a time of immanence will come, vomiting up the tyranny of the Hand. There is no metadispatcher, it's as simple as that. That God, at least, does not exist; no one has ever been able to occupy that position, whether it is called Market or State. No one has ever had that kind of knowledge, that prescience. There is no Providence.

CAN THE MODERNS BECOME AGNOSTIC IN MATTERS OF THE ECONOMY ⊙

The question becomes the following: "Can the Moderns finally become *agnostic* where The Economy is concerned?" We can understand the temptation to believe in a providential metadispatcher, of course. The history of that temptation, of that fall into facility, has been written many times. There is, first, the desire to shelter riches from the greed of princes: how better to protect them than by constructing the strongbox of an economic matter with which they must not be allowed to interfere? As Michel Foucault showed so well, it is first of all from those who govern that the effectiveness of the invisible hand must be hidden. They can be told: "Hands off!" Then, at just the right moment, examples such as clocks, scales, and Watt's regulator came along to offer proof that one could automate even the equilibrium that had until then always been a matter of simple practical wisdom on the part of "well-balanced" people with whom "dealings were pleasant." How could anyone have resisted assimilating the economic machine to the gigantic automated machinery that was filling up the industrial landscape so fast? Didn't the economic machinery infinitely surpass the forces of poor human intelligence? Was it not in the hands of specialists? Then Darwin appeared, with his new violent, bloody, and pitilessly just nature, and economics didn't lose any time collecting the legacy (whereas Darwin's theory of evolution ruined all the plans of the providential second Nature as much as those of the first). A self-regulating machine, constructed on the model of the most prestigious physical sciences, imitating the eternal harshness of Nature, protecting the horn of plenty from interference by the powerful, economics was the preserve of cold, serious experts; moreover, it was flooded with mathematical results owing to the very scope of the numerical data. Let us be honest here: this secular religion could not fail to spread!

To designate this subversion that makes the scripts of organization invisible and calculations of the optimum automatic, the term CAPITALISM has come into use. Unfortunately, the critical power of this term wears out very quickly if it is used to change only the name of the metadispatcher, the one who "has in hand" the destinies of all those who consider themselves with delight to be slaves of the immense forces to whose service they devote themselves wholly. The atrocious irony of this particular religion is that it is constantly reinforced by the very actors who wanted our eyes to stop turning Heavenward. A proof if ever there were one that from the good faith of martyrs one cannot draw any proof as to the quality of the Faith for which they sacrificed themselves. How many people have died in the name of the struggle against the "opium of the people"? It is The Economy that ought to say: "My Kingdom is not of this world." All the more so because their class enemies, terrified, would have committed any crime to avoid the reign of the commissars, even that of believing in a "self-regulated" market. Anything rather than entrusting one's destiny to the cold monster called the State, the supposed repository of the "common good" without calculation, without experimentation, without groping.

If it took so many wars of religion before anyone could dream of separating the State from religion, how many "wars over The Economy" will we have to endure before we decide to separate ourselves from both the Providence of the State and that of the Market? When will we put an end to this long infantilizing, this situation of dependency, by becoming true materialists? Is it conceivable that those who thought they were teaching unbelief to the rest of the world and who boasted of their supposed "secularization" may finally learn economic freedom? Not content to have soiled, perhaps irreversibly, the planet that has given them shelter, they have gone on to degrade the very word "LIBERALISM." Just as the meaning of the word "republic" has been reversed by the addition of the adjective "Islamic," the meaning of the word "liberalism" has been perverted by the addition of the prefix "neo." They think that the expression *laisser faire, laisser passer* translates faithfully the admirable injunction "Don't let anything go, don't let anything pass!"

⊙ AND PROVIDE A NEW
FOUNDATION FOR THE
DISCIPLINE OF ECONOMICS?

And yet nothing in the arrangements of ECONOMICS as a discipline warrants these excesses of belief or of critique. It is only a matter of calculation matrices that make it possible to trace the intermingling of scripts and to note clearly and comprehensibly who possesses what until when and up to what amount, in order to have a protocol that can be used to bring together those who must constantly recalculate the optimum. Everything in the formalization practiced by economics aims exclusively at permitting the local renegotiation of scripts. Nothing requires the occult presence of a metadispatcher. Without economization, we would never know when it is "our turn" to play, or what is "ours," nor would we know how to get back on the paths of optimization or on what timetable to balance the accounts.

If you have doubts about this simple, humble truth, imagine that, through the action of some mysterious virus, all value meters are erased from all hard disks and all their backups. The scope, the multiplicity, the variety of attachments are such that we cannot find our way around in them without constantly refining the tools that make them, if not entirely calculable, at least legible on the numerous screens that the new intellectual technologies keep multiplying [ATT · ORG]. It is owing to the scale of the overflow that the framings become so invasive. Just as the development of the sciences is making the equipment needed for paving the paths of reference [REF] more and more visible, the artificiality of the tools of economization ought to make it easier every day to grasp the cleavage between the disciplines of economics and the metaphysics of The Economy—the latter being only the phantom image left behind by the development of the former. And yet we never manage to fix our gaze on the techniques of visualization, counting, statistics, modeling, dissemination, and metrology that allow the circulation of the accounting disciplines deep into our own heart of hearts.

Curiously, the Whites are prepared to reconnect the thread of experience only when they encounter other cultures whose imbroglios do not strike them as obeying "economic rationality"! For in fact, everywhere in the world, from the time of the great discoveries on, the Moderns have encountered peoples strange enough not to share their blindness as to the respective roles of what lies underneath and what is on the surface where imbroglios of goods, bads, and people are concerned. It was the genius of Karl Polanyi, in *The Great Transformation*, to reinstitute this contrast for us while avoiding the double exoticism. The Whites see perfectly well among the others—whom they wrongly take to be exotic—what they do not see in themselves: the proliferation of what overflows the *framing* of the calculation devices that nourish economics, a discipline radically distinct from what constitutes the imbroglios themselves. OCCIDENTALISM consists in believing that the Whites are economized up to their eyebrows, while the "others" (whether we regret this or admire it) are still at the stage of "confusing" the symbolic play of nonequivalences with the rules for transforming market exchanges. "They" would be all mixed up, entangled in their goods and their bads; "we" alone would be capable of calculation, of reason, of rationality (or at least that is how we saw things before the Others, "they" too, began to take on these capabilities and exercise them better than we do . . .). Now, if there is one case in which *symmetrical* anthropology makes sense, it is surely this one: "We" are exactly like "Them," in the sense that the discipline of economics counts *a great deal* but at the same time *counts for very little*, counts *clumsily*, and its calculations never come out right.

Our anthropologist is obviously naïve to an unhealthy degree— this is her vocation and also her charm—and she really believes it is possible to turn The Economy around by making visible the three modes of existence that the stress on the "rational" aimed, quite deliberately, to keep out of sight. As The Economy is a recent domain—Tim Mitchell even situates its coagulation in the postwar period—she imagines that it can become more fluid again. In her eyes, nothing ought to prevent us from deploying the chains that link passionate interests or from discerning the allocation keys and the cadencing of scripts that underlie the calculations. She even dreams—a waking dream—that the abandoned agora is filling up again with all those who are called to take up the calculations of optimization anew. She envisions an assembly that would at last have the equipment, the technology, the politics, and the morality to proclaim, without any God intervening to misdirect its immanence: "And now, let's calculate." In short, a civilization.

What could be more utopian than putting an end to the utopia of The Economy, whose worldwide success seems definitive? Still, it would be quite cruel to reproach our ethnologist for dreaming. Has she not done her work well? How can we fail to sense, in the author and perhaps in the reader, the suspicion, perhaps even the hope, that she has prepared the ground for another event. After all, The Economy has the signal weakness of living off the ground, away from the Earth, out of this world. How could it survive the return of the world? How could it withstand the reminder from a Creditor who does not have the ability to cancel debts but who, on the contrary, waves before our dumbfounded eyes the powerful threat of canceling all the quittances? "You are no longer quits at all!" We have to start all over again. Let's get back to our calculations. *Calculemus.*

CAN WE PRAISE THE CIVILIZATION TO COME?

To avoid failure, we must use a series of tests to define the trial that the inquiry must undergo:

First test: can the experiences detected be shared?

Second test: does the detection of one mode allow us to respect the other modes?

Third test: can accounts other than the author's be proposed?

Fourth test: can the inquiry mutate into a diplomatic arrangement ⊙ so that institutions adjusted to the modes can be designed ⊙ while a new space is opened up for comparative anthropology ⊙ by a series of negotiations over values?

For new wars, a new peace.

WHEN HE WAKES UP IN THE MORNING, AT THE END OF HIS LABORS, THE AUTHOR, UNEASY, WONDERS WHETHER WHAT he has just put together from bits and pieces, gathered over many years without ever being shown to the public, looks like a gingerbread house, or a painting by Le Douanier Rousseau: a hodgepodge of curiosities that says a lot about the odd tastes of the autodidact who collected them, but very little about the world he claims to be describing. Try as he might to reassure himself by telling himself that the questionnaire that is the basis for the study has "held up" for a quarter of a century, that he has never "let go" of it along the way, that he has always drawn from it clarifying effects that have often enchanted him, he knows how fragile this testimony is and how many ruses the Sphinx is capable of deploying to deceive the one it places before the enigma of the "work of art to be done."

I shall never manage to reassure myself except by taking this work as a "provisional report" on a collective inquiry that can now begin, at last. So that it will be more than just one more whim on my part, I have to try to explain how subsequent research, once equipped and made collective, might extend this book, which is only, let me say it once again, a summary of documentation whose dimensions have become excessive. To do this, I must first spell out the various tests to which the inquiry has to be prepared to respond.

The first reading test, as I announced in the introduction, is to determine whether the experience of the modes is *shareable* with others, despite a rendering that differs greatly from good sense—but not, and this is what is really at stake, from COMMON SENSE. The method or, to use a more modest term, the navigation advice I have offered along the way is actually very simple: we choose, among the Moderns, the DOMAINS to which they seem to hold the most; we shift attention from these domains to NETWORKS; then we look at the way the networks expand, in order to detect the distinct tonalities that we gradually extract by comparing each network with the other modes of extension, two at a time; finally, and this is the hardest part, we try to entrust ourselves exclusively to the often fragile guidance of these discontinuous trajectories, abandoning the reassuring but vacuous help of a transcendent level. This last is a delicate operation that presupposes a long-cultivated skepticism toward the aggregates inherited from the history of the Bifurcation— Nature, Language, Society, Economy, without forgetting the all-purpose pincers Object and Subject. A few surprises, many strange encounters, of course, but nothing that can surprise common sense, much less shock it.

FIRST TEST: CAN THE EXPERIENCES DETECTED BE SHARED?

The first test can thus be formulated this way: by following these navigation procedures, have I been able to make perceptible to readers a certain number of tonalities or wavelengths that modern institutions in their shambles have made it impossible to capture? Have I really earned the right, following William James, to use the name *second* empiricism for this way of following the beings bearing relationships that the first empiricism situated in the human mind? Can I then take pride in having brought an end to the BIFURCATION between PRIMARY QUALITIES and SECONDARY QUALITIES? If this test fails, if my readers do not feel better equipped to become sensitive to the experiences assembled here, if their attention is not directed toward the beings whose specifications differ in each case, then the affair is over. They will have merely visited, with a blend of amusement and irritation, a scale model of modernism, a kind of ideal palace like the famous one built over decades in a mix of different styles by a French mailman, Ferdinand Cheval—full of fantasy, to be sure, but about as useful as a reconstruction of Paris in matchsticks or Beijing in wine corks...

If, on the contrary, this first test succeeds, at least partially, if readers, profiting from the gap our method has established between the thread of experience and its institutional rendering, begin to register documents parallel to mine, concerning one mode or several, we can go on to the second test: that of the comparison between modes. Do we gain in quality by crossing several ontological templates in order to evaluate, little by little, what is distinctive about each one? And, an even more daunting subtest: do we gain in verisimilitude by treating all the modes at once in such a move of envelopment?

I am well aware that I passed over each mode too quickly, and that each crossing would require volumes of erudition, even if the modes and crossings are more fully developed in the digital environment that accompanies this text. But it would be unfair to ask me to go more deeply into each domain, each institution, each period, each counter-move of one mode on another, at the level of detail required by specialists. So this is not the way this inquiry asks to be judged: the only usable criterion is whether the light shed on a given mode makes it possible, through successive crossings, to reinstitute the *other modes*. Specialists obviously don't have to ask themselves this question. The anthropologist does. This is an operation that the Moderns have never attempted, in my opinion, and their failure to do so has led them to multiply their misunderstandings about the other collectives, first of all, but also about themselves. They were too busy forging ahead; they never consented to look at themselves straight on and acknowledge all their trump cards. Strangely enough, in the history of anthropology, there haven't been any "first contacts" with the Whites. The mythical stage that I propose to set is thus this one: by unfolding their entire set of values all at once, do we do them justice at last?

I know perfectly well that the attempt is far-fetched. But still, if you were to inherit from a relative who was both wealthy and eccentric, who had died intestate, wouldn't you try to draw up a list of all his properties and all his papers? Since we have never been modern and never will be, we still really need to know what has happened, no? And another factor justifies being somewhat vague on details so as to have a clearer view of the whole: the fact that our goal is to make the collectives more readily comparable while using a system of coordinates different from those of the MODERNIZATION FRONT, with all its advances and its lags, for that front measures movements by distinguishing between the objective and the subjective, reason and unreason, the archaic and the progressive, science and culture, the local and the global. I believe I have shown to what extent that system is incapable of continuing to offer adequate bearings. If, instead of modernizing, it now behooves us to ecologize, it is perfectly normal to change *operating systems*, to use a metaphor from computer science.

The second test can be formulated as follows: have the readers found it an advantage to grasp all the values strung together, as it were, an advantage that would compensate in part, in their eyes, for the rapidity of the exercise? If this second test is a failure, the specialists will have no trouble tearing our fragile assembly apart: it will never be anything but one more system, a Master Narrative that will slip without difficulty into the shredder's maw. I will be forgiven, I hope, for betting on the success of this test: restoring the gleam to a value is one thing, but allowing that gleam to shed light on another, one that has been left heretofore in the shadows, and then to another, and yet another—this is what seems much more promising to me. To encounter the Moderns in order to ask them: "But finally, what do you really care about?"—is it not to prepare ourselves better for the DIPLOMACY to come?

Before moving on to the other tests, I need to address a likely objection to the number of modes envisaged here. I have no answer to this question, and I am actually quite embarrassed, I confess, by the number twelve (plus three) and by these five groups of three that line up too well in an overly tidy table of categories. To say that it has held up robustly for a quarter of a century is not to say anything particularly reassuring about the mental health of its author. As I believe in the modest advantage of a systematic approach, even though I am skeptical of the systematic spirit, I consider this number of modes as the fortuitous effect of a historical contingency among those whom I study as well as in the inquirer. So I seek no justification other than a lovely image from Souriau: the colors of the Lascaux cave are quite simply those that the painter found underfoot; "yellow ochre, red ochre; green clay, black smoke. He has to make do." To adopt a more recent metaphor, let us say that with these fifteen colors, we can get an image of the modern experience taken as a whole with a satisfactory *resolution*. It would be better with 103,404 colors? No doubt, but then you're in charge of setting up the Pivot Table; that's not something I'm capable of doing. In any case, the question of the number of modes needed will resolve itself in practical terms, like all the rest, when other inquirers propose other candidates—but only after a serious examination.

THIRD TEST: CAN ACCOUNTS OTHER THAN THE AUTHOR'S BE PROPOSED?

Which leads us to the third test, the one that has to do with diplomatic preparation. If the text had already passed the first two tests, I would approach the third with less trepidation, for this one basically concerns just my own presuppositions. They are idiosyncratic? Well, yes, you have to be *idiotic* to throw yourself into something like this, no question, but this sort of idiocy is a stage that may not be superfluous, if it serves to open up a space between experience and its institutional rendering. In the end, the name of the volunteer oblivious enough to stick with it matters little; what counts is the opening he or she has created, or not. Nothing prevents readers who have now become COINVESTIGATORS from proposing to restitute experiences and link values in ways that differ completely from my own.

In other words, this test amounts to asking whether I have made enough effort in this provisional report to distinguish between (1) experience; (2) the metaphysics in which it is sheltered—as I see it, almost always inopportunely; and finally (3) another metaphysics, my own, provisional and disposable. But before my formulations are torn to shreds, the reader mustn't deal with them too unjustly. Their weirdness should be counterbalanced by the fact that they are supposed to allow each mode to enter into resonance with all the others, but also to be differentiated from the institution that has often betrayed it, as well as from the domain that encloses it, sometimes very badly; and, finally—the hardest part—to open up a conduit for comparative anthropology. I am the first to recognize the weakness of my formulations, but if someone claims to dismiss them, the challenger needs to commit to responding to these three constraints.

FOURTH TEST: CAN THE INQUIRY MUTATE INTO A DIPLOMATIC ARRANGEMENT ⊙

The main difficulty, as I am well aware, comes from the fact that the inquiry has to be maintained under the auspices of prepositions [PRE]. I have shown this several times; each mode takes all the others under its protection. For the beings of fiction, everything, even nature, even law, even science, is an occasion to aestheticize. But for religion, everything, even organization, even morality, even nature, has no goal but to "sing the glory of God." And, of course, for knowledge, everything must bend to the requirements of its chains of reference, everything, even habit, even religion, even metamorphoses, even politics—and even the beings of reproduction, which nevertheless follow an entirely different path. And so forth, across the whole Pivot Table. It isn't easy to multiply the comparisons, especially when one refuses to give in to the facility of the template to end all templates offered by Double Click. Now, this inquiry does not proceed under the reign of any of the domains but under that of the prepositions, which offer an unfounded FOUNDATION, a very fragile one since it says nothing about what follows—only the meticulous description of situations can do this—even as it specifies in what key everything that follows must be taken.

Thus while I have spoken all along of an inquiry and even of a questionnaire, it is not in the mode of knowledge that I claim to be working. The term "inquiry" has to be taken in a plurimodal sense whose object is to preserve the diversity of modes. Can we call this approach "empirical philosophy"? I am not sure, given how indifferent philosophy has become to the tasks of description. Experimental metaphysics? Cosmopolitics? Comparative anthropology? Practical ontology? Mailman Cheval will surely have made more than one discovery along his way that can serve as an emblem for this enterprise. To situate this reprise of the rationalist adventure, but to mark clearly that it will not take place under the auspices of Double Click, I have entrusted it to the term *diplomacy*. Values, if they are not to disappear, have to be diplomatically negotiated. A practical RELATIONISM that seeks, in a protocol of relationship-building and benchmarking, to avoid the ravages of RELATIVISM—that absolutism of a single point of view.

I know that I myself am a terrible diplomat and that here we have a clear-cut case of miscasting, since I have said all sorts of bad things about the INSTITUTIONS charged with taking in the famous values that I have presented, conversely, in the best possible light. In the final analysis, no institution really stays in place; no domain is really solid, not even law, although law has been less disturbed by the ravages of modernism than the other modes. Go ahead, try to start diplomatic talks after such overtures to peace! The diplomat who shows up in the agora saying that the institutions all those folks cherish are corrupt will meet with a rousing reception! We might just as well entrust the role of head of protocol to Diogenes and his dog . . . And yet these are the stakes: we risk getting expelled from the agora, at first, but we will be *called back* there as soon as the participants begin to feel that there is more to be gained than lost.

⊙ SO THAT INSTITUTIONS ADJUSTED TO THE MODES CAN BE DESIGNED ⊙   Let us recall what is at stake: the institutions have no justification but to take in values; without institutions that are maintained and cherished, only fundamentalism can win. If institutions start to express the thread of values, why continue to critique them? But if they interrupt the movement of this thread by relying on foundations that are too distant or too transcendent, why respect them? An institution that is sure of itself is dreadfully worrying, but an institution that is afraid of losing its treasure is not reassuring either. What we want is an institution that follows the trajectory of its own mode of existence without prejudging the rest, without insulting the others and without believing, either, that it is going to be able to last in the absence of any REPRISE, through simple inertia.

Wouldn't sciences delivered from Science, finally capable of deploying their chains of reference without insults on their lips, be lovelier—I mean more objective, more respectable? Wouldn't a God of salvation wholly linked to the quality of conversion of those who invoke him, freed from the obligation of insulting fetishes and making laws concerning public and private behavior, be more capable of finding a place in a finally reassembled Church? Since we can't live an instant without the help and the menace of the beings of metamorphosis, couldn't we finally recognize them in all the arrangements charged with taking them in rather than feeling obliged to insult invisible beings and explore the inner depths of the ego?

⊙ WHILE A NEW SPACE IS OPENED UP FOR COMPARATIVE ANTHROPOLOGY ⊙   All that appears absurd? Yes, unless we are using it to seek the help of the other collectives whose competencies we had rejected, in the belief that our first duty was to bring them out of their archaism, by modernizing them. What an absurd idea, to sell them objective, scientific reason when we ourselves didn't have the slightest idea of what we ourselves were doing in the name of science! How often have we been disappointed in purporting to universalize democracy and the State based on the rule of law, when nothing in our official definition of politics and law corresponded to our own experience? What should we call the operation through which we have claimed to extend economic rationality, without being capable ourselves of defining what our own market organizations might look like and on what multiform values they might depend? What a shame, when we think about the fetishes burned by missionaries who were incapable of renewing, at home, the holy Church they had inherited! But imagine, now, before it is too late, before modernization has struck equally everywhere, what a planetary discussion might be like with Moderns who finally know what they care about and can at last share the secrets of their institutions with the others!

I have often wondered, contemplating the mutilated frieze of the Parthenon through the black clouds of pollution in Athens or in the room in the British Museum housing the marbles stolen by Lord Elgin, what a contemporary Panathenaic procession would look like. Who would be our representatives? How many genres and species would be included? Under what label would they be arrayed? Toward what vast enclosure would they be heading? How many of them would have human form? If they had to speak, swear, or sacrifice in common, what civic or religious rites would be capable of assembling them, and in what agora? If a song had to accompany their march, or a rhythm were to punctuate their long undulations, what sounds would they make, and on what instruments? Can we imagine such Panathenaics?

If this text were capable of passing that fourth test, things would become truly interesting; a negotiation could take shape. An internal negotiation, ⊙ BY A SERIES OF NEGOTIATIONS OVER VALUES? first, and then an external one. I recalled at the outset that the notion of CATEGORY engages the one who is speaking to move into the agora in order to restore to those to whom these values are of primary concern a version entirely or partially different from their ideals. Now, as the history of diplomacy can attest, diplomats often have more trouble with their principals than with the other parties. How many laboriously concluded treaties have not been ratified because the principals back home had not been convinced of their true interests? If those interests are too narrowly conceived, or if they appear threatened, it is useless to start talks; the diplomat will immediately be accused of betrayal. The operation in which I would like to engage consists in testing a series of formulas for peace, by proposing a sort of bargain, a deal, a *combinazione*: "Yes, of course, in defining Science, or The Economy, or Society this way you lose a defense that you took to be essential, but on the one hand this defense was broken down long ago, and on the other hand you gain other solid agreements with values that have been extinguished or scorned up to now. Isn't the game worth the candle?"

You have to be naïve to believe in such talks? Well, yes, but the diplomat is a hybrid figure, as devious as he is naïve. I maintain that the only way to tell whether what I am proposing is illusory or not is to carry out negotiations for real, with those who are directly interested in formulating other versions of their ideals; this is what we are going to be able to do with our collaborative research project, the third phase of which presupposes just such talks, focused on the "hottest" zones of conflicts among values.

FOR NEW WARS, A NEW PEACE. Identifying such zones and bringing together principals who are worried enough to trust themselves to the tribulations of diplomacy obviously presupposes a situation that is no longer that of war, or at least no longer a situation of mindless confidence in the rapid success of a war of conquest. The modernization front is such a state of war, not declared, perhaps, but all the more violent in that it has made its warriors more apt to shoot as traitors those who didn't show up for roll call. How much clarification we would introduce into all these questions of modernity and modernization and globalization and universalization if we finally treated that front as a front in an explicitly declared conflict, with its "war objectives" spelled out at last! If it isn't a war, let's try to say who is the sovereign arbiter; if it is indeed a war, let's declare it for real and define its fronts and its goals. Yes, let's draw up the "fourteen points," so the belligerents will at least know why we are fighting them.

It will be said that no one can be interested in such a clarification of the stakes as long as they believe they are at peace, or at least as long as they are too sure of winning. Unless we are at war on two new fronts that modernism has been totally unable to anticipate, launched as it has been into its utopias of the future. Universalization, globalization, and modernization pleased Westerners as long as they were the unique source of such movements of irreversible conquest. Now that others, everywhere, in the East and in the South, are globalizing these movements, universalizing them, modernizing them, the Westerners find the affair much less amusing! A good opportunity, among all these new ways of globalizing and modernizing, to single out the ones that correspond to local histories contrasting with those of the aforementioned Moderns.

But it is the second front that has kept us awake from the beginning and that is going to redistribute the cards so thoroughly—I have little doubt about this—that the most confident modernizers will inevitably make their way to these strange diplomatic talks. If there is only one Earth and it is against us, what are we going to do? No polemology prepares us for conflicts so asymmetrical that we are helpless in the face of Gaia, which is helpless before us, but which can nevertheless rid itself of us Earthlings, or so we are told. A strange war that we can only lose: if we win, we lose; if we lose, we still lose . . . And it's all the stranger in that we don't even know for sure whether there is any being in existence capable of summoning up these retroactive loops of the Earth system, nor even if such a being would be hostile to us.

The modernizers knew how to survive a nature indifferent to their projects; but when Nature ceases to be indifferent, when the Nature of the ANTHROPOCENE becomes sensitive, even hypersensitive, to their weight, how is anyone to define what it is looking for, when in fact it is not even interested in us, but in itself? Go ahead, try to talk about mastery and possession to something that can master and possess us without even attaching any importance to our survival. Now there's a real "phony war." What we are fearful of hearing is, as Sloterdijk says, the announcement—a terrifying one for those who have always lived under the tension of transcendence—of this "monogeism": there is no God, there is only one Earth. Our predecessors got quite frightened at the "death of God," which seems to have deprived them of all their bearings and all their limits: without God, they used to say, "everything is permitted." The return of narrow limits obliges us to seek quite different bearings: if Gaia is against us, then *not much* is permitted any longer. While we wait for Gaia, it isn't the sense of the absurd that threatens us now, but rather our lack of adequate preparation for the civilization to come. It is that civilization that our inquiry seeks to praise in advance, in order to ward off the worst.

NAME	HIATUS	TRAJECTORY
[REP]RODUCTION	Risks of reproduction	Prolonging existents
[MET]AMORPHOSIS	Crises, shocks	Mutations, emotions, transformations
[HAB]IT	Hesitations and adjustments	Uninterrupted courses of action
[TEC]HNOLOGY	Obstacles, detours	Zigzags of ingenuity and invention
[FIC]TION	Vacillation between material and form	Triple shifting: time, space, actant
[REF]ERENCE	Distance and dissemblances of forms	Paving with inscriptions
[POL]ITICS	Impossibility of being represented or obeyed	Circle productive of continuity
[LAW]	Dispersal of cases and actions	Linking of cases and actions via means
[REL]IGION	Break in times	Engendering of persons
[ATT]ACHMENT	Desires and lacks	Multiplication of goods and bads
[ORG]ANIZATION	Disorders	Production and following of scripts
[MOR]ALITY	Anxiety about means and ends	Exploration of the links between ends and means
[NET]WORK	Surprise of association	Following heterogeneous connections
[PRE]POSITION	Category mistakes	Detection of crossings
[DC] DOUBLE CLICK	Horror of hiatuses	Displacement without translation

This table summarizes the state of the inquiry presented in the report.

The fifteen modes recognized up to now are listed in rows; the columns give the four canonical questions asked of each mode: by what hiatus and what trajectory are they

FELICITY/INFELICITY CONDITIONS	BEINGS TO INSTITUTE	ALTERATION	NAME
Continue, inherit, disappear	Lines of force, lineages, societies	Explore continuities	[REP]
Make (something) pass, install, protect/alienate, destroy	Influences, divinities, psyches	Explore differences	[MET]
Pay attention/lose attention	Veil over prepositions	Obtain essences	[HAB]
Rearrange, set up, adjust/ fail, destroy, imitate	Delegations, arrangements, inventions	Fold and redistribute resistances	[TEC]
Make (something) hold up, make believe/cause to fail, lose	Dispatches, figurations, forms, works of art	Multiply worlds	[FIC]
Bring back/lose information	Constants through transformations	Reach remote entities	[REF]
Start over and extend/ suspend or reduce the Circle	Groups and figures of assemblies	Circumscribe and regroup	[POL]
Reconnect/break levels of enunciation	Safety-bearers	Ensure the continuity of actions and actors	[LAW]
Save, bring into presence/ lose, take away	Presence-bearers	Achieve the end times	[REL]
Undertake, interest/ stop transactions	Passionate interests	Multiply goods and bads	[ATT]
Master scripts/lose scripts from view	Framings, organizations, empires	Change the size or extension of frames	[ORG]
Renew calculations/ suspend scruples	The "kingdom of ends"	Calculate the impossible optimum	[MOR]
Traverse domains/lose freedom of inquiry	Networks of irreductions	Extend associations	[NET]
Give each mode its template/ crush the modes	Interpretive keys	Ensure ontological pluralism	[PRE]
Speak literally/speak through figures and tropes	Reign of indisputable Reason	Maintain the same despite the other	[DC]

distinguished? (columns 1 and 2); what are their felicity and infelicity conditions? (column 3); what beings must they be prepared to institute? (column 4); finally, to what alteration is being-as-other subjected in each case? (column 5).